MW01673088

UWorld | GRAD SCHOOL

MCAT® Critical Analysis and Reasoning Skills (CARS)

2026–2027 Prep Book

2nd Edition: An Annotated Passage Guide

Acknowledgments for the 2nd Edition

The accuracy and up-to-date content of the course materials in this book result from the combined efforts of our team of content experts, editors, illustrators, software developers, and dedicated support staff. UWorld's passion for education drives every product we create, and it is supported by our commitment to quality and student success.

How To Use This Booklet

In the following pages, you will find copies of the passages used in the UWorld MCAT Critical Analysis and Reasoning Skills (CARS) book. Each passage appears twice in this booklet: once with annotations focusing on the passage's structural features, and a second time with annotations focusing on its content. The annotations are divided into categories for organization and ease of reference as shown below.

Structural Annotations	
Paragraph Summary	A brief synopsis of each paragraph
Topic Shift	When a passage switches its subject of discussion
Addressing Opposing View	Raising and responding to potential counterpoints to passage claims
Connected Ideas	Information found in a later part of the passage that relates back to a previous idea

Content Annotations: Passage Observations	
Claim	A statement that seems notable or important for understanding the passage
Focus	A statement that helps to indicate the passage's main idea
Author's View	Information that suggests or reveals the author's opinion
Emphasis	A point that the author particularly stresses or draws extra attention to
Likely Question Topic	Information that could easily form the basis for a question
Outside Source	A reference to ideas other than the author's
Elaboration	Information that provides further details about a previous claim

Content Annotations: Potential Pitfalls	
Misreading	Cases where it would be easy to make a mistake about what is being said
Details	Information containing many nuances or complexities to keep track of
Missed Distinction	Cases where it would be easy to overlook a point the author makes, or to assume the author makes a point that they actually do not
Trap Answer	Claims that could make an incorrect answer to a question seem attractive

You can think of these annotations as representing the kind of mental notes that a CARS expert would make while reading. Accordingly, reviewing them in context will give you a sense of the types of things that are important to notice when analyzing a passage. As you continue to practice and become more experienced with CARS, you may find yourself naturally making similar (but less formal) mental notes.

In the main UWorld CARS book, you can read and analyze the passages as they would appear on the exam (without annotations). Then, in this booklet, you can review the annotated versions of each passage.

Table of Contents

Passage A: The Knights Templar

This passage is associated with 5 questions analyzed in the UWorld MCAT CARS book.

- Subskill 1a. Main Idea or Purpose Question 1
- Subskill 2d. Connecting Claims With Evidence Question 1
- Subskill 2e. Determining Passage Perspectives Question 1
- Subskill 3d. External Scenario Support or Challenge Question 5
- Subskill 3f. Additional Conclusions From New Information Question 2

Passage A: The Knights Templar

Structural Annotations

Passage Structure: 6 Annotations

[P1] The seal of the Knights Templar depicts two knights astride a single horse, a visual testament of the order's poverty at its inception in 1119. Nevertheless, these Knights of the Temple—who swore oaths not only of poverty but of chastity, loyalty, and bravery—would eventually become one of the wealthiest and most powerful organizations in the medieval world. So far-reaching was their strength and acclaim that their destruction must have seemed as sudden and surprising as it was utter and irrevocable. The signs of danger could not have been wholly invisible, however, as the Templars' growing influence became perceived as a threat to European rulers.

Paragraph 1 Summary

The Templars rose from humble beginnings to become extremely wealthy and powerful. Nevertheless, they came to a sudden and surprising end due to conflict with European rulers.

[P2] The first Templars were nine knights who took an oath to defend the Holy Land and any pilgrims who journeyed there after the First Crusade. Having secured a small benefice from Jerusalem's King Baldwin II, the knights inaugurated their mission at the site of the great Temple of Solomon. They quickly attracted widespread admiration as well as many recruits from crusaders and other knights. Within a year of its founding, the order received a financial endowment from the deeply impressed Count of Anjou, whose example was soon followed by other nobles and monarchs. As early as 1128, the Templars even gained official papal recognition, and their wealth, holdings, and numbers swelled both in the Holy Land and throughout Europe, especially in France and England.

Paragraph 2 Summary

Because of the great admiration they inspired, the Templars rapidly gained members, wealth, and influence.

[P3] [1] However, this growing power contained the seeds of the order's downfall. Although the Templars were generally held in high esteem, the passage of time saw censure and suspicion directed toward them. The failure of the disastrous siege of Ascalon in 1153 was attributed by some to Templar greed. Similarly, in 1208 Pope Innocent III condemned the wickedness he believed to exist within their ranks. Moreover, their increasingly elevated status brought them into conflict with established authorities. One revealing example occurred in 1252 when, because of the Templars' "many liberties" as well as their "pride and haughtiness," England's King Henry III proposed to curb the order's strength by reclaiming some of its

[1] Connected Ideas

The author revisits the idea from the introduction that the Templars' power would eventually lead to their downfall. This relationship is likely a major theme of the passage.

possessions. The Templars' response was unambiguous: "So long as thou dost exercise justice thou wilt reign; but if thou infringe it, thou wilt cease to be King!"

Paragraph 3 Summary

The Templars' growing power led to suspicion and tension with rulers, whose authority they were strong enough to challenge.

[P4] Ultimately, the impoverished Philip the Fair of France joined forces with Pope Clement V to engineer the Templars' downfall beginning in 1307. Conspiring to seize the order's wealth, Clement invited the Templar Master, Jacques de Molay, to meet with him on the pretext of organizing a new crusade to retake the Holy Land. Shortly thereafter, Philip's forces arrested de Molay and his knights on charges the preponderance of which were almost certainly fabricated. Ranging from the mundane to the unspeakably perverse, the accusations even included an incredible entry citing "every crime and abomination that can be committed." Suffering tortures nearly as horrific as the acts of which they were accused, many Templars confessed.

Paragraph 4 Summary

The Templars met their end after being tricked by the pope and arrested on false charges by the king of France.

[P5] Not everyone believed these charges. Despite Philip's urging, Edward II of England remained convinced that the accusations were false, a view seemingly shared by other rulers. Nevertheless, Clement ordered Edward to extract confessions, a task the king tried to carry out with some measure of mercy. By 1313, the Templar Order had been dissolved by papal decree, and many of its members were dead. The following year, Jacques de Molay was burned at the stake after declaring that the Templars' confessions were lies obtained under torture. Stories would spread that as he died, he condemned Clement and Philip to join him within a year. Although it is impossible to say whether he truly called divine vengeance down upon them, within a few months' time both pope and king had gone to their graves.

Paragraph 5 Summary

Many rulers believed the accusations were false, but the pope forced them to comply with the Templars' condemnation. Thus, the conflict between the Templars and established authorities ended with the destruction of the Templars by Pope Clement and King Philip of France.

Passage A: The Knights Templar

Content Annotations

Passage Observations: 10 Annotations

[P1] The seal of the Knights Templar depicts two knights astride a single horse, [1] a visual testament of the order's poverty at its inception in 1119. Nevertheless, these Knights of the Temple—who swore oaths not only of poverty but of chastity, loyalty, and bravery—would eventually become one of the wealthiest and most powerful organizations in the medieval world. [2] So far-reaching was their strength and acclaim that their destruction must have seemed as sudden and surprising as it was utter and irrevocable. The signs of danger could not have been wholly invisible, however, as the Templars' growing influence became perceived as a threat to European rulers.

[1] Claim

The passage sets up a contrast between the Templars' initial poverty and their eventual wealth and power.

[2] Focus

The passage will likely center on explaining the rise and fall of the Knights Templar.

[P2] The first Templars were nine knights who took an oath to defend the Holy Land and any pilgrims who journeyed there after the First Crusade. Having secured a small benefice from Jerusalem's King Baldwin II, the knights inaugurated their mission at the site of the great Temple of Solomon. [3] They quickly attracted widespread admiration as well as many recruits from crusaders and other knights. Within a year of its founding, the order received a financial endowment from the deeply impressed Count of Anjou, whose example was soon followed by other nobles and monarchs. [4] As early as 1128, the Templars even gained official papal recognition, and their wealth, holdings, and numbers swelled both in the Holy Land and throughout Europe, especially in France and England.

[3] Claim

Admiration for the Templars led to voluntary financial support and enthusiastic recruits.

[4] Elaboration

Not only did the Templars grow in wealth and power, they did so quickly: their resources and membership swelled less than ten years after their founding.

[P3] [5] However, this growing power contained the seeds of the order's downfall. Although the Templars were generally held in high esteem, the passage of time saw censure and suspicion directed toward them. The failure of the disastrous siege of Ascalon in 1153 was attributed by some to Templar greed. Similarly, in 1208 Pope Innocent III condemned the wickedness he believed to exist within their ranks. Moreover, their increasingly elevated status brought them into conflict with established authorities. [6] One revealing example occurred in 1252 when, because of the Templars' "many liberties" as well as their "pride and haughtiness," England's King Henry III proposed to curb the order's strength by reclaiming some of its

[5] Claim

The author connects the Templars' eventual downfall with their own great power.

[6] Emphasis

The author highlights this example as especially representative of how the Templars came into conflict with established rulers. Henry will "cease to be king" if he tries to reclaim the Templars' possessions.

possessions. The Templars' response was unambiguous: "So long as thou dost exercise justice thou wilt reign; but if thou infringe it, thou wilt cease to be King!"

[P4] [7] Ultimately, the impoverished Philip the Fair of France joined forces with Pope Clement V to engineer the Templars' downfall beginning in 1307. Conspiring to seize the order's wealth, Clement invited the Templar Master, Jacques de Molay, to meet with him on the pretext of organizing a new crusade to retake the Holy Land. Shortly thereafter, Philip's forces arrested de Molay and his knights on [8] charges the preponderance of which were almost certainly fabricated. Ranging from the mundane to the unspeakably perverse, the accusations even included an incredible entry citing "every crime and abomination that can be committed." Suffering tortures nearly as horrific as the acts of which they were accused, many Templars confessed.

[7] Claim

The Templars' destruction was caused by the pope and the king of France teaming up to ruin them and take their wealth.

[8] Author's View

The author believes that most of the charges against the Templars were false, and that their confessions were forced.

[P5] [9] Not everyone believed these charges. Despite Philip's urging, Edward II of England remained convinced that the accusations were false, a view seemingly shared by other rulers. Nevertheless, Clement ordered Edward to extract confessions, a task the king tried to carry out with some measure of mercy. By 1313, the Templar Order had been dissolved by papal decree, and many of its members were dead. The following year, Jacques de Molay was burned at the stake after declaring that the Templars' confessions were lies obtained under torture. Stories would spread that as he died, he condemned Clement and Philip to join him within a year. [10] Although it is impossible to say whether he truly called divine vengeance down upon them, within a few months' time both pope and king had gone to their graves.

[9] Claim

The Templars' arrest was controversial, as many rulers didn't believe the accusations. Nevertheless, Pope Clement forced these rulers to comply with the arrests and convictions.

[10] Claim

The Templars may have achieved some vengeance for their destruction as both Pope Clement and King Philip soon died.

Passage B: Jackie in 500 Words

This passage is associated with 5 questions analyzed in the UWorld MCAT CARS book.

- Subskill 1a. Main Idea or Purpose Question 2
- Subskill 1b. Meaning of Term Question 3
- Subskill 1c. Direct Passage Claims Question 4
- Subskill 1d. Implicit Claims or Assumptions Question 1
- Subskill 3a. Exemplar Scenario for Passage Claims Question 1

Passage B: Jackie in 500 Words

Structural Annotations

Passage Structure: 5 Annotations

[P1] Born in New York in 1929, Jacqueline Bouvier first came into the public eye as the wife of the 35th president of the United States, John F. Kennedy. The president was assassinated in 1963, and by the end of what turned out to be a turbulent decade, Mrs. Kennedy had transformed herself into the enigmatic Jackie O., wife of Greek shipping magnate Aristotle Onassis. Multifaceted and always elusive, the former first lady never ceased to fascinate; however, people had to be satisfied with only glimpses of this fashion icon, culture advocate, historic preservationist, polyglot, equestrienne, and book editor. Indeed, upon her death in 1994, Jacqueline Kennedy Onassis was described as "the most intriguing woman in the world." Often topping lists of the most admired individuals of the second half of the 20th century, this celebrated woman is likely someone many wish they had known. Barring such a possibility, the best way to fully appreciate Jackie's exceptional nature might be to consider the people she wished she had known.

Paragraph 1 Summary

We can come to understand the "multifaceted" and "fascinating" Jackie better by looking at the people she wished she could have known.

[P2] It is unsurprising that this woman who captured the public's imagination for decades distinguished herself from her peers early on. Notably, in 1951, Ms. Bouvier entered a scholarship contest sponsored by *Vogue* and open to young women in their final undergraduate year, the annual Prix de Paris. Among other assignments, applicants were asked to compose a 500-word essay, "People I Wish I Had Known," spotlighting three individuals influential in art, literature, or culture. The future first lady chose an iconoclastic trio from the Victorian era: the French symbolist poet Charles Baudelaire, the Irish wit Oscar Wilde, and the innovative Ballets Russes dance company founder Sergei Diaghilev.

Paragraph 2 Summary

The author expands on the previous idea of knowing Jackie better by introducing her scholarship essay about people she wished she had known.

[P3] In a brief composition, Jackie provided deep insights into this bohemian threesome of poet, aesthete, and impresario with whom she strongly identified. She concluded that Baudelaire deployed "venom and despair" as "weapons" in his poetry. She idolized Wilde for being able "with the flash of an epigram to bring about what serious reformers had for years been trying to accomplish." Diaghilev she defined as an artist of a different sort, someone who "possessed what is rarer than artistic genius in any one field—the sensitivity to take the best of each man and incorporate it into a masterpiece." As Jackie poignantly observed, such a work is "all the more precious because it lives only in the minds of those who have seen it," dissipating soon after. Furthermore, although these men espoused different disciplines, she discerned that "a common theory runs through their work, a certain concept of the interrelation of the arts." Finally, foreshadowing her self-assumed role in the White House as the nation's unofficial minister of the arts, Jackie paid homage with her vision: "If I could be a sort of Overall Art Director of the Twentieth Century, watching everything from a chair hanging in space, it is their theories of art that I would apply to my period."

Paragraph 3 Summary

Jackie's scholarship essay expressed her admiration for Baudelaire, Wilde, and Diaghilev. All three saw the arts as interconnected, and Jackie would apply their theories to art direction if she had the chance.

[P4] The contest committee judged Jackie's essay to have exhibited a profound appreciation for the arts combined with a truly outstanding level of intellectual maturity and originality of thought. Similarly, biographer Donald Spoto deemed Jackie "remarkably unorthodox," not unlike the men about whom she wrote in her unusual composition which he pronounced "a masterpiece of perceptive improvisation." Thus, from a pool of 1,279 applicants representing 224 colleges, Jacqueline Bouvier was declared the winner.

Paragraph 4 Summary

The author describes why Jackie won the scholarship contest and notes how she and the men she wrote about shared an unorthodox perspective.

[P5] Although Ms. Bouvier went on to decline the prestigious award, which would have involved living and working in Paris, she never gave up her dream of being the century's art director. As first lady, she tirelessly promoted the arts and culture. Today, the John F. Kennedy Center for the Performing Arts in Washington, DC, is a legacy of Jackie's vision.

Paragraph 5 Summary

The vision expressed in Jackie's winning essay informed her adult life and motivated her activities as first lady.

Passage B: Jackie in 500 Words

Content Annotations

Passage Observations: 11 Annotations

[P1] Born in New York in 1929, Jacqueline Bouvier first came into the public eye as the wife of the 35th president of the United States, John F. Kennedy. The president was assassinated in 1963, and by the end of what turned out to be a turbulent decade, Mrs. Kennedy had transformed herself into the enigmatic Jackie O., wife of Greek shipping magnate Aristotle Onassis. [1] Multifaceted and always elusive, the former first lady never ceased to fascinate; however, people had to be satisfied with only glimpses of this fashion icon, culture advocate, historic preservationist, polyglot, equestrienne, and book editor. Indeed, upon her death in 1994, Jacqueline Kennedy Onassis was described as [2] "the most intriguing woman in the world." Often topping lists of the most admired individuals of the second half of the 20th century, this celebrated woman is likely someone many wish they had known. Barring such a possibility, [3] the best way to fully appreciate Jackie's exceptional nature might be to consider the people she wished she had known.

[1] Author's View

The author portrays Jackie as being fascinating because of her mysterious nature and many diverse roles in life.

[2] Author's View

By referring to "Jackie's exceptional nature," the author implies that they share the positive attitude many people have toward the "intriguing" and "admired" Jackie.

[3] Focus

The author's goal seems to be to discuss what they see as Jackie's admirable qualities by considering the qualities that Jackie herself admired in others.

[P2] It is unsurprising that this woman who captured the public's imagination for decades distinguished herself from her peers early on. Notably, in 1951, Ms. Bouvier entered a scholarship contest sponsored by *Vogue* and open to young women in their final undergraduate year, the annual Prix de Paris. Among other assignments, applicants were asked to compose a 500-word essay, "People I Wish I Had Known," spotlighting three individuals influential in art, literature, or culture. [4] The future first lady chose an iconoclastic trio from the Victorian era: the French symbolist poet Charles Baudelaire, the Irish wit Oscar Wilde, and the innovative Ballets Russes dance company founder Sergei Diaghilev.

[4] Claim

Jackie's essay centered on three men from the Victorian era: a poet (Baudelaire), a "wit" (Wilde), and the founder of a ballet company (Diaghilev).

[P3] In a brief composition, Jackie provided deep insights into this bohemian threesome of poet, aesthete, and impresario with whom she strongly identified. [5] She concluded that Baudelaire deployed "venom and despair" as "weapons" in his poetry. She idolized Wilde for being able "with the flash of an epigram to bring about what serious reformers had for years been trying to accomplish." Diaghilev she defined as an artist of a different

[5] Outside Source

The passage cites Jackie's own words to describe the specific reasons she admired the men about whom she wrote.

sort, someone who "possessed what is rarer than artistic genius in any one field—the sensitivity to take the best of each man and incorporate it into a masterpiece." As [6] Jackie poignantly observed, such a work is "all the more precious because it lives only in the minds of those who have seen it," dissipating soon after. Furthermore, [7] although these men espoused different disciplines, she discerned that "a common theory runs through their work, a certain concept of the interrelation of the arts." Finally, foreshadowing her self-assumed role in the White House as the nation's unofficial minister of the arts, Jackie paid homage with her vision: [8] "If I could be a sort of Overall Art Director of the Twentieth Century, watching everything from a chair hanging in space, it is their theories of art that I would apply to my period."

[6] Author's View

The positive term "poignantly" suggests that the author agrees with Jackie's sentiment.

[7] Likely Question Topic

The interrelation of the arts is presented as a unifying theme between the different men Jackie admired. Thus, this connection seems important for understanding Jackie's views and could easily serve as the basis for a question.

[8] Outside Source

The author again cites Jackie's own words to describe her essay and her overall view of art.

[P4] [9] The contest committee judged Jackie's essay to have exhibited a profound appreciation for the arts combined with a truly outstanding level of intellectual maturity and originality of thought. [10] Similarly, biographer Donald Spoto deemed Jackie "remarkably unorthodox," not unlike the men about whom she wrote in her unusual composition which he pronounced "a masterpiece of perceptive improvisation." Thus, from a pool of 1,279 applicants representing 224 colleges, Jacqueline Bouvier was declared the winner.

[9] Claim

Jackie won the contest because of her maturity, originality, and appreciation of art.

[10] Outside Source

The quotations from Donald Spoto largely echo the contest committee's perspective, reinforcing the idea that Jackie's essay displayed unusual insight and quality.

[P5] Although Ms. Bouvier went on to decline the prestigious award, which would have involved living and working in Paris, [11] she never gave up her dream of being the century's art director. As first lady, she tirelessly promoted the arts and culture. Today, the John F. Kennedy Center for the Performing Arts in Washington, DC, is a legacy of Jackie's vision.

[11] Claim

The passage seems to imply that the values expressed in Jackie's essay informed her goals and activities as a promoter of the arts throughout her life.

Passage C: Probability and The Universe

This passage is associated with 6 questions analyzed in the UWorld MCAT CARS book.

- Subskill 1a. Main Idea or Purpose Question 3
- Subskill 1e. Identifying Passage Perspectives Question 1
- Subskill 2a. Logical Relationships Within Passage Question 1
- Subskill 2b. Function of Passage Claim Question 2
- Subskill 2e. Determining Passage Perspectives Question 5
- Subskill 3a. Exemplar Scenario for Passage Claims Question 3

Passage C: Probability and The Universe

Structural Annotations

Passage Structure: 11 Annotations

[P1] The idea of probability is frequently misunderstood, in large part because of a conceptual confusion between objective probability and subjective probability. The failure to make this distinction leads to an erroneous conflation of genuine possibility with what is in fact merely personal ignorance of outcome. An example will clarify.

Paragraph 1 Summary

The passage distinguishes between objective and subjective probability and will go on to illustrate how they are often misunderstood.

[P2] A standard die is rolled on a table, but the outcome of the roll is concealed. Should an observer be asked the chance that a particular number was rolled—five, say—the natural response is 1/6. However, this answer is incorrect. To say there is a one-in-six chance that a five was rolled implies there is an equal chance that any of the other numbers were rolled. But there is no equal chance, because the roll has already occurred. Hence, the probability that the result of the roll is a five is either 100% or 0%, and the same is true for each of the other numbers.

Paragraph 2 Summary

An example of a die roll illustrates the difference between objective and subjective probability: the actual chance of something occurring is unaffected by what a person knows (or doesn't know) about its occurrence.

[P3] [1] It might be objected that such an analysis is an issue of semantics rather than a substantive claim. For declaring the probability to be 1/6 is merely an expression that, for all we know, any number from 1 through 6 might have been rolled. But the difference between *for all we know* and what *is* remains crucially important, because it forestalls a tendency to make scientific assertions from a perspective biased by human perception....

[1] Addressing Opposing View

The author anticipates a possible objection to their position so they can refute that objection.

Paragraph 3 Summary

The objective/subjective probability distinction is not just a matter of wording but an important understanding to prevent human bias from influencing scientific assertions.

[P4] [2] [T]hus, it should be clear that claims about the purpose of the universe rest on shaky ground. In particular, we must be wary of inferences drawn from juxtaposing the existence of intelligent life with the genuinely improbable cosmological conditions that make such life possible. Joseph Zycinski provides an instructive account of those conditions: "If twenty billion years ago the rate of expansion were minimally smaller, the universe would have entered a stage of contraction at a time when temperatures of thousands of degrees were prevalent and life could not have appeared. If the distribution of matter were more uniform the galaxies could not have formed. If it were less uniform, instead of the presently

[2] Topic Shift

The author switches topics here, building on the "crucially important" distinction in the preceding paragraph to discuss claims about the purpose of the universe.

observed stars, black holes would have formed from the collapsing matter." In short, the existence of life (let alone intelligent life) depends upon the initial conditions of the universe having conformed to an extremely narrow range of possible values.

[P5] [3] It is tempting to go beyond Zycinski's factual point to draw deep cosmological, teleological, and even theological conclusions. But no such conclusions follow. The reason why a life-sustaining universe exists is that if it did not, there would be no one to wonder why a life-sustaining universe exists. This fact is a function not of purpose, but of pre-requisite. [4] For instance, suppose again that a five was rolled on a die. We now observe a five on its face, not because the five was "meant to be" but because one side of the die had to land face up. Even if we imagine that the die has not six but six *billion* sides, that analysis is unaffected. Yes, it was unlikely that any given value would be rolled, but *some* value had to be rolled. That "initial condition" then set the parameters for what kind of events could possibly follow; in this case, the observation of the five.

[P6] Similarly, the existence of intelligent beings is evidence that certain physical laws obtained in the universe. However, it is not evidence that those beings or laws were necessary or intended rather than essentially random. [5] The subjective probability of that outcome is irrelevant, because in this universe the objective probability of its occurrence is 100%. Thus, the seemingly low initial chance that such a universe would exist is not in itself indicative of a purpose to its existence.

Paragraph 4 Summary

The author applies the previous points about probability and human bias to the existence of the universe. Even though the conditions for life were extremely improbable, that fact does not justify inferences about the universe's purpose.

[3] Addressing Opposing View

The author raises a "tempting" view in order to argue against it.

[4] Connected Ideas

The connection between the original die-rolling example and this new, more extensive example parallels the topic shift from probability to the existence of the universe. The original example illustrated the right way to think about probability; this second example illustrates the right way to think about whether the universe has a purpose.

Paragraph 5 Summary

The conditions necessary for intelligent life were improbable, but that doesn't mean they were "meant to be." A die with 6 billion sides is an analogy—any individual roll is unlikely, but some side has to land face up.

[5] Connected Ideas

The author directly links their claims about the purpose of the universe to the distinction in Paragraph 1 between objective and subjective probability.

Paragraph 6 Summary

The distinction between objective and subjective probability parallels the difference between facts and ideas about purpose. Just because a universe with intelligent life was improbable does not mean there is a purpose to its existence.

Passage C: Probability and The Universe

Content Annotations

Passage Observations: 9 Annotations
Potential Pitfalls: 2 Annotations

[P1] [1] The idea of probability is frequently misunderstood, in large part because of a conceptual confusion between objective probability and subjective probability. The failure to make this distinction leads to an erroneous conflation of genuine possibility with what is in fact merely personal ignorance of outcome. An example will clarify.

[1] Claim

The passage distinguishes between "objective" and "subjective" probability, which parallels the contrast between "genuine possibility" and "merely personal ignorance." Thus, the implication seems to be that only objective probability is genuine.

[P2] A standard die is rolled on a table, but the outcome of the roll is concealed. Should an observer be asked the chance that a particular number was rolled—five, say—[1] the natural response is 1/6. However, this answer is incorrect. [2] To say there is a one-in-six chance that a five was rolled implies there is an equal chance that any of the other numbers were rolled. But there is no equal chance, because the roll has already occurred. Hence, the probability that the result of the roll is a five is either 100% or 0%, and the same is true for each of the other numbers.

[1] Potential Pitfall: Trap Answer

Since the author says this "natural response" is wrong, questions and answer choices might play on the tendency to think in this way in order to trick you.

[2] Elaboration

According to the author, this is the correct (but not obvious) way of understanding the probability of the roll. We now have the information that was meant to clarify the author's points from the previous paragraph.

[P3] It might be objected that such an analysis is an issue of semantics rather than a substantive claim. For declaring the probability to be 1/6 is merely an expression that, for all we know, any number from 1 through 6 might have been rolled. But [3] the difference between *for all we know* and what *is* remains crucially important, because it forestalls a tendency to make scientific assertions from a perspective biased by human perception....

[3] Emphasis

The author highlights this difference as crucially important, suggesting this point is a key part of their argument.

[P4] [T]hus, it should be clear that claims about the purpose of the universe rest on shaky ground. In particular, we must be wary of inferences drawn from juxtaposing the existence of intelligent life with the genuinely improbable cosmological conditions that make such life possible. [4] Joseph Zycinski provides an instructive account of those conditions: "If twenty billion years ago the rate of expansion were minimally smaller, the universe would have entered a stage of contraction at a time when temperatures of thousands of degrees were prevalent and life could not have appeared. If the distribution of matter were more uniform the galaxies could not have

[4] Outside Source

The author cites Zycinski's words to explain the specific conditions necessary for a universe to support intelligent life.

formed. If it were less uniform, instead of the presently observed stars, black holes would have formed from the collapsing matter." [5] In short, the existence of life (let alone intelligent life) depends upon the initial conditions of the universe having conformed to an extremely narrow range of possible values.

[P5] It is tempting to go beyond [6] Zycinski's factual point to draw deep cosmological, teleological, and even theological conclusions. [2] But no such conclusions follow. [7] The reason why a life-sustaining universe exists is that if it did not, there would be no one to wonder why a life-sustaining universe exists. This fact is a function not of purpose, but of pre-requisite. For instance, suppose again that a five was rolled on a die. We now observe a five on its face, not because the five was "meant to be" but because one side of the die had to land face up. Even if we imagine that the die has not six but six *billion* sides, that analysis is unaffected. Yes, it was unlikely that any given value would be rolled, but *some* value had to be rolled. That "initial condition" then set the parameters for what kind of events could possibly follow; in this case, the observation of the five.

[P6] [8] Similarly, the existence of intelligent beings is evidence that certain physical laws obtained in the universe. However, it is not evidence that those beings or laws were necessary or intended rather than essentially random. The subjective probability of that outcome is irrelevant, because in this universe the objective probability of its occurrence is 100%. [9] Thus, the seemingly low initial chance that such a universe would exist is not in itself indicative of a purpose to its existence.

Probability and The Universe ©UWorld

[5] Elaboration

The passage first provides and then summarizes Zycinski's quote, which explains that the existence of life depends on "extremely narrow" conditions.

[6] Author's View

The author agrees with Zycinski's view, describing it as "factual."

[2] Potential Pitfall: Misreading

It would be easy to think that "no such conclusions follow" means the author is disagreeing with Zycinski. But attentive reading shows the opposite is true—the author calls Zycinski's point "factual" and is cautioning against going *beyond* what he said.

[7] Claim

As a counterpoint to the "tempting" conclusions just mentioned, the author asserts that nothing about purpose can be inferred from the mere fact that intelligent life exists. We can ask why intelligent life exists only because the necessary conditions for it did occur, however improbable those conditions might seem.

[8] Author's View

The author further stresses that the low probability of intelligent life existing is not evidence that such life was meant to exist rather than a random occurrence.

[9] Focus

This concluding sentence likely represents the main point of the passage. The author's argument is meant to establish that: the low subjective probability that a universe with intelligent life would exist is not evidence that there is a purpose to the universe's existence.

Passage D: American Local Motives

This passage is associated with 7 questions analyzed in the UWorld MCAT CARS book.

- Subskill 1b. Meaning of Term Question 1
- Subskill 2a. Logical Relationships Within Passage Question 2
- Subskill 2c. Extent of Passage Evidence Question 1
- Subskill 2d. Connecting Claims With Evidence Question 3
- Subskill 3c. New Claim Support or Challenge Question 4
- Subskill 3d. External Scenario Support or Challenge Question 1
- Subskill 3g. Identifying Analogies Question 3

Passage D: American Local Motives

Structural Annotations

Passage Structure: 9 Annotations

[P1] Locomotives were invented in England, with the first major railroad connecting Liverpool and Manchester in 1830. However, it was in America that railroads would be put to the greatest use in the nineteenth century. On May 10, 1869, the Union Pacific and Central Pacific lines met at Promontory Point, Utah, joining from opposite directions to complete a years-long project—the Transcontinental Railroad. This momentous event connected the eastern half of the United States with its western frontier and facilitated the construction of additional lines in between. As a result, journeys that had previously taken several months by horse and carriage now required less than a week's travel. By 1887 there were nearly 164,000 miles of railroad tracks in America, and by 1916 that number had swelled to over 254,000.

Paragraph 1 Summary

Trains were invented in nineteenth-century England but used more extensively in America. The completion of the Transcontinental Railroad in 1869 dramatically transformed travel in the United States, and the amount of railroad tracks continued to increase over the next few decades.

[P2] While the United States still has the largest railroad network in the world, it operates largely in the background of American life, and citizens no longer view trains with the sense of importance those machines once commanded. Nevertheless, the economic and industrial advantages those citizens enjoy today would not have been possible without America's history of trains; as Tom Zoellner reminds us, "Under the skin of modernity lies a skeleton of railroad tracks." [1] Although airplanes and automobiles have now assumed greater prominence, the time has arrived for the resurgence of railroads. A revitalized and advanced railway system would confer numerous essential benefits on both the United States and the globe.

[1] Topic Shift

While so far the passage has provided historical information about American railroads, the author now switches to arguing that railroads should be increasingly used in the future.

Paragraph 2 Summary

Trains were essential to creating modern American society but have fallen out of prominence in favor of cars and planes. In the author's view, however, they should once again be used to transform American life.

[P3] [2] The chief obstacles to garnering support for such a project are the current dominance of the automobile and the languishing technology of existing railroads. In a sense these two obstacles are one, as American dependence on personal automobiles is partially due to the paucity of rapid public transportation. The railroads of Europe and Japan, by comparison, have vastly outpaced their American counterparts. Japan has operated high-speed rail lines continuously since 1964, and in 2007, a French train set a record of 357 miles per

[2] Addressing Opposing View

The author discusses two factors that could inhibit people from supporting a revitalized train system.

hour. While that speed was achieved under tightly controlled conditions, it still speaks to the great disparity in railroad development between the United States and other countries since the mid-twentieth century. British trains travel at speeds much higher than those in America, where both the trains themselves and the infrastructure to support them have simply been allowed to fall behind. In much of Europe it is common for trains to travel at close to 200 miles per hour.

Paragraph 3 Summary

Americans today generally prefer automobiles to trains, but this can be attributed to the poor state of American railways compared to other parts of the world. Japan, England, and much of Europe have developed vastly superior modern trains.

[P4] [3] To invest in a modern network of railroads would improve the United States in much the same way that the first railroads did in the nineteenth and early twentieth centuries. A high-speed passenger rail system would dramatically transform American life as travel between cities and states became quicker and more convenient, encouraging commerce, business, and tourism. Such a system would also make important strides in environmental preservation. According to a 2007 British study, "CO2 emissions from aircraft operations are...at least five times greater" than those from high-speed trains. For similar reasons, Osaka, Japan, was ranked as "the best...green transportation city in Asia" by the *2011 Green City Index*. As Lee-in Chen Chiu notes in *The Kyoto Economic Review*, Osakans travel by railway more than twice as much as they travel by car.

[3] Connected Ideas

The author draws an explicit comparison between the proposed modern rail system and the dramatic impact that railroads have historically had in the U.S.

Paragraph 4 Summary

The author argues that high-speed trains would create improvements in many areas of American life. Such trains would also be good for the environment, as indicated by data from multiple studies.

[P5] [4] It is true that developing a countrywide high-speed rail system would come with significant costs. However, that was also true of the original Transcontinental Railroad, as indeed it is with virtually any great project undertaken for the public good. We should thus move ahead with confidence that the rewards will outweigh the expenditure as citizens increasingly choose to travel by train. Both for society's gain and the crucial well-being of the planet, our path forward should proceed upon rails.

[4] Connected Ideas

As in the previous paragraph, the author compares the historical impact of the first railroads with the potential impact of a new, high-speed rail system. The dramatic transformation such a modern system would create seems to be an overarching theme of the passage.

Paragraph 5 Summary

The author argues that although a high-speed rail system would be expensive, the benefits to society and the environment would outweigh those costs. Just as the Transcontinental Railroad was enormously beneficial in the nineteenth century, we should embrace a modern rail system today.

Passage D: American Local Motives

Content Annotations

Passage Observations: 9 Annotations

[P1] [1] Locomotives were invented in England, with the first major railroad connecting Liverpool and Manchester in 1830. However, it was in America that railroads would be put to the greatest use in the nineteenth century. On May 10, 1869, the Union Pacific and Central Pacific lines met at Promontory Point, Utah, joining from opposite directions to complete a years-long project—the Transcontinental Railroad. [2] This momentous event connected the eastern half of the United States with its western frontier and facilitated the construction of additional lines in between. As a result, journeys that had previously taken several months by horse and carriage now required less than a week's travel. By 1887 there were nearly 164,000 miles of railroad tracks in America, and by 1916 that number had swelled to over 254,000.

[P2] While the United States still has the largest railroad network in the world, it operates largely in the background of American life, and citizens no longer view trains with the sense of importance those machines once commanded. [3] Nevertheless, the economic and industrial advantages those citizens enjoy today would not have been possible without America's history of trains; as Tom Zoellner reminds us, "Under the skin of modernity lies a skeleton of railroad tracks." Although airplanes and automobiles have now assumed greater prominence, the time has arrived for the resurgence of railroads. [4] A revitalized and advanced railway system would confer numerous essential benefits on both the United States and the globe.

[P3] The chief obstacles to garnering support for such a project are the current dominance of the automobile and the languishing technology of existing railroads. [5] In a sense these two obstacles are one, as American dependence on personal automobiles is partially due to the paucity of rapid public transportation. The railroads of Europe and Japan, by comparison, have vastly outpaced their American counterparts. [6] Japan has operated high-speed rail lines continuously since 1964, and in 2007, a French train set a record of 357 miles per

[1] Claim

While trains were an English invention, they had more effect on life in America.

[2] Claim

The completion of the Transcontinental Railroad is described as "momentous," with a comparison illustrating how drastically it changed the speed of travel.

[3] Outside Source

The author quotes Zoellner to support the assertion that trains were essential to creating modern American society.

[4] Focus

Based on this sentence and the previous one, it seems the author's main purpose is to argue for the creation of a revitalized American railway system. In the author's view, the restored prominence of trains would produce both domestic and global benefits.

[5] Author's View

According to the author, the outdated technology of current railroads and the American preference for automobiles are two sides of the same coin. Thus, the implication seems to be that this preference would change if better railroads were available.

[6] Elaboration

Data is given to illustrate the superiority of trains in Europe and Japan compared to the "languishing technology" of trains in America.

hour. While that speed was achieved under tightly controlled conditions, it still speaks to the great disparity in railroad development between the United States and other countries since the mid-twentieth century. British trains travel at speeds much higher than those in America, where both the trains themselves and the infrastructure to support them have simply been allowed to fall behind. In much of Europe it is common for trains to travel at close to 200 miles per hour.

[P4] To invest in a modern network of railroads would improve the United States in much the same way that the first railroads did in the nineteenth and early twentieth centuries. [7] A high-speed passenger rail system would dramatically transform American life as travel between cities and states became quicker and more convenient, encouraging commerce, business, and tourism. Such a system would also make important strides in environmental preservation. [8] According to a 2007 British study, "CO2 emissions from aircraft operations are...at least five times greater" than those from high-speed trains. For similar reasons, Osaka, Japan, was ranked as "the best...green transportation city in Asia" by the *2011 Green City Index*. As Lee-in Chen Chiu notes in *The Kyoto Economic Review*, Osakans travel by railway more than twice as much as they travel by car.

[7] Author's View

Although the author has previously mentioned that high-speed trains would be beneficial, they only now state specifically what those benefits would be: improvements in commerce, business, tourism, and the environment.

[8] Outside Source

Data from British and Japanese sources support the author's claim that high-speed rail is good for the environment.

[P5] It is true that developing a countrywide high-speed rail system would come with significant costs. However, that was also true of the original Transcontinental Railroad, as indeed it is with virtually any great project undertaken for the public good. We should thus move ahead with [9] confidence that the rewards will outweigh the expenditure as citizens increasingly choose to travel by train. Both for society's gain and the crucial well-being of the planet, our path forward should proceed upon rails.

[9] Author's View

According to the author, we can count on people wanting to travel by train once a modern railway is available.

Passage E: The Divine Sign of Socrates

This passage is associated with 6 questions analyzed in the UWorld MCAT CARS book.

- Subskill 1b. Meaning of Term Question 2
- Subskill 1d. Implicit Claims or Assumptions Question 2
- Subskill 2e. Determining Passage Perspectives Question 3
- Subskill 2e. Determining Passage Perspectives Question 4
- Subskill 2f. Drawing Additional Inferences Question 2
- Subskill 3b. Passage Applications to New Context Question 2

Passage E: The Divine Sign of Socrates

Structural Annotations

Passage Structure: 5 Annotations

[P1] From his bare feet to his bald pate, the potentially shapeshifting figure of Socrates found in the literary tradition that arose after his controversial trial and death presents an intriguing array of oddities and unorthodoxies. Most conspicuously, his unshod and shabby sartorial state flaunted poverty at a time when the city of Athens had become obsessed with wealth and its trappings. Yet the philosopher's peculiar appearance was but a hint of the strange new calling he embraced. Inspired perhaps by the famous Delphic dictum "Know thyself," he embarked on a mission devoted to finding truth through dialogue. In what struck some as a dangerous new method of inquiry, he subjected nearly everyone he encountered to intense cross-examination, mercilessly exposing the ignorance of his interlocutors. Moreover, in a culture that still put stock in magic, the highly charismatic, entertaining, and at times infuriating Socrates appeared to be a sorcerer bewitching the aristocratic young men of Athens who followed him fanatically about the agora.

Paragraph 1 Summary

Socrates was a strange figure whose odd behavior, constant questioning, and bewitching influence over the youth caused some to view him with alarm.

[P2] By all credible accounts, this exceedingly eccentric, self-styled radical truth-seeker had more than a whiff of the uncanny about him. As Socrates himself explains in Plato's *Republic*, he was both blessed and burdened with a supernatural phenomenon in the form of a *daimonion* or inner spirit that always guided him: "This began when I was a child. It is a voice, and whenever it speaks, it turns me away from something I am about to do, but it never encourages me to do anything." An overtly rational thinker, Socrates nonetheless considered these warnings—or, in James Miller's words, "the audible interdictions he experienced as irresistible"—to be infallible. Such oracular injunctions were highly anomalous as tutelary spirits were thought to assume a more nuanced presence. Some scholars have dismissed Socrates' recurring sign as a hallucination or psychological aberration. Others have conjectured that the internal voice might be attributable to the cataleptic or trancelike episodes from which the philosopher purportedly suffered. Indeed, as Miller notes, "Socrates was

Paragraph 2 Summary

A particularly odd feature of Socrates was the supposed inner voice that he claimed guided him. Such a voice did not fit with typical storied for the abstracted states that overtook him"; not infrequently, his companions would see him stop in his tracks and stand still for hours, completely lost in thought.

Athenian belief, and some modern scholars attribute it to psychological causes. However, Socrates viewed it as an infallible spirit that gave him warnings.

[P3] As Socrates further insisted, it was only the protestations of this apotreptic voice that held him back from entering the political arena. Even so, its personal admonitions could not spare him persecution. Despite the political amnesty extended by the resurgent democracy that succeeded the interim pro-Spartan oligarchy, the thinker's notoriety and ambiguous allegiances aroused suspicions. In 399 BCE, Socrates was brought before the court on trumped-up charges of impiety; these included [1] willfully neglecting the traditional divinities, flagrantly introducing new gods to the city, and wittingly corrupting the youth. Athenian society recognized no division between religious and civic duties, and capricious gods demanded constant appeasement through sacrifices and rituals. Consequently, belief in a purely private deity—particularly a wholly benevolent deity conveying unequivocal messages—was inadmissible. Worse, as Socrates' own testimony revealed, he honored this personal god's authority above even the laws of the city. Hence, the philosopher's *daimonion* loomed over his indictment, conviction, and sentencing.

[1] Connected Ideas

Here we can see a thread connecting the first three paragraphs. Socrates seemed to "bewitch" the young men of Athens, and he spoke of a strange spirit that guided him. Thus, he was accused of corrupting the youth and introducing new gods to the city.

Paragraph 3 Summary

Socrates' *daimonion* was used as an excuse to bring him to trial on politically motivated charges. Nevertheless, Socrates admitted he saw his inner spirit as a higher authority than the city's laws. Thus, the *daimonion* contributed to his conviction for impiety and corrupting the youth.

[P4] Nevertheless, in his defense speech as reconstructed by Plato in the *Apology*, Socrates maintained confidence in the protective nature and prophetic powers of his inner monitor. He never questioned its affirmatory silence toward his predicament, remarking, "The divine faculty would surely have opposed me had I been going to evil and not to good." Thus, Socrates acknowledged that his *daimonion* had its reasons, however inscrutable. Variously described as malcontent and martyr, public nuisance and prophet, laughingstock and hero, the mercurial Athenian, like the sign that guided him, was difficult to fathom yet impossible to ignore.

Paragraph 4 Summary

Socrates continued to believe his *daimonion* was infallible even after he was convicted and sentenced. Thus, this guiding spirit represented the remarkable strangeness of Socrates' life and character.

The Divine Sign of Socrates ©UWorld

Passage E: The Divine Sign of Socrates

Content Annotations

Passage Observations: 12 Annotations

Potential Pitfalls: 1 Annotation

[P1] From his bare feet to his bald pate, [1] the potentially shapeshifting figure of Socrates found in the literary tradition that arose after his controversial trial and death presents an intriguing array of oddities and unorthodoxies. Most conspicuously, his unshod and shabby sartorial state flaunted poverty at a time when the city of Athens had become obsessed with wealth and its trappings. [2] Yet the philosopher's peculiar appearance was but a hint of the strange new calling he embraced. Inspired perhaps by the famous Delphic dictum "Know thyself," he embarked on a mission devoted to finding truth through dialogue. In what struck some as a dangerous new method of inquiry, he subjected nearly everyone he encountered to intense cross-examination, mercilessly exposing the ignorance of his interlocutors. Moreover, [3] in a culture that still put stock in magic, the highly charismatic, entertaining, and at times infuriating Socrates appeared to be a sorcerer bewitching the aristocratic young men of Athens who followed him fanatically about the agora.

[P2] By all credible accounts, [4] this exceedingly eccentric, self-styled radical truth-seeker had more than a whiff of the uncanny about him. As Socrates himself explains in Plato's *Republic*, he was both blessed and burdened with a supernatural phenomenon in the form of a *daimonion* or inner spirit that always guided him: [5] "This began when I was a child. It is a voice, and whenever it speaks, it turns me away from something I am about to do, but it never encourages me to do anything." An overtly rational thinker, Socrates nonetheless considered these warnings—or, in James Miller's words, "the audible interdictions he experienced as irresistible"—to be infallible. Such oracular injunctions were [6] highly anomalous as tutelary spirits were thought to assume a more nuanced presence. Some scholars have dismissed Socrates' recurring sign as a hallucination or psychological aberration. Others have conjectured that the internal voice might be attributable to [1] the cataleptic or trancelike episodes from which the

[1] Author's View

The author portrays Socrates as mysterious and strange.

[2] Claim

Socrates' behavior was even more disconcerting than his appearance, as he habitually questioned everyone he met in an attempt to discover truth.

[3] Claim

Socrates drew people to himself easily, but some found his behavior and sway over the youth to be a cause for anger or concern.

[4] Focus

The author moves from an overview of Socrates' strangeness to discussing a specific oddity in depth. Thus, Socrates' *daimonion* is likely the main subject of the passage. (This inference is also suggested by the passage's title.)

[5] Outside Source

Both Socrates himself and James Miller are quoted to support the passage description of Socrates' *daimonion*.

[6] Claim

Although Socrates saw his *daimonion*'s warnings as both clear and infallible, this view diverged from typical Athenian belief about spirits.

[1] Potential Pitfall: Misreading

The author affirms that Socrates experienced trancelike episodes but does not otherwise philosopher purportedly suffered. Indeed, as Miller notes, "Socrates was storied for the abstracted states that overtook him"; not infrequently, his companions would see him stop in his tracks and stand still for hours, completely lost in thought.

express agreement or disagreement with the scholars mentioned. Thus, we cannot necessarily infer the author's own view about Socrates' *daimonion*.

[P3] [7] As Socrates further insisted, it was only the protestations of this apotreptic voice that held him back from entering the political arena. Even so, its personal admonitions could not spare him persecution. Despite the political amnesty extended by the resurgent democracy that succeeded the interim pro-Spartan oligarchy, [8] the thinker's notoriety and ambiguous allegiances aroused suspicions. In 399 BCE, Socrates was brought before the court on trumped-up charges of impiety; these included willfully neglecting the traditional divinities, flagrantly introducing new gods to the city, and wittingly corrupting the youth. [9] Athenian society recognized no division between religious and civic duties, and capricious gods demanded constant appeasement through sacrifices and rituals. Consequently, belief in a purely private deity—particularly a wholly benevolent deity conveying unequivocal messages—was inadmissible. [10] Worse, as Socrates' own testimony revealed, he honored this personal god's authority above even the laws of the city. Hence, the philosopher's *daimonion* loomed over his indictment, conviction, and sentencing.

[7] Claim

The voice warned Socrates not to get involved in politics. Despite following this advice, Socrates was eventually persecuted anyway.

[8] Author's View

According to the author, the charges against Socrates were unjust and politically motivated.

[9] Likely Question Topic

Here we have specific information about why Socrates and his *daimonion* seemed troubling to Athenian society. It would not be surprising to see a question asking about the reasons for Socrates' trial or why his actions were viewed as disruptive.

[10] Emphasis

There were many reasons why the Athenians viewed Socrates' *daimonion* with suspicion, but especially concerning was that Socrates respected it more than he did the city laws.

[P4] Nevertheless, in his defense speech as reconstructed by Plato in the *Apology*, [11] Socrates maintained confidence in the protective nature and prophetic powers of his inner monitor. He never questioned its affirmatory silence toward his predicament, remarking, "The divine faculty would surely have opposed me had I been going to evil and not to good." [12] Thus, Socrates acknowledged that his *daimonion* had its reasons, however inscrutable. Variously described as malcontent and martyr, public nuisance and prophet, laughingstock and hero, the mercurial Athenian, like the sign that guided him, was difficult to fathom yet impossible to ignore.

[11] Claim

Socrates had faith in his *daimonion* even after his conviction, believing that things must have turned out for the best.

[12] Focus

The main purpose of the passage seems to be to describe Socrates' *daimonion* and its role in his unusual life.

The Divine Sign of Socrates ©UWorld

Passage F: The Victorian Internet

This passage is associated with 5 questions analyzed in the UWorld MCAT CARS book.

- Subskill 1c. Direct Passage Claims Question 1
- Subskill 1d. Implicit Claims or Assumptions Question 4
- Subskill 1e. Identifying Passage Perspectives Question 2
- Subskill 1f. Further Implications of Passage Claims Question 2
- Subskill 2b. Function of Passage Claim Question 1

Passage F: The Victorian Internet

Structural Annotations

Passage Structure: 8 Annotations

[P1] Lasting roughly from 1820 to 1914, the Victorian Era is often defined by its many distinctive sociological conditions, including industrialization, urbanization, railroad travel, imperialism, territorial expansionism, and the frictions sparked by Darwinism and democratic reform. However, descriptors that may not come as readily to mind are "the Information Age" and "the Age of Communication"; nor would we likely associate this era with something called the "highway of thought," which emanated from the electric telegraph—an invention that, in retrospect, could be renamed "the Victorian Internet." Yet the rise of the electric telegraph in the mid-1800s constituted the greatest revolution in communication since the invention of the printing press in the fifteenth century and until the launch of the World Wide Web at the end of the twentieth century.

Paragraph 1 Summary

The Victorian Era is not usually thought of as an "Age of Communication" but it should be; the invention of the electric telegraph was a revolutionary event in human history.

[P2] Historically, messages could be conveyed only as fast as a person could travel from one location to another. However, by the end of the eighteenth century, the Chappe brothers had constructed a rudimentary optical telegraph on a hilltop tower in France. Using the large, jointed arms attached to the roof, operators could form various configurations to communicate a message to a similar tower farther away, which would then relay the message to a third tower, and so on, in a long-distance chain. Nevertheless, even with telescopes, visibility severely hampered the efficacy of this semaphore system. After countless attempts by scientists to use electricity to transmit messages, innovators in both Britain and America, including the painter and polymath Samuel F.B. Morse, worked at harnessing electromagnetic forces to send communications via cable. But while in Britain the technology was initially reserved for railway signaling, in the U.S. it culminated in the transmission of a message along a 40-mile telegraphic wire from Washington, D.C., to Baltimore in 1844. Using a binary code of short and long electrical impulses, or "dots and dashes," Morse dispatched the words: "What hath God wrought?"

Paragraph 2 Summary

An earlier optical messaging system had been used in France, but it was limited and impractical. By the mid-eighteenth century, British and American inventors like Morse succeeded in sending messages by electrical wire.

[P3] Within 20 years, telegraph cables crisscrossed the continental U.S. and much of Europe. The telegraph office became ubiquitous, and [1] telegrams—often bouncing from one office to another like emails tossed from server to server—reached ever more remote destinations. Following some spectacular failures, a durable cable was stretched across the floor of the Atlantic Ocean, enabling Europe and the U.S. to exchange messages within minutes. In the 1870s, the British Empire connected London with outposts in India and Australia. While individual telegram speed remained relatively constant, an endless deluge of information began to pour through the wires. There was, as Tom Standage observes, "an irreversible acceleration in the pace of business life," reflecting how "telegraphy and commerce thrived in a virtuous circle." It was as though "rapid long-distance communication had effectively obliterated time and space," begetting the phenomenon known as "globalization."

[1] Connected Ideas

This comparison extends the analogy presented in Paragraph 1 between the telegraph and the modern internet.

Paragraph 3 Summary

Telegraphic communication increased rapidly both domestically and intercontinentally. Communication and business mutually thrived, giving rise to globalization.

[P4] A new type of skilled worker, the telegrapher, was born. He or she belonged to a vast, online community, whose semi-anonymous members shared a unique intermediary role as well as a language of dots and dashes—[2] vaguely prefiguring the exchange of bits and bytes along modern computer networks. Like today's online communities, these telegraphers fostered their own subculture: jokes and anecdotes flew over the wires; some operators invented systems to play games, and occasional romances blossomed. Regrettably, hackers, scammers, and shady entrepreneurs also frequented the byways of this early internet.

[2] Connected Ideas

The passage draws a further analogy between the "Victorian Internet" and the modern internet.

Paragraph 4 Summary

Telegraphers were an important new type of worker, and they took part in online communities that were similar in many ways to today's internet.

[P5] Predictably, newer forms of the original telegraphic technology—first the telephone and, much later, the fax machine—eventually encroached. The last telegram departed from India in 2013, but [3] the twenty-first century has arguably seen a revival of this communication form in the text message. As Standage claims, texting has not only "reincarnate[d] a defunct nineteenth-century technology" but reinforced "the democratization of telecommunications" inaugurated by the miraculous Victorian Internet.

[3] Connected Ideas

The passage again describes how the telegraph was like the modern internet, here directly comparing telegrams with text messages. This extended comparison is clearly an overarching theme of the passage.

Paragraph 5 Summary

Although there are no more telegrams now, text messages are arguably a modern incarnation of them. Thus, the communication revolution begun by the telegraph continues today.

The Victorian Internet ©UWorld

Passage F: The Victorian Internet

Content Annotations

Passage Observations: 7 Annotations
Potential Pitfalls: 1 Annotation

[P1] Lasting roughly from 1820 to 1914, the Victorian Era is often defined by its many distinctive sociological conditions, including industrialization, urbanization, railroad travel, imperialism, territorial expansionism, and the frictions sparked by Darwinism and democratic reform. [1] However, descriptors that may not come as readily to mind are "the Information Age" and "the Age of Communication"; nor would we likely associate this era with something called the "highway of thought," which emanated from the electric telegraph—[2] an invention that, in retrospect, could be renamed "the Victorian Internet." Yet the rise of the electric telegraph in the mid-1800s constituted the greatest revolution in communication since the invention of the printing press in the fifteenth century and until the launch of the World Wide Web at the end of the twentieth century.

[1] Author's View

While the Victorian Era is known for many things, the author believes the significance of the electric telegraph is underacknowledged.

[2] Focus

Given the term "revolution in communication," the passage will likely center on explaining the telegraph's immense societal impact.

[P2] Historically, messages could be conveyed only as fast as a person could travel from one location to another. However, by the end of the eighteenth century, the Chappe brothers had constructed a rudimentary optical telegraph on a hilltop tower in France. Using the large, jointed arms attached to the roof, operators could form various configurations to communicate a message to a similar tower farther away, which would then relay the message to a third tower, and so on, in a long-distance chain. [3] Nevertheless, even with telescopes, visibility severely hampered the efficacy of this semaphore system. [1] After countless attempts by scientists to use electricity to transmit messages, innovators in both Britain and America, including the painter and polymath Samuel F.B. Morse, worked at harnessing electromagnetic forces to send communications via cable. But while in Britain the technology was initially reserved for railway signaling, in the U.S. it culminated in the transmission of a message along a 40-mile telegraphic wire from Washington, D.C., to Baltimore in 1844. Using a binary code of short and long electrical impulses, or "dots and dashes," Morse dispatched the words: "What hath God wrought?"

[3] Claim

Previous attempts at long-distance communication were still limited by dependence on proximity and vision.

[1] Potential Pitfall: Details

These sentences contain nuances that could lead to wrong answers. For instance, the passage says the technology was "initially reserved" for railway signaling in Britain, but it doesn't say for how long, nor whether the U.S. also used it for railways in addition to messages.

[P3] Within 20 years, telegraph cables crisscrossed the continental U.S. and much of Europe. The telegraph office became ubiquitous, and telegrams—often bouncing from one office to another like emails tossed from server to server—reached ever more remote destinations. Following some spectacular failures, a durable cable was stretched across the floor of the Atlantic Ocean, enabling Europe and the U.S. to exchange messages within minutes. In the 1870s, the British Empire connected London with outposts in India and Australia. While individual telegram speed remained relatively constant, [4] an endless deluge of information began to pour through the wires. There was, as Tom Standage observes, "an irreversible acceleration in the pace of business life," reflecting how "telegraphy and commerce thrived in a virtuous circle." It was as though [5] "rapid long-distance communication had effectively obliterated time and space," begetting the phenomenon known as "globalization."

[4] Outside Source

Standage's quote describes a mutually reinforcing relationship between telegraphy and commerce in which each contributed to the rapid growth of the other.

[5] Likely Question Topic

Given how dramatically the electric telegraph changed the world, a question could easily ask about its effects or how life differed before and after its introduction.

[P4] A new type of skilled worker, the telegrapher, was born. He or she belonged to a vast, online community, whose semi-anonymous members shared a unique intermediary role as well as a language of dots and dashes—vaguely prefiguring the exchange of bits and bytes along modern computer networks. Like today's online communities, these telegraphers fostered their own subculture: jokes and anecdotes flew over the wires; some operators invented systems to play games, and occasional romances blossomed. Regrettably, hackers, scammers, and shady entrepreneurs also frequented the byways of this early internet.

[P5] Predictably, newer forms of the original telegraphic technology—first the telephone and, much later, the fax machine—eventually encroached. The last telegram departed from India in 2013, but the twenty-first century has arguably seen a revival of this communication form in the text message. [6] As Standage claims, texting has not only "reincarnate[d] a defunct nineteenth-century technology" but reinforced "the democratization of telecommunications" inaugurated by [7] the miraculous Victorian Internet.

[6] Outside Source

The author again cites Standage's words to support claims about the impact and importance of the telegraph.

[7] Author's View

The author calls the telegraph "miraculous." Along with the previous description of it as the "greatest revolution in communication" between the printing press and the modern internet, the author appears to view the telegraph as both extremely influential and positive.

Passage G: Food Costs and Disease

This passage is associated with 6 questions analyzed in the UWorld MCAT CARS book.

- Subskill 1c. Direct Passage Claims Question 2
- Subskill 1e. Identifying Passage Perspectives Question 4
- Subskill 2b. Function of Passage Claim Question 3
- Subskill 3b. Passage Applications to New Context Question 1
- Subskill 3d. External Scenario Support or Challenge Question 2
- Subskill 3e. Applying Passage Perspectives Question 3

Passage G: Food Costs and Disease

Structural Annotations

Passage Structure: 10 Annotations

[P1] Because frequent consumption of unhealthy foods is strongly linked with cardiometabolic diseases, one way for governments to combat those afflictions may be to modify the eating habits of the general public. Applying economic incentives or disincentives to various types of foods could potentially alter people's diets, leading to more positive health outcomes.

Paragraph 1 Summary

Unhealthy food is linked to disease incidence, so government interventions on food prices could help prevent disease.

[P2] Utilizing national data from 2012 regarding food consumption, health, and economic status, Peñalvo et al. concluded that such price adjustments would help to prevent deaths related to cardiometabolic diseases. According to their analysis, increasing the prices of unhealthy foods such as processed meats and sugary sodas by 10%, while reducing the prices of healthy foods such as fruit and vegetables by 10%, would prevent an estimated 3.4% of yearly deaths in the U.S. Changing prices by 30% would have an even stronger effect, preventing an estimated 9.2% of yearly deaths. This data comports with that found in other countries, such as "previous modeling studies in South Africa and India, where a 20% SSB [sugar-sweetened beverage] tax was estimated to reduce diabetes prevalence by 4% over 20 years." The effects of price adjustments would be most pronounced on persons of lower socioeconomic status, as the researchers "found an overall 18.2% higher price-responsiveness for low versus high SES [socioeconomic status] groups."

Paragraph 2 Summary

Multiple data sets support the previous hypothesis: governments can prevent disease by making healthy food more affordable and unhealthy food less affordable.

[P3] [1] This differential effect based on socioeconomic status contributes to concerns about such interventions, however. In *Harvard Public Health Review*, Kates and Hayward ask: "Well intentioned though they may be, at what point do these taxes overstep government influence on an individual's right to autonomy in decision-making? On whom does the increased financial burden of this taxation fall?" They note that taxes on sugar-sweetened beverages, for instance, "are likely to have a greater impact on low-income individuals...because individuals in those settings are more likely to be beholden to cost when making decisions about food."

[1] Topic Shift

Up until this point, the passage has focused on whether adjusting food prices is *effective* in preventing disease. But now the author is discussing whether such interventions *ought* to be implemented.

Paragraph 3 Summary

Some worry that government price interventions may violate consumers' autonomy or unfairly burden the poor.

[P4] [2] However, "well intentioned though they may be," the worries that Kates and Hayward express are to some extent misguided. In particular, the idea that taxing unhealthy foods would burden those least able to afford it misses the point. Although the increased taxes would affect anyone who continued to purchase the items despite the higher prices, the goal of raising prices on unhealthy foods is precisely to dissuade people from buying them. As Kates and Hayward themselves remark, "Those in low-income environments may also be the largest consumers of obesogenic foods and therefore most likely to benefit from such a lifestyle change indirectly posed by SSB taxes." As the goal of the taxes is to promote those lifestyle changes, the financial burden objection is a non-starter.

[P5] Given this recognition, the question regarding autonomy constitutes a more substantial issue. [3] Nevertheless, that concern also rests on a dubious assumption, as people's autonomy is not necessarily respected in the current situation either. The fact that those of lower socioeconomic status are more likely to have poorer diets suggests that such persons' food choices are the result of financial constraint, not fully autonomous, rational deliberation. Hence, by making healthy foods more affordable relative to unhealthy ones, government intervention might actually *facilitate* autonomous choices rather than hindering them.

[P6] [4] On the other hand, suppose that the disproportionate consumption of unhealthy foods—and associated higher incidence of disease—among certain groups is not the result of financial hardship but rather the result of those persons' perceived self-interest. If so, that would suggest that members of these groups are being encouraged to persist in harmful dietary habits for the sake of corporate profits. In that case, violating autonomy for the sake of health may be permissible, as that practice would be morally preferable to the present system of corporate exploitation.

Food Costs and Disease ©UWorld

[2] Addressing Opposing View

The author argues against the objection that price interventions might unethically burden low-income individuals.

Paragraph 4 Summary

The author argues that Kates and Hayward's concern about burdening the poor misses the point of the price interventions.

[3] Addressing Opposing View

The author argues against the objection that price interventions might unethically violate people's autonomy.

Paragraph 5 Summary

The author agrees that autonomy is a legitimate concern, but suggests that price interventions might not violate it after all.

[4] Connected Ideas

The author previously argued that price interventions are justified if poor diet is caused by financial constraint. Now they consider whether those interventions might still be justified even if poor diet is *not* caused by financial constraint.

Paragraph 6 Summary

The author argues that autonomy is violated whether or not there are price interventions, so it's better to violate it for improved health than for corporate profits.

Passage G: Food Costs and Disease

Content Annotations

Passage Observations: 8 Annotations
Potential Pitfalls: 3 Annotations

[P1] Because frequent consumption of unhealthy foods is strongly linked with cardiometabolic diseases, one way for governments to combat those afflictions may be to modify the eating habits of the general public. [1] Applying economic incentives or disincentives to various types of foods could potentially alter people's diets, leading to more positive health outcomes.

[1] Focus

The passage seems to be exploring two relationships: how people's diets affect their incidence of disease; and how government price interventions affect people's diets.

[P2] Utilizing national data from 2012 regarding food consumption, health, and economic status, [2] Peñalvo et al. concluded that such price adjustments would help to prevent deaths related to cardiometabolic diseases. According to their analysis, increasing the prices of unhealthy foods such as processed meats and sugary sodas by 10%, while reducing the prices of healthy foods such as fruit and vegetables by 10%, would prevent an estimated 3.4% of yearly deaths in the U.S. Changing prices by 30% would have an even stronger effect, preventing an estimated 9.2% of yearly deaths. [3] This data comports with that found in other countries, such as "previous modeling studies in South Africa and India, where a 20% SSB [sugar-sweetened beverage] tax was estimated to reduce diabetes prevalence by 4% over 20 years." [4] The effects of price adjustments would be most pronounced on persons of lower socioeconomic status, as the researchers "found an overall 18.2% higher price-responsiveness for low versus high SES [socioeconomic status] groups."

[2] Outside Source

Peñalvo et al.'s research supports the claim that adjustments to food prices have a positive effect on disease prevention.

[3] Outside Source

Research from other countries supports the same relationship between food prices and disease prevention.

[4] Likely Question Topic

Socioeconomic status adds another variable to the relationship between price adjustments and health outcomes. Thus, it is likely to play a role in questions about price interventions.

[P3] This differential effect based on socioeconomic status contributes to concerns about such interventions, however. [5] In *Harvard Public Health Review*, Kates and Hayward ask: "Well intentioned though they may be, at what point do these taxes overstep government influence on an individual's right to autonomy in decision-making? On whom does the increased financial burden of this taxation fall?" [1] They note that taxes on sugar-sweetened beverages, for instance, "are likely to have a greater impact on low-income individuals…because individuals

[5] Outside Source

Kates and Hayward are cited to raise ethical concerns about food price interventions.

[1] Potential Pitfall: Missed Distinction

Kates and Hayward are not disputing the data about price interventions, only whether such interventions are ethical. Thus, thinking of them as just "disagreeing with price interventions" could make a wrong answer appear correct.

in those settings are more likely to be beholden to cost when making decisions about food."

[P4] However, "well intentioned though they may be," the worries that Kates and Hayward express are to some extent misguided. In particular, the idea that taxing unhealthy foods would burden those least able to afford it misses the point. Although the increased taxes would affect anyone who continued to purchase the items despite the higher prices, the goal of raising prices on unhealthy foods is precisely to dissuade people from buying them. [2] As Kates and Hayward themselves remark, "Those in low-income environments may also be the largest consumers of obesogenic foods and therefore most likely to benefit from such a lifestyle change indirectly posed by SSB taxes." [6] As the goal of the taxes is to promote those lifestyle changes, the financial burden objection is a non-starter.

[2] Potential Pitfall: Missed Distinction

There is no disagreement about the facts of price interventions, only their ethical implications. Thus, thinking of the author as just "disagreeing with Kates and Hayward" could make a wrong answer appear correct.

[6] Author's View

According to the author, price interventions do disproportionately impact the poor, but this is an intended feature of the interventions rather than a drawback.

[P5] [7] Given this recognition, the question regarding autonomy constitutes a more substantial issue. Nevertheless, that concern also rests on a dubious assumption, as people's autonomy is not necessarily respected in the current situation either. The fact that those of lower socioeconomic status are more likely to have poorer diets suggests that such persons' food choices are the result of financial constraint, not fully autonomous, rational deliberation. [3] Hence, by making healthy foods more affordable relative to unhealthy ones, government intervention might actually *facilitate* autonomous choices rather than hindering them.

[7] Author's View

The author believes that concerns about autonomy are significant, unlike the previous objection to price interventions (about the impact on low-income people).

[3] Potential Pitfall: Missed Distinction

Note that the author is not arguing that it is okay to violate autonomy here. Rather, they state that price interventions might increase autonomy instead. Overlooking this distinction could lead to mistakes about the author's position.

[P6] On the other hand, suppose that the disproportionate consumption of unhealthy foods—and associated higher incidence of disease—among certain groups is not the result of financial hardship but rather the result of those persons' perceived self-interest. If so, that would suggest that members of these groups are being encouraged to persist in harmful dietary habits for the sake of corporate profits. [8] In that case, violating autonomy for the sake of health may be permissible, as that practice would be morally preferable to the present system of corporate exploitation.

[8] Focus

The author agrees that autonomy is valuable but argues that violating it for health is more ethical than allowing corporate exploitation. Especially given the second half of the passage, this conclusion suggests that the author's main purpose is to defend the use of government price interventions to improve public health.

Food Costs and Disease ©UWorld

Passage H: For Whom the Bell Toils

This passage is associated with 5 questions analyzed in the UWorld MCAT CARS book.

- Subskill 1c. Direct Passage Claims Question 3
- Subskill 1f. Further Implications of Passage Claims Question 3
- Subskill 2f. Drawing Additional Inferences Question 1
- Subskill 3c. New Claim Support or Challenge Question 2
- Subskill 3f. Additional Conclusions From New Information Question 1

Passage H: For Whom the Bell Toils

Structural Annotations

Passage Structure: 6 Annotations

[P1] In nineteenth-century America, most people dismissed the notion that someone might assassinate the president. The presumption was based not only on ethics but practicality: a president's term is inherently limited, and an unpopular one could be voted out of office. Therefore, it was reasoned, there would be no need to consider removal through violence. This belief persisted even after the shocking murder of Abraham Lincoln in 1865, which was viewed as an aberration. Thus it was that on July 2, 1881, Charles Guiteau could simply walk up to President James A. Garfield and shoot him in broad daylight. As Richard Menke portrays events, "Guiteau was in fact a madman who had come to identify with a disgruntled wing of the Republican Party after his deranged fantasies of winning a post from the new administration had come to nothing." Believing that God had told him to kill the president, Guiteau thought this act would garner fame for his religious ideas and thereby help to usher in the Apocalypse.

Paragraph 1 Summary

People used to believe that no one would consider assassinating a president, so Charles Guiteau could easily shoot President Garfield in 1881. Guiteau was motivated both by his political frustration and his radical religious views.

[P2] [1] In an interesting parallel, Garfield had felt a sense of divine purpose for his own life after surviving a near-drowning as a young man. Unlike Guiteau's fanatical ravings, however, Garfield's vision worked to the betterment of himself and the world. Candice Millard describes his ascent from extreme poverty to incredible excellence in college, where "by his second year...they made him a professor of literature, mathematics, and ancient languages." Garfield would go on to join the Union Army, where he attained the rank of major general and argued that black soldiers should receive the same pay as their white compatriots. While serving in the war he was nominated for the House of Representatives but accepted the seat only after President Lincoln declared that the country had more need of him as a congressman than as a general. The reluctant politician would later himself become president under similar circumstances, after multiple factions of a deadlocked Republican convention unexpectedly nominated him instead of their original candidates in 1880. An honest man who opposed corruption within the party, Garfield strove both to heal the

[1] Topic Shift

Here the passage switches from discussing the assassination to providing biographical details about Garfield.

Paragraph 2 Summary

The author contrasts Guiteau's delusion with Garfield's aspiration. In addition to extreme academic achievement, Garfield served in the Union Army and argued for more racial fairness.

fractures of the Civil War and to uphold the aims for which it was fought, until "the equal sunlight of liberty shall shine upon every man, black or white, in the Union."

He became both a congressman and eventually president almost by accident, using those offices to fight for honesty and equality.

[P3] [2] Although Guiteau's bullet would ultimately dim this light for Garfield, the president actually survived the initial attack and for a time appeared headed for recovery. Tragically, however, the hubris shown by his main physician, Dr. Willard Bliss, would lead instead to weeks of prolonged suffering. None of the doctors who examined Garfield were able to locate the bullet, and its lingering presence—along with the unwashed hands of the doctors who probed for it—led to an infection. As the president's condition worsened, inventor Alexander Graham Bell attempted to adapt his patented telephone technology to locate foreign metal in the human body. Inspired by speculation that the bullet's electromagnetic properties might be detectable, Bell used his newly developed "Induction Balance" device to listen for the sounds of electrical interference he hoped would isolate the site of the bullet.

[2] Topic Shift

The passage now returns to the original topic of Garfield's assassination.

Paragraph 3 Summary

Garfield seemed like he might recover, but Dr. Bliss' pride caused him to keep suffering instead. Because doctors were unable to locate the bullet, Alexander Graham Bell tried to make a device to find it.

[P4] Unfortunately, Bell's searches were unsuccessful. Like Garfield's doctors, he had been looking for the bullet in the wrong area. Menke asserts that Bell's efforts "would probably have fallen short" regardless. However, other historians suggest that Dr. Bliss, unwilling to consider challenges to his original assessment, prevented Bell from more thoroughly searching the president's body. Certainly, Bliss ignored the advice and protestations of other physicians, even as Garfield continued to decline. With death imminent, Garfield asked to be taken to his seaside cottage, where he died on the 19th of September.

Paragraph 4 Summary

Bell failed to find the bullet, possibly in part due to Dr. Bliss' interference. Dr. Bliss also refused to listen to other doctors despite Garfield's continued decline, and Garfield died under his care.

For Whom the Bell Toils ©UWorld

Passage H: For Whom the Bell Toils

Content Annotations

Passage Observations: 11 Annotations
Potential Pitfalls: 1 Annotation

[P1] [1] In nineteenth-century America, most people dismissed the notion that someone might assassinate the president. The presumption was based not only on ethics but practicality: a president's term is inherently limited, and an unpopular one could be voted out of office. Therefore, it was reasoned, there would be no need to consider removal through violence. This belief persisted even after the shocking murder of Abraham Lincoln in 1865, which was viewed as an aberration. [2] Thus it was that on July 2, 1881, Charles Guiteau could simply walk up to President James A. Garfield and shoot him in broad daylight. [3] As Richard Menke portrays events, "Guiteau was in fact a madman who had come to identify with a disgruntled wing of the Republican Party after his deranged fantasies of winning a post from the new administration had come to nothing." Believing that God had told him to kill the president, Guiteau thought this act would garner fame for his religious ideas and thereby help to usher in the Apocalypse.

[P2] [4] In an interesting parallel, Garfield had felt a sense of divine purpose for his own life after surviving a near-drowning as a young man. Unlike Guiteau's fanatical ravings, however, Garfield's vision worked to the betterment of himself and the world. [5] Candice Millard describes his ascent from extreme poverty to incredible excellence in college, where "by his second year…they made him a professor of literature, mathematics, and ancient languages." Garfield would go on to join the Union Army, where he attained the rank of major general and argued that black soldiers should receive the same pay as their white compatriots. [6] While serving in the war he was nominated for the House of Representatives but accepted the seat only after President Lincoln declared that the country had more need of him as a congressman than as a general. The reluctant politician would later himself become president under similar circumstances, after multiple factions of a deadlocked Republican convention unexpectedly nominated him instead of their original candidates in 1880. An honest man who opposed corruption within the party, [7] Garfield strove

[1] Claim

Part of the historical context surrounding Garfield's assassination was the widespread belief that no one would consider such a thing.

[2] Focus

Based on this statement and how the preceding information leads up to it, the passage will likely center on describing the circumstances of Garfield's assassination.

[3] Outside Source

The author cites Menke's words to explain Guiteau's motivations: political revenge and religious delusions.

[4] Author's View

The author depicts Garfield and Guiteau as mirror images of each other: one embodied intelligence and moral courage, the other lunacy and destructive violence.

[5] Outside Source

Millard's words illustrate Garfield's extraordinary mind—he was "promoted" from college student to professor.

[6] Claim

Garfield became both a congressman and eventually the president despite not intentionally running for either office.

[7] Likely Question Topic

This is a dense paragraph in which the author highlights several unusual aspects of Garfield's both to heal the fractures of the Civil War and to uphold the aims for which it was fought, until "the equal sunlight of liberty shall shine upon every man, black or white, in the Union."

[P3] Although Guiteau's bullet would ultimately dim this light for Garfield, [8] the president actually survived the initial attack and for a time appeared headed for recovery. Tragically, however, the hubris shown by his main physician, Dr. Willard Bliss, would lead instead to weeks of prolonged suffering. [1] None of the doctors who examined Garfield were able to locate the bullet, and its lingering presence—along with the unwashed hands of the doctors who probed for it—led to an infection. As the president's condition worsened, [9] inventor Alexander Graham Bell attempted to adapt his patented telephone technology to locate foreign metal in the human body. Inspired by speculation that the bullet's electromagnetic properties might be detectable, Bell used his newly developed "Induction Balance" device to listen for the sounds of electrical interference he hoped would isolate the site of the bullet.

[P4] Unfortunately, Bell's searches were unsuccessful. Like Garfield's doctors, he had been looking for the bullet in the wrong area. [10] Menke asserts that Bell's efforts "would probably have fallen short" regardless. However, other historians suggest that Dr. Bliss, unwilling to consider challenges to his original assessment, prevented Bell from more thoroughly searching the president's body. [11] Certainly, Bliss ignored the advice and protestations of other physicians, even as Garfield continued to decline. With death imminent, Garfield asked to be taken to his seaside cottage, where he died on the 19th of September.

For Whom the Bell Toils ©UWorld

life. It would be reasonable to expect a question about Garfield's unconventional political career or his impressive accomplishments and character.

[8] Author's View

The author implies that Dr. Bliss was partially to blame for Garfield's suffering and death.

[1] Potential Pitfall: Misreading

Since the author just said Dr. Bliss' hubris caused Garfield to suffer, it would be natural to expect this next sentence to provide details. But it doesn't actually tell us anything more about Dr. Bliss or his hubris; it explains only why Garfield's wound became infected. Thus, the proximity of the sentences and our natural expectations could lead to mistakes about the point of this claim.

[9] Claim

Alexander Graham Bell attempted to develop a device that could locate the bullet and thereby help save Garfield.

[10] Outside Source

Based on the disagreement between Menke and other historians, it is unclear whether Bell might have found the bullet without Bliss' interference.

[11] Elaboration

The previous paragraph stated that Dr. Bliss' hubris caused Garfield to keep suffering instead of recovering. However, it is only now that the author specifies: Bliss might have kept Bell from finding the bullet, and he definitely refused to listen to other doctors' dissenting opinions even though Garfield was dying.

Passage I: Meaning: Readers or Authors?

This passage is associated with 5 questions analyzed in the UWorld MCAT CARS book.

- Subskill 1c. Direct Passage Claims Question 5
- Subskill 2f. Drawing Additional Inferences Question 3
- Subskill 3d. External Scenario Support or Challenge Question 4
- Subskill 3e. Applying Passage Perspectives Question 2
- Subskill 3g. Identifying Analogies Question 2

Passage I: Meaning: Readers or Authors?

Structural Annotations

Passage Structure: 9 Annotations

[P1] Of late it has become popular among linguists and literary theorists to assert that a work's *meaning* depends upon the individual reader. It is readers, we are told, not authors, who create meaning, by interacting with a text rather than simply receiving it. Thus, a reader transcends the aims of the author, producing their own reading of the text. Indeed, on this line of thinking, even to speak of "the" text is to commit a conceptual error; every text is in fact many texts, a plurality of interpretations that resist comparative evaluation. This view is nonsense. That many otherwise sensible scholars should be attracted to it can perhaps be readily explained, but we should first delineate why the theory goes so far astray....

Paragraph 1 Summary

The passage author strongly opposes the view of literary interpretation that a work's meaning depends on readers rather than authors, and that those readers' different interpretations resist evaluation. The passage will go on to explain why that view should be considered nonsense.

[P2] The absurdity of the view can be demonstrated by a practical analogy. Suppose Smith is conveying his ideas to Jones in conversation (the particular topic is of no consequence). Afterward, we discover that the men differ in their accounts of what Smith had expressed. At this point, Jones may decide that he misunderstood Smith, or perhaps that Smith was unclear. A more complex supposition might be that Smith misused some key term, so his words did not fully match his intentions. Any of these possibilities would reasonably describe why Jones and Smith possessed different opinions about what Smith had said.

Paragraph 2 Summary

The passage author describes a situation where one man misunderstands another, as part of an analogy meant to show that the reader-based view of meaning is absurd.

[P3] [3] What Jones may *not* justifiably conclude is that his own interpretation is what Smith *really meant*. He may not, in effect, say: "Yes, I admit that Smith honestly claims to have been saying something different, but I have formed my own equally correct understanding." Someone who made such an assertion would be suspected of making a joke; if he proved to be serious, we could only conclude that he was deeply confused or else being deliberately quarrelsome. For in questions about what Smith meant, it is surely Smith whose answer must be accepted.... [T]his is not a matter of *agreeing* with a speaker; Jones might judge Smith's ideas to be wrong, unfounded, etc. But whether Smith's ideas are right or wrong is a different

[3] Connected Ideas

The passage author began describing the analogy in the previous paragraph, but it is here that the argument about it actually starts.

Paragraph 3 Summary

In the passage author's analogy, a listener knows he misunderstood what a speaker meant but still claims his own interpretation is "equally correct." The listener's position is portrayed as matter from what those ideas *are*. On that count, Smith must be the authority.

[P4] [5] However, this observation is in no way changed if Smith's ideas are written rather than spoken—sent by letter, for instance. Regardless of any interpretation Jones may produce, the letter's true meaning is whatever Smith intended to convey. Likewise it is, then, with a book, poem, or whatsoever object of literature a scholar (or ordinary reader) encounters. The writing down of ideas does not magically imbue them with malleability or render their content amorphous. From the loftiest tomes of Shakespeare or Milton to the lowliest of yellowed paperbacks, authors produce works with a particular message in mind. [6] It is readers' task to discern that message, not to superimpose their own volitional perspectives.

[P5] To think otherwise is to undermine the foundation of literary scholarship. For what is the purpose of such scholarship, if not to seek understanding of an author's creation? One examines the text, taking note of style, historical context, allusions to other works, and other factors, in addition, of course, to the surface sense of the words themselves. [7] If such an enterprise is to be reasonable, it must presume the existence of standards for success: accuracy and inaccuracy, depth or shallowness of analysis, grounds for preferring one interpretation to another. Different readers may come to different conclusions about a text, it is true. But to excise authorial intent from the evaluation of those conclusions does a disservice both to individual works and to literary study as a discipline.

Meaning: Readers or Authors? ©UWorld

ridiculous, since it is clearly the speaker himself who knows what he really meant.

[5] Connected Ideas

The passage author's analogy and associated argument extend from the second to the fourth paragraphs, forming the bulk of the passage.

[6] Connected Ideas

This claim refers back to the first paragraph, again rejecting the idea that the reader "transcends the aims of the author."

Paragraph 4 Summary

The passage author completes their argument by analogy: Just as it is absurd to think a listener determines the meaning of what a speaker says, it is also absurd to think a reader determines the meaning of what an author writes.

[7] Connected Ideas

The author again refers back to the first paragraph, rejecting the idea that different interpretations "resist comparative evaluation."

Paragraph 5 Summary

The passage author concludes that literary meaning is determined by an author's intention. To accept the alternative, reader-based theory of meaning is a misguided and harmful approach to literary scholarship.

Passage I: Meaning: Readers or Authors?

Content Annotations

Passage Observations: 9 Annotations
Potential Pitfalls: 1 Annotation

[P1] [1] Of late it has become popular among linguists and literary theorists to assert that a work's *meaning* depends upon the individual reader. It is readers, we are told, not authors, who create meaning, by interacting with a text rather than simply receiving it. Thus, a reader transcends the aims of the author, producing their own reading of the text. Indeed, on this line of thinking, even to speak of "the" text is to commit a conceptual error; every text is in fact many texts, a plurality of interpretations that resist comparative evaluation. [2] This view is nonsense. That many otherwise sensible scholars should be attracted to it can perhaps be readily explained, but [3] we should first delineate why the theory goes so far astray....

[1] Claim

According to one popular view, the meaning of a literary work is highly subjective and varies based on each individual reader.

[2] Author's View

The passage author not only disagrees with the view being described but expresses a strongly negative attitude toward it.

[3] Focus

The purpose of the passage seems to be to argue against the reader-based view of meaning just described.

[P2] [4] The absurdity of the view can be demonstrated by a practical analogy. Suppose Smith is conveying his ideas to Jones in conversation (the particular topic is of no consequence). Afterward, we discover that the men differ in their accounts of what Smith had expressed. At this point, Jones may decide that he misunderstood Smith, or perhaps that Smith was unclear. A more complex supposition might be that Smith misused some key term, so his words did not fully match his intentions. Any of these possibilities would reasonably describe why Jones and Smith possessed different opinions about what Smith had said.

[4] Author's View

The passage author again criticizes the view being discussed, referring to its "absurdity."

[P3] What Jones may *not* justifiably conclude is that his own interpretation is what Smith *really meant*. He may not, in effect, say: "Yes, I admit that Smith honestly claims to have been saying something different, but I have formed my own equally correct understanding." Someone who made such an assertion would be suspected of making a joke; if he proved to be serious, we could only conclude that he was deeply confused or else being deliberately quarrelsome. [5] For in questions about what Smith meant, it is surely Smith whose answer must be accepted.... [6] [T]his is not a matter of *agreeing* with a speaker; Jones might judge Smith's ideas to be wrong,

[5] Author's View

The passage author seems to take this point for granted (it is "surely" correct). Thus, this claim—that it is the speaker, not the listener, who knows what the speaker really meant—is likely a key point in the passage author's overall argument.

[6] Elaboration

The passage author distinguishes between knowing what a speaker means and whether unfounded, etc. But whether Smith's ideas are right or wrong is a different matter from what those ideas *are*. On that count, Smith must be the authority.

that speaker is correct. People always know what their own ideas are, even if those ideas are misguided or wrong.

[P4] However, this observation is in no way changed if Smith's ideas are written rather than spoken—sent by letter, for instance. Regardless of any interpretation Jones may produce, the letter's true meaning is whatever Smith intended to convey. [7] Likewise it is, then, with a book, poem, or whatsoever object of literature a scholar (or ordinary reader) encounters. [8] The writing down of ideas does not magically imbue them with malleability or render their content amorphous. From the loftiest tomes of Shakespeare or Milton to the lowliest of yellowed paperbacks, authors produce works with a particular message in mind. It is readers' task to discern that message, not to superimpose their own volitional perspectives.

[7] Elaboration

According to the passage author, their conclusion about meaning applies to all forms of writing and all types of readers.

[8] Author's View

The use of the word "magically" further characterizes the reader-based view of meaning as nonsensical.

[P5] To think otherwise is to undermine the foundation of literary scholarship. For what is the purpose of such scholarship, if not to seek understanding of an author's creation? [9] One examines the text, taking note of style, historical context, allusions to other works, and other factors, in addition, of course, to the surface sense of the words themselves. If such an enterprise is to be reasonable, it must presume the existence of standards for success: accuracy and inaccuracy, depth or shallowness of analysis, grounds for preferring one interpretation to another. [1] Different readers may come to different conclusions about a text, it is true. But to excise authorial intent from the evaluation of those conclusions does a disservice both to individual works and to literary study as a discipline.

[9] Author's View

Here the passage author describes how they think literary interpretation ought to work (rather than the reader-based view of meaning they have been arguing against).

[1] Potential Pitfall: Missed Distinction

Note that the passage author never disputes the idea that interpretations of a work may differ; they simply reject the idea that such interpretations are equally correct. Missing this point could lead to mistakes about the passage author's position.

Meaning: Readers or Authors? ©UWorld

Passage J: The Inkblots

This passage is associated with 5 questions analyzed in the UWorld MCAT CARS book.

- Subskill 1d. Implicit Claims or Assumptions Question 3
- Subskill 1e. Identifying Passage Perspectives Question 3
- Subskill 2c. Extent of Passage Evidence Question 2
- Subskill 2e. Determining Passage Perspectives Question 2
- Subskill 3a. Exemplar Scenario for Passage Claims Question 2

Passage J: The Inkblots

Structural Annotations

Passage Structure: 7 Annotations

[P1] For almost 100 years now, the psychological evaluation known as the Rorschach Inkblot Test has engendered much controversy, including skepticism about its value, questions about its scoring, and, especially, criticism of its interpretive methods as too subjective. Thus, the Rorschach test, which emerged from the same early twentieth-century zeitgeist that produced Einstein's physics, Freudian psychoanalysis, and abstract art, seems one of modernity's most misbegotten children. Destined never to be completely accepted or discredited, the test remains a perennial outlier in its field. Nevertheless, the inkblots' mystery and aesthetic appeal have caused them to be indelibly printed on our cultural fabric.

Paragraph 1 Summary

The Rorschach test has long been controversial, especially because it is criticized as too subjective. Nevertheless, it continues to have great cultural impact.

[P2] The now iconic inkblots were introduced to the world by Swiss psychiatrist Hermann Rorschach in his 1921 book *Psychodiagnostics*. As both director of the Herisau Asylum in Switzerland and an amateur artist, Rorschach was uniquely positioned to wed the new practice of psychoanalysis to the budding phenomenon of abstract art. For instance, reading Freud's work on dream symbolism prompted him to recall his childhood passion for a game based on inkblot art called *Klecksographie*. He was also cognizant that in a recently published dissertation, his colleague Szymon Hens had used inkblots to try to probe the imagination of research subjects; moreover, a few years earlier, the French psychologist and father of intelligence testing Alfred Binet had used them to measure creativity.

Paragraph 2 Summary

Rorschach had many influences in creating his test: his love of art, Freud's dream symbolism, and other scientists' use of inkblots in research

[P3] Motivated by these developments, the Herisau director decided to revisit that childhood pastime that had awakened his curiosity about how visual information is processed. In particular, he wondered why different people saw different things in the same image. Traditionally, psychoanalysts had relied on language for insights; however, as biographer Damion Searls reports, Rorschach's theories would exemplify the principle that "who we are is a matter less of what we say than of what we see." Indeed, through a process of perception termed pareidolia, the mind projects meaning onto images,

Paragraph 3 Summary

Most psychoanalysis had been based on language, but the inkblot test was instead based detecting in them familiar objects or shapes. Consequently, what a person sees in an image reveals more about that person than about the image itself.

on visual interpretation. Each person's mind may impart a different meaning to the same image.

[P4] Rorschach experimented with countless inkblots, eventually selecting ten—five black on white, two also featuring some red, and three pastel-colored—to use with research subjects. For these perfectly symmetrical images—each of which he was said to have "meticulously designed to be as ambiguous and 'conflicted' as possible"—the primary question was always "What do *you* see?" Rorschach was especially careful to note how much attention individuals paid to various components of each inkblot (such as form, color, and a sense of movement) and whether they concentrated on details or the whole image. Having observed that his patients with schizophrenia gave distinctly different responses from the control group, Rorschach envisioned his experiment as a diagnostic tool for the disease. Nevertheless, he resisted the notion that its results could be used to assess personality. In fact, until his untimely death from a ruptured appendix in 1922, Rorschach referred to his project as an "interpretive form experiment" rather than a test. [1] Ironically, however, by the 1960s, the Rorschach Inkblot Test was known chiefly as a personality assessment and had become the most frequently administered projective personality test in the US.

[1] Connected Ideas

The test became primarily used in a way that Rorschach never intended. Thus, this fact may relate to the inkblots' controversial aspects as mentioned in Paragraph 1.

Paragraph 4 Summary

The inkblots are ambiguous and elicit varying interpretations, and Rorschach envisioned them as helping to diagnose schizophrenia. Ironically, however, they became primarily used as a general personality assessment, something that Rorschach himself resisted.

[P5] [2] Rorschach's test has survived nearly incessant scrutiny, including a 2013 comprehensive study of all Rorschach test data and repeated revisions to its scoring, yet doubts about its validity and reliability persist. Much like the inkblots themselves—which tantalize us with the possibility of divulging the secrets of who we are and how we see the world—the test has (for better or worse) defied attempts to fix its meaning. Thus, what has been called "the twentieth century's most visionary synthesis of art and science" stands tempered by harsh criticism.

[2] Topic Shift

Although most of the passage has discussed the Rorschach test's development, the author now returns to the controversy mentioned in the first paragraph.

Paragraph 5 Summary

The Rorschach test remains controversial, continuing both to hold great appeal and to receive strong criticism.

Passage J: The Inkblots

Content Annotations

Passage Observations: 11 Annotations
Potential Pitfalls: 1 Annotation

[P1] For almost 100 years now, the psychological evaluation known as [1] the Rorschach Inkblot Test has engendered much controversy, including skepticism about its value, questions about its scoring, and, especially, criticism of its interpretive methods as too subjective. Thus, the Rorschach test, which emerged from the same early twentieth-century zeitgeist that produced Einstein's physics, Freudian psychoanalysis, and abstract art, seems one of modernity's most misbegotten children. [2] Destined never to be completely accepted or discredited, the test remains a perennial outlier in its field. Nevertheless, the inkblots' mystery and aesthetic appeal have caused them to be indelibly printed on our cultural fabric.

[1] Claim

The Rorschach test has been considered controversial for several reasons, especially its subjective nature.

[2] Focus

The introduction seemed to be emphasizing the controversy around the Rorschach test, but it ends by talking about the inkblots' effect on our culture. Thus, the passage is likely a more descriptive account of the inkblots, rather than an argument about their controversial aspects.

[P2] The now iconic inkblots were introduced to the world by Swiss psychiatrist Hermann Rorschach in his 1921 book *Psychodiagnostics*. As both director of the Herisau Asylum in Switzerland and an amateur artist, [3] Rorschach was uniquely positioned to wed the new practice of psychoanalysis to the budding phenomenon of abstract art. For instance, [4] reading Freud's work on dream symbolism prompted him to recall his childhood passion for a game based on inkblot art called *Klecksographie*. He was also cognizant that in a recently published dissertation, his colleague Szymon Hens had used inkblots to try to probe the imagination of research subjects; moreover, a few years earlier, the French psychologist and father of intelligence testing Alfred Binet had used them to measure creativity.

[3] Claim

The Rorschach test combines psychoanalysis and art, which was one of Rorschach's interests.

[4] Likely Question Topic

This list of specific individuals and works could easily generate questions about Rorschach's inspirations and the scholarship that led up to the test's development.

[P3] Motivated by these developments, the Herisau director decided to revisit that childhood pastime that had awakened his curiosity about how visual information is processed. In particular, he wondered why different people saw different things in the same image. [5] Traditionally, psychoanalysts had relied on language for insights; however, as biographer Damion Searls reports, Rorschach's theories would exemplify the principle that "who we are is a matter less of what we say than of what we see." [6] Indeed, through a process of perception

[5] Outside Source

Searls is quoted to explain how Rorschach's approach was atypical. While most psychoanalysis had been based on language, the Rorschach test instead focused on the interpretation of visual stimuli.

[6] Likely Question Topic

By connecting the Rorschach test to a specific psychological phenomenon, this information termed pareidolia, the mind projects meaning onto images, detecting in them familiar objects or shapes. Consequently, what a person sees in an image reveals more about that person than about the image itself.

could generate questions about pareidolia, the functioning of the Rorschach test, or both.

[P4] [7] Rorschach experimented with countless inkblots, eventually selecting ten—five black on white, two also featuring some red, and three pastel-colored—to use with research subjects. For these perfectly symmetrical images—each of which he was said to have [8] "meticulously designed to be as ambiguous and 'conflicted' as possible"—the primary question was always "What do *you* see?" Rorschach was especially careful to note how much attention individuals paid to various components of each inkblot (such as form, color, and a sense of movement) and whether they concentrated on details or the whole image. Having observed that his [9] patients with schizophrenia gave distinctly different responses from the control group, Rorschach envisioned his experiment as a diagnostic tool for the disease. Nevertheless, he resisted the notion that its results could be used to assess personality. In fact, until his untimely death from a ruptured appendix in 1922, Rorschach referred to his project as an "interpretive form experiment" rather than a test. [1] Ironically, however, by the 1960s, the Rorschach Inkblot Test was known chiefly as a personality assessment and had become the most frequently administered projective personality test in the US.

[7] Elaboration

This is the first (and only) physical description of the inkblots in the passage.

[8] Elaboration

The test functions through the inkblots' ambiguity: people project their own meaning onto part or all of a given image.

[9] Claim

Rorschach thought his inkblots could help diagnose schizophrenia, but he specifically resisted the idea that they represented a test for assessing personality. The term "resisted" implies that others did believe the inkblots should be used in that way, revealing a conflict over their purpose.

[1] Potential Pitfall: Details

This paragraph presents a fair amount of overlapping information, including distinctions concerning how the Rorschach test functioned, how Rorschach envisioned its purpose, and how it has mostly been used. It would be easy to mix up some of this information when approaching questions.

[P5] Rorschach's test has survived nearly incessant scrutiny, including a 2013 comprehensive study of all Rorschach test data and repeated revisions to its scoring, yet doubts about its validity and reliability persist. [10] Much like the inkblots themselves—which tantalize us with the possibility of divulging the secrets of who we are and how we see the world—the test has (for better or worse) defied attempts to fix its meaning. [11] Thus, what has been called "the twentieth century's most visionary synthesis of art and science" stands tempered by harsh criticism.

[10] Author's View

Just as the inkblots are known for their ambiguity, the author characterizes the Rorschach test as ambiguous overall.

[11] Focus

As in the first paragraph, the passage does not argue for or against the validity or reliability of the Rorschach test, but simply affirms that such criticism persists. Thus, the purpose of the passage as a whole seems to be to describe the test's development and the controversy that surrounds it.

Passage K: Lengthening the School Day

This passage is associated with 7 questions analyzed in the UWorld MCAT CARS book.

- Subskill 1f. Further Implications of Passage Claims Question 1
- Subskill 2a. Logical Relationships Within Passage Question 3
- Subskill 2a. Logical Relationships Within Passage Question 4
- Subskill 2c. Extent of Passage Evidence Question 3
- Subskill 3c. New Claim Support or Challenge Question 1
- Subskill 3d. External Scenario Support or Challenge Question 3
- Subskill 3e. Applying Passage Perspectives Question 1

Passage K: Lengthening the School Day

Structural Annotations

Passage Structure: 8 Annotations

[P1] There may be reasons to reject the idea of lengthening the school day. None of them, however, are *good* reasons. Rather, the supposed demerits of such a proposal fall easily in the face of its numerous financial and social benefits for families.

Paragraph 1 Summary

The author will argue that lengthening the school day has many benefits, and that any supposed drawbacks can be dismissed.

[P2] The greatest of these benefits lies in reducing the need for childcare. It is a curious fact of American life that the adult's work schedule and the child's school schedule are misaligned. Children rise with the sun to head to classes, only to be sent home again hours before parents return from their jobs. In a society where, more often than not, both parents work, this discordance creates the need for an expensive arrangement to fill the gap in families' routines. For instance, studies show that in 2016, childcare costs accounted for 9.5 to 17.5 percent of median family income, depending on the state. Today, 40 percent of families nationwide spend over 15 percent of their income on childcare. Transportation to and from care sites only adds to that expense.

Paragraph 2 Summary

Based on the cited statistics about childcare expenses, reducing these would be the greatest benefit of a lengthened school day.

[P3] An additional advantage of an extended school day would be to allow for greater diversity and depth in curricula. Schools across the country have increasingly cut instruction in arts, music, and physical education (as well as recess) in order to meet objectives in reading and math. While this unfortunate state of affairs can be partially blamed on overzealous attention to standardized tests, it points to the larger deleterious trend of narrowing students' instruction. With a longer school day, such eliminated subjects can be restored, enriching students with a more well-rounded education.

Paragraph 3 Summary

Another benefit of a longer school day is enabling schools to offer a broader range of subjects and programs. In the author's view, this change would enrich students and improve their overall education.

[P4] [1] To this proposal, however, critics may object that the added time would impose strain on educators. Can we truly ask schoolteachers—already among the most overworked individuals in society—to endure even more hours in the classroom? The answer is that a lengthened school day need not distress teachers nor add to their already cumbersome workload. By providing for additional areas of study in the arts and humanities, the extension would give schools cause to hire new, perhaps specialized, faculty to offer these courses.

[1] Addressing Opposing View

The author raises an objection in order to address it: the worry that lengthening the school day would overburden teachers.

Paragraph 4 Summary

Lengthening the school day would not burden educators as some critics suggest, because additional teachers would be hired and some of

Moreover, the time could also be allocated to sports, academic clubs, and other extracurricular activities.

the added time could go to extracurricular activities.

[P5] [2] However, this point speaks to another objection, namely, the cost of adjusting the school day. Whether through paying current teachers more or hiring new ones, implementing such a proposal would entail a significant financial expenditure. There are at least two responses to this line of thinking. First is that this increase in the cost of schooling would be offset and likely surpassed by the aforementioned savings in childcare. Thus, while it is true that schools would require greater funding (likely necessitating higher property taxes), parents would ultimately pay the same or less overall, with greater educational opportunities for their children and fewer transportational burdens. Second is that schools should be better funded regardless. Recently, some schools—especially those in rural areas—have even reduced school weeks to only four days as a cost-saving measure. It is beyond dispute that schools across the board both need and deserve a radically increased investment from citizens. Lengthening the school day is simply one manifestation of how such funding should be utilized.

[2] Addressing Opposing View

A second objection is addressed: that lengthening the school day would be expensive.

Paragraph 5 Summary

The author agrees that lengthening the school day comes at a financial cost. However, they argue that these costs will be offset by savings in childcare and that schools should be better funded anyway.

[P6] With this one change, states can coordinate the lives of parents and children, reduce the need for costly childcare, and expand curricular offerings. These worthy and desirable aims provide a clear justification for extending the school day.

Paragraph 6 Summary

Having explored the benefits of and potential objections to lengthening the school day (as promised in the introduction), the author concludes that the case has been made in favor of this proposal.

Passage K: Lengthening the School Day

Content Annotations

Passage Observations: 11 Annotations

[P1] There may be reasons to reject the idea of lengthening the school day. None of them, however, are *good* reasons. Rather, [1] the supposed demerits of such a proposal fall easily in the face of its numerous financial and social benefits for families.

[1] Focus

The purpose of the passage seems to be to defend the claim that school days should be made longer.

[P2] [2] The greatest of these benefits lies in reducing the need for childcare. It is a curious fact of American life that the adult's work schedule and the child's school schedule are misaligned. Children rise with the sun to head to classes, only to be sent home again hours before parents return from their jobs. In a society where, more often than not, both parents work, this discordance creates the need for an expensive arrangement to fill the gap in families' routines. [3] For instance, studies show that in 2016, childcare costs accounted for 9.5 to 17.5 percent of median family income, depending on the state. Today, 40 percent of families nationwide spend over 15 percent of their income on childcare. Transportation to and from care sites only adds to that expense.

[2] Author's View

According to the author, addressing childcare needs is the most important reason for lengthening the school day.

[3] Outside Source

The author cites data supporting the claims about the expense of childcare.

[P3] [4] An additional advantage of an extended school day would be to allow for greater diversity and depth in curricula. Schools across the country have increasingly cut instruction in arts, music, and physical education (as well as recess) in order to meet objectives in reading and math. [5] While this unfortunate state of affairs can be partially blamed on overzealous attention to standardized tests, it points to the larger deleterious trend of narrowing students' instruction. With a longer school day, such eliminated subjects can be restored, enriching students with a more well-rounded education.

[4] Author's View

The author's second reason for favoring a longer school day is increasing the variety of subjects that can be taught.

[5] Author's View

The author believes education should be broader and include more instruction in the arts and humanities, and that standardized tests have been overemphasized.

[P4] To this proposal, however, critics may object that the added time would impose strain on educators. [6] Can we truly ask schoolteachers—already among the most overworked individuals in society—to endure even more hours in the classroom? The answer is that a lengthened school day need not distress teachers nor add to their already cumbersome workload. By providing for additional areas of study in the arts and humanities, the extension would give schools cause to hire new, perhaps specialized, faculty to offer these courses.

[6] Author's View

The author seems to take for granted that teachers are overworked but argues that lengthening the school day would not add to this problem.

Moreover, the time could also be allocated to sports, academic clubs, and other extracurricular activities.

[P5] However, this point speaks to another objection, namely, the cost of adjusting the school day. Whether through paying current teachers more or hiring new ones, implementing such a proposal would entail a significant financial expenditure. [7] There are at least two responses to this line of thinking. [8] First is that this increase in the cost of schooling would be offset and likely surpassed by the aforementioned savings in childcare. Thus, while it is true that schools would require greater funding (likely necessitating higher property taxes), parents would ultimately pay the same or less overall, with greater educational opportunities for their children and fewer transportational burdens. [9] Second is that schools should be better funded regardless. Recently, some schools—especially those in rural areas—have even reduced school weeks to only four days as a cost-saving measure. [10] It is beyond dispute that schools across the board both need and deserve a radically increased investment from citizens. Lengthening the school day is simply one manifestation of how such funding should be utilized.

[P6] With this one change, states can [11] coordinate the lives of parents and children, reduce the need for costly childcare, and expand curricular offerings. These worthy and desirable aims provide a clear justification for extending the school day.

Lengthening the School Day ©UWorld

[7] Likely Question Topic

The phrase "at least two responses" suggests that what follows will be a list of reasons for the author's position. Such information could easily be the basis for questions about the author's view or the evidence for passage claims.

[8] Claim

The author argues that savings in childcare would help offset the added expense of a longer school day.

[9] Likely Question Topic

While the author's first response was about the effects of lengthening the school day, this second response is based on an independent claim that schools should receive more funding in general. This difference could be relevant to distinguishing a correct from an incorrect answer.

[10] Author's View

Claiming the point "is beyond dispute" suggests the author is taking it for granted and assuming the reader's agreement.

[11] Author's View

Here the author gives a concise summation of the benefits of lengthening the school day.

Passage L: When Defense Is Indefensible

This passage is associated with 5 questions analyzed in the UWorld MCAT CARS book.

- Subskill 2d. Connecting Claims With Evidence Question 2
- Subskill 3b. Passage Applications to New Context Question 3
- Subskill 3c. New Claim Support or Challenge Question 3
- Subskill 3f. Additional Conclusions From New Information Question 3
- Subskill 3g. Identifying Analogies Question 1

Passage L: When Defense Is Indefensible

Structural Annotations

Passage Structure: 8 Annotations

[P1] Suppose a prosecutor is considering whether to bring a case to trial. He is not sure that the suspect is guilty—in fact, based on the evidence, it's more likely that the suspect is *not* guilty. Nevertheless, he feels confident he can secure a guilty verdict. His powers of persuasion are considerable, and there's a good chance he could trick a jury into believing the evidence is strong instead of weak. In addition, the case is high profile and could be very lucrative; winning would likely lead to a substantial raise or promotion. He decides to charge the suspect, and ultimately succeeds in persuading the jury to convict.

Paragraph 1 Summary

In a hypothetical scenario, a prosecutor tricks a jury into convicting a man who is likely innocent so the prosecutor himself will personally gain.

[P2] Looking at this situation, most of us would easily judge the prosecutor as extremely unethical. His conduct is outrageous and wrong—he clearly acted with corrupt intent, perpetrating injustice in order to profit financially. Why is it not shocking, then, that we tolerate the mirror image of this behavior from defense attorneys? For they engage in the same outrageous conduct, only on the other side. Paid handsomely to represent even the vilest of clients, they apply their oratorical prowess to manipulating jury perception, keeping the guilty free and unpunished in exchange for money and status. To the extent that this behavior takes place, are some defense attorneys as unethical as our hypothetical prosecutor?

Paragraph 2 Summary

According to the author, some defense attorneys act just like the hypothetical prosecutor, which suggests they are just as unethical. So, the author questions why we don't seem to judge such attorneys in the same way.

[P3] It is worth distinguishing two senses of the word "ethical" here. For there are standards of *professional* ethics to which any attorney must conform, including standards particular to the defense. Most relevant to our purposes, a lawyer is obligated to provide their client with a "zealous defense." In other words, once an attorney takes on a client, they are ethically bound to promote that client's rights, interests, or innocence—in fact, *not* to do so would be *unethical*. Thus, one might try to suggest that this obligation undermines the claim that some defense attorneys act unethically.

Paragraph 3 Summary

The word "ethical" can mean two different things, and in particular there are *professional* ethics that defense attorneys must follow. However, the author hints that this fact still might not justify defense attorneys' behavior.

[P4] [1] However, meeting that professional standard is not the same as being ethical in the general sense of the word. The standard depends on the condition: *once an attorney takes on a client*. With the exception of court-appointed attorneys or public defenders, who are assigned to provide representation to those who would otherwise lack it, an attorney is never required to represent a defendant. Therefore, meeting one's obligations as a defense attorney does not necessarily make one ethical, because the choice to accept a specific case (and thus to incur those obligations in the first place) may itself be an unethical act.

[P5] Moreover, the role of court-appointed attorneys is to help protect the rights of citizens who cannot secure their own representation, usually for financial reasons. Although preserving those rights is necessary to uphold justice, this situation highlights how wealth and class contribute to *injustice*. While some defendants possess the means to hire top-level private lawyers, others must depend on public servants—frequently less experienced lawyers from overloaded, understaffed agencies. As a result, the rich are more likely to escape conviction even when they are guilty, and the poor are more likely to be convicted even when they are innocent.

[P6] It is doubtful that private defense attorneys could be somehow forbidden from representing guilty clients. Hence, the needed reforms to the system must come from individual attorneys committing to work for the right reasons. For those who strive to ensure citizens' rights, or who truly believe their clients are innocent, providing a defense is a noble undertaking. But for those whose overriding motivation is greed, [2] that legally "zealous defense" is ethically indefensible.

When Defense Is Indefensible ©UWorld

[1] Connected Ideas

The author now discusses the second half of the distinction made in the previous paragraph. There is a general sense of the word "ethical" that may differ from professional ethics, and upholding one is not the same as upholding the other.

Paragraph 4 Summary

The author distinguishes between public attorneys who cannot choose their clients and private attorneys who can. Since the private attorneys are not obligated to defend certain clients, their choice to do so may be unethical.

Paragraph 5 Summary

The author argues that the differences between private and public attorneys ensure that rich people are more likely to escape justice while poor people are more likely to be wrongly convicted.

[2] Connected Ideas

The author revisits defense attorneys' professional obligation to provide clients with a "zealous defense." As the author first suggested in Paragraph 4, private defense attorneys can still be unethical for choosing to take on this obligation in the first place.

Paragraph 6 Summary

Attorneys who truly believe their clients are innocent or who serve to ensure people's rights are ethical. But private defense attorneys who defend the guilty for money are still unethical. The system can be reformed only if such attorneys choose to give up that behavior.

Passage L: When Defense Is Indefensible

Content Annotations

Passage Observations: 9 Annotations

[P1] Suppose a prosecutor is considering whether to bring a case to trial. He is not sure that the suspect is guilty—in fact, based on the evidence, it's more likely that the suspect is *not* guilty. Nevertheless, he feels confident he can secure a guilty verdict. His powers of persuasion are considerable, and there's a good chance he could trick a jury into believing the evidence is strong instead of weak. In addition, the case is high profile and could be very lucrative; winning would likely lead to a substantial raise or promotion. He decides to charge the suspect, and ultimately succeeds in persuading the jury to convict.

[P2] [1] Looking at this situation, most of us would easily judge the prosecutor as extremely unethical. His conduct is outrageous and wrong—he clearly acted with corrupt intent, perpetrating injustice in order to profit financially. [2] Why is it not shocking, then, that we tolerate the mirror image of this behavior from defense attorneys? For they engage in the same outrageous conduct, only on the other side. Paid handsomely to represent even the vilest of clients, they apply their oratorical prowess to manipulating jury perception, keeping the guilty free and unpunished in exchange for money and status. To the extent that this behavior takes place, [3] are some defense attorneys as unethical as our hypothetical prosecutor?

[P3] [4] It is worth distinguishing two senses of the word "ethical" here. For there are standards of *professional* ethics to which any attorney must conform, including standards particular to the defense. Most relevant to our purposes, a lawyer is obligated to provide their client with a "zealous defense." In other words, once an attorney takes on a client, they are ethically bound to promote that client's rights, interests, or innocence—in fact, *not* to do so would be *unethical*. [5] Thus, one might try to suggest that this obligation undermines the claim that some defense attorneys act unethically.

[1] Author's View

The author believes there is wide agreement that the prosecutor's conduct is unethical.

[2] Author's View

The author argues that the behavior of many defense attorneys is just like that of the hypothetical prosecutor. Therefore, we ought to view these defense attorneys as being equally unethical.

[3] Focus

This question about defense attorneys seems to be the author's real point in talking about the hypothetical prosecutor. Thus, the passage's main idea likely concerns whether defense attorneys are unethical.

[4] Claim

The author distinguishes between two meanings of "ethical." In particular, the notion of *professional* ethics may affect how defense attorneys should be judged.

[5] Author's View

The author raises the possibility that defense attorneys' behavior might not be unethical after all because of their professional obligations. However, the phrase "one might try to suggest" indicates that the author may not find that position convincing.

[P4] However, meeting that professional standard is not the same as being ethical in the general sense of the word. The standard depends on the condition: *once an attorney takes on a client*. [6] With the exception of court-appointed attorneys or public defenders, who are assigned to provide representation to those who would otherwise lack it, an attorney is never required to represent a defendant. Therefore, [7] meeting one's obligations as a defense attorney does not necessarily make one ethical, because the choice to accept a specific case (and thus to incur those obligations in the first place) may itself be an unethical act.

[6] Claim

The author makes another distinction: public attorneys must represent whomever they are assigned to, while private attorneys can choose their clients.

[7] Author's View

Building on the distinction between public and private attorneys, the author argues that *the choice to represent* a client may be unethical regardless of lawyers' professional obligations.

[P5] Moreover, the role of court-appointed attorneys is to help protect the rights of citizens who cannot secure their own representation, usually for financial reasons. Although preserving those rights is necessary to uphold justice, [8] this situation highlights how wealth and class contribute to *injustice*. While some defendants possess the means to hire top-level private lawyers, others must depend on public servants—frequently less experienced lawyers from overloaded, understaffed agencies. As a result, the rich are more likely to escape conviction even when they are guilty, and the poor are more likely to be convicted even when they are innocent.

[8] Claim

The public vs. private attorney distinction can also contribute to class-based injustice. The rich hire top-level private lawyers and so are more likely to go free, while the poor depend on less-experienced public attorneys and so are more likely to be convicted.

[P6] [9] It is doubtful that private defense attorneys could be somehow forbidden from representing guilty clients. Hence, the needed reforms to the system must come from individual attorneys committing to work for the right reasons. For those who strive to ensure citizens' rights, or who truly believe their clients are innocent, providing a defense is a noble undertaking. But for those whose overriding motivation is greed, that legally "zealous defense" is ethically indefensible.

[9] Author's View

According to the author, there is probably no way to force private attorneys to represent only clients they truly believe are innocent. Therefore, the system can be reformed only if attorneys themselves decide to uphold that ethical standard.

The Science of Shrimp Keeping

A Fact-Based Approach to Flourishing Neocaridina & Caridina Shrimp Tanks

Print v1.0

By Shrimply Explained

Dedicated to my wife and mother – two different people, to be clear.

Thank you both for supporting my obsessions.

ISBN 979-8-234-04369-6

Table of Contents

Notes

- In this book, "Shrimp" refers to dwarf shrimp including Neocaridina and soft-water Caridina (Crystals, Tigers, Taiwan Bees) covered in the Shrimp of Interest chapter.
    - "Shrimps" technically refers to multiple shrimp, but we think it sounds silly so "shrimp" is used to refer to one or more shrimp based on the context.
- There are many research papers cited in this book but not all of them cover testing done on dwarf shrimp like Neocaridina or Caridina. These species are a relatively small sub-section of the aquarium hobby so there is not much money to fund research. As such, some conclusions are extrapolated from studies on other species. We look for patterns across multiple research papers and species that are closely related to our Shrimp Of Interest (p5) to find ways to optimize their care with the available science.

> Roman numeral superscript indicates there is a link to a source or product available at www.shrimplyexplained.com/book-links/.

- Information specifically related to Caridina is indicated by the Caridina graphic seen to the right.

- The following chapters include information based on articles from The Shrimp School[i], along with additional photos and sections:
    - Parameters – pH, GH, KH, and TDS
    - Parameter Control – pH, GH, KH
    - Cycling
    - The Best Water for Shrimp Tanks
    - How To Acclimate Shrimp
    - Differences Between Male and Female Shrimp
    - Treating Shrimp Diseases
        - Quarantine Tank Setup
        - Salt Dip Procedure
        - Hydrogen Peroxide Treatment
- Say hi to Shrimply if you haven't been introduced yet! Things get serious when he puts on his lab coat.

Introduction

The purpose of this book is to help understand the "why" of shrimp keeping – the science behind different aspects of our setup and how our actions affect aquatic life in our tank. This background knowledge lets us build the perfect tank to achieve our goal of breeding for profit, breeding for show (yes, there are shrimp shows, just like dog shows), creating art through aquascaping, or massively reducing maintenance time.

Understanding the science underpinning a tank isn't just useful – it's necessary for a shrimp keeper. We need to know what an additive will do to our water parameters; how to perform maintenance safely; what precautions to take when feeding different types of food and why; when and how to treat diseases. We don't want to create extra work for ourselves or potentially harm our shrimp by blindly following advice online, which is why this book has 150+ citations. Our goal is to save you hundreds of hours of research by providing all the science of shrimp keeping in one well-organized book.

There are a lot of scientific concepts in this guide, but don't worry - We keep things shrimple, so nothing is outside of your ability to understand. **If you struggle with a concept, then that's our failure in explaining it**. Please don't hesitate to reach out (contact@shrimplyexplained.com) and ask for clarification. We're happy to help! Doing so also lets us know how to edit the book in the future to reduce confusion for others.

That being said, shrimp keeping is also about having fun! Once a tank is set up properly and we know what signs to look for in a healthy tank, it becomes much more relaxing to sit back and watch the shrimp go about their daily lives. What are they doing over there? How are these ones interacting? Is this one on drugs? Is that one cosplaying as a Skyrim merchant? Who knows!

Enjoy it 😊

Mother?

Are you... me?
Yes, Henrietta... If I was a million times dumber. Lay off the drugs.

Our Responsibility as Pet Owners

Prepare to start on the incredible journey of building a healthy shrimp tank! It is a fantastic opportunity to learn about these awesome creatures and the environment they live in. It isn't all fun and games though. Doing so comes with some risks and responsibilities to be aware of:

First, we are taking care of living creatures. Please treat them with respect and care for them to the best of your ability.

Second, plants and animals from a tank must not enter local aquatic environments, even if they appear to be dead. If some of these species end up in local waterways, they may become invasive and destroy the balance of that ecosystem. This has happened in many places with the introduction of plants like duckweed. Neocaridina are already invasive in areas like Hawaii and may be causing the decline of native species (Andrea D. Shluker, 2003). **Even seemingly dead organisms may recover, or bodies may contaminate the water with microorganisms or diseases** that local populations are not used to.

There may come a time when we must put an unhealthy pet out of its misery because it is an ethical decision (the Humane Euthanasia section [p298]). It is terrible to see an animal suffer and can feel easier to simply release it and let nature take its course but please do not do this. It's difficult to make the decision to end a life and should never be taken lightly but *not doing so can lead to the suffering of many more organisms*. This is a part of responsible pet ownership.

Okay, now that's out of the way, let's learn some background about the shrimp species this book focuses on!

Shrimp Of Interest

Neocaridina AKA "Neos"

Cherry Shrimp (Neocaridina davidi) from Shrimply Explained HQ

Species Name: *Neocaridina davidi* (Formerly *Neocaridina heteropoda*)

Common Breeds: Wild Type, Red Cherry, Bloody Mary, Red Rili, Blue Dream, Green Jade, Orange Sunkist, Pumpkin Orange, Orange Rili, Carbon Rili, Blue Jelly, Blue Velvet, Chocolate, Black Rose, 24k Yellow, Red Onyx, and others (breeders are constantly making new ones)

Species Name: *Neocaridina palmata* (Formerly *Neocaridina zhangjiajiensis*)

Common Breeds: Snowball (or White Pearl), Blue Pearl, Amber Pearl

Overview

Originally found in the rivers of China, Taiwan, Japan, and Korea, Neocaridina (Neos) are now present on at least 6 continents and are the most common variety in the hobby for two reasons (Han C.-C. , Hsu, Fang, & et al., 2019):

- They are resilient over a wide range of *stable* parameters.
- They are easily bred.

Due to these traits, Neos were some of the earliest freshwater shrimp species to be kept in aquariums around the world, although they look very different now than their ancestors from even 20 years ago. When first pulled from the wild, Neos were typically clear with brown pigmentation on their shell (wild type coloration). Years ago, a breeder found shrimp that had more red pigments and selectively bred them with others that had the same trait, eventually leading to the cherry shrimp we see today. Since then, a rainbow of colors has been discovered so just about anyone can find their favorite colored Neos

(except purple – so far. You might get lucky and find it!) In addition, there are various "grades" of each color depending on hue and shell opacity, with the highest grades having the most vibrant and solid colors.

One of the reasons we like this hobby so much is that we never know exactly what the offspring will look like - especially if there is a strange mutation. We could have all red cherry shrimp and then one day a blue-colored baby pops up! With more than one tank, it becomes possible to separate the offspring with characteristics that we like to create a higher-grade line of shrimp, or we can experiment and make our own line. There are actually shrimp shows to compete in using homebred shrimp, if you want to get deeper into the shrimp breeding hobby. This book also covers those shows on p248!

A few other characteristics worth mentioning for Neos:

1. Due to their popularity, high-grade (brightly colored) Neos are common and often available for cheap from local hobbyists, fish stores, or online.
2. Different colors will breed together and produce a variety of offspring. If we're not careful about mixing colors, a shrimp colony could lose its distinctive hue and turn into the original brown (wild type, or WT) Neos within a year or two. Some hobbyists like this variety of offspring.
3. Neos are ornamental shrimp that are typically kept for their color rather than their ability to remove algae. In fact, they have almost no impact on algae at all because, while they may be seen picking at algae, they are actually scraping off biofilm found on the algae's surface. Neocaridina are still good for cleaning up waste though.
4. *N. palmata* and *N. davidi* can breed together.
5. Neocaridina cannot breed with any Caridina shrimp.

Caridina

Tiger Shrimp

*Tangerine Tiger (*Caridina serrata*) from Shrimply Explained HQ*

Species Name: *Caridina serrata*

Common Breeds: Tangerine Tiger, Aurora Blue/Red, Cheetah, and others

Species Name: *Caridina mariae*

Common Breeds: Wild Tigers, Super Tigers, Orange-Eye Blue Tigers, Orange-Eye Black Tigers, Red Tigers, and others

Overview

These are typically the next step up from Neos because tiger shrimp are also ornamental but slightly more challenging to keep. They don't breed quite as quickly or easily as Neos but do breed faster than some of the more difficult Caridina species.

Tiger shrimp can be kept with Neos, but they evolved in more acidic freshwater rivers with fewer minerals and a smaller range of parameters. *C. mariae* typically do better in Neo parameters than *C. serrata* in terms of hardiness and breeding success. That being said, softer, more acidic water matching *C. cantonensis* parameters is still best for both Tiger Shrimp species. **This book covers everything a shrimp keeper needs to know to control parameters and get the best water for each species**.

In exchange for the increased difficulty (and price) compared to Neos, Tiger Caridina have striking black, blue, or red stripes running across their shells, which explains the name. In addition, breeds like the orange-eyed blue tiger have bright orange eyes that really stand out against their shell.

Because they are different species, Neos and Tiger shrimp cannot breed together so there is no risk of diluting the colors and patterns. Tiger shrimp can breed with *C. cantonensis*,

however, leading to breeds like Galaxy and Snowflake Caridina. *C. mariae* carry the traits for strong blue/red stripes and orange eyes, while *Caridina serrata* carry traits for strong backlines (fishbone pattern), faster breeding, and more dots on the head (preferred by breeders). The traits make tiger shrimp extremely important for crossbreeding projects. More on that in the Caridina Shrimp Breeding chapter (p251).

Crystal Shrimp

Crystal Red Shrimp (Caridina Cantonensis) Credit: 神谷純平, CC BY-SA 4.0 via Wikimedia Commons

Species Name: *Caridina cantonensis*

Common Breeds: Crystal Black, Crystal Red

Overview

Crystal shrimp (Crystals) have bright splotches of red or black on their shell (like koi fish) and are of medium difficulty to keep. While they are the same species as Tiger shrimp, Crystals require more acidic--or softer—water and have an even smaller range of acceptable parameters. They are also generally more expensive to keep because **they require active substrates** and **remineralized RO/DI water** to maintain healthy conditions. Once the right conditions are provided, crystals breed relatively easily and can overrun the tank with eye-catching bright colors that stand out against any plant or substrate. This explains why they are the second most popular breed of ornamental shrimp after Neos.

Active Substrate - A term used for substrate that covers the bottom of the tank AND affects water parameters. For example, Fluval Stratum[ii] is an active substrate that lowers pH and KH, while crushed coral is an active substrate that increases pH, GH, and KH.
Remineralized RO/DI water - Reverse Osmosis/Deionized water - Purified water that must have minerals added to it to make it suitable for an aquarium. Remineralized water is often used to avoid potential contaminants found in tap water and ensure healthy water parameters.

As with Tiger shrimp, *C. cantonensis* cannot breed with Neos but can breed with *C. serrata* and *C. mariae*. *C. cantonensis* carry genes for white shells and more opaque shell coloration so many crossbreeding projects start with them. See the Caridina Shrimp Breeding section (p251) for more information.

Taiwan Bee Shrimp

*Blue Bolt (*Caridina Cantonensis*) from Shrimply Explained HQ*

Species Name: *Caridina cantonensis*

Common Breeds: Blue Bolt, Red Bolt, Green Hulk, Black King Kong, Shadow Panda, Panda, Wine Red, and others

Overview

Taiwan Bee shrimp are essentially Crystal shrimp that have mutated or been **selectively bred** for specific colors and patterns. All this selective breeding has led to shrimp that are

more expensive, sensitive, and slow-breeding. Due to their sensitivity, Taiwan bees need an active substrate, remineralized RO/DI water, and a well-established tank that has *very* stable water parameters with particularly low pH. They generally thrive at an even lower pH than crystal shrimp. It can be difficult to set up and care for a tank at such low pH (waste isn't broken down as quickly) which makes them the most difficult breed of Caridina shrimp to care for.

> **Selective Breeding**: The process of breeding animals and selecting desired traits over generations. This process has been used since at least 200 BC to develop crops and livestock with traits like better disease resistance or flavor[iii].

Black Fancy Tiger - Credit: Master Breeder Kerric Peyton

Once we have a bit of experience with shrimp keeping, keeping Taiwan Bees opens up some of the most interesting colorations (like the Fancy Tiger, Galaxy Pinto or Metallic Blue Boa) which are typically created by breeding Taiwan Bee and Tiger shrimp – creating TaiTiBees. These genetic crosses create fascinating patterns - and the Tiger shrimp genetics make the shrimp hardier so even beginners may be able to keep these shrimp in the future. Different types of Caridina are discussed later on (p251).

Now that we're on the same page about what kinds of shrimp we're discussing, let's move onto arguably the most important part of the book – understanding water parameters.

Water Parameters

For humans, air quality is extremely important to our survival. If we inhale smoke-filled air for too long, we develop breathing problems like asthma. If we travel to a high altitude where there's less oxygen, then we struggle to exercise and may feel lightheaded.

Water is to shrimp like air is to humans. Achieving healthy water conditions in a tank is the most critical factor for success in shrimp keeping.

But what does "healthy" water mean? How do we know whether water is "healthy" or not?

That's where water parameters like pH, GH, and KH come into play. A good shrimp keeper needs to know what water parameters matter and how to test them. We cover these first because **all decisions for tank setup hinge on water parameters**. There are eight important parameters to consider in a tank:

- Temperature (p14)
- pH (p17)
- Carbonate Hardness (KH) (p21)
- General Hardness (GH) (p23)
- Ammonia, Nitrite, Nitrate (p30)
- Dissolved Oxygen (DO) (p31)

Honorable Mentions:

- Total Dissolved Solids (TDS) (p32)

The rest of this section covers the Need-To-Know information for these parameters, including their importance, testing, and control methods. Recommended ranges for each type of shrimp are covered at the end of the chapter (p46).

If we get across nothing else in this book, ***please remember that keeping stable parameters is more important than having the perfect parameters***. Many shrimp keepers fail because they constantly fiddle with parameters to try making them perfect and end up doing more harm than good. If shrimp are doing well in a tank with parameters that are "out of range", then it's often best to leave the parameters as is rather than change them. That being said, there are two times when we recommend making changes if the parameters are out of range:

1. If there's a problem (stressed/dead shrimp)
2. If the tank does not have shrimp in it yet

SIDE NOTE: Understanding these parameters is not only important for shrimp keeping - It also matters for our own health. If we measure our tap water and find it's extremely high in nitrate or ammonia or copper, then it might be affecting our health. In this case, purchasing a filtration system may benefit both us and our shrimp.

While we're on the topic of water - please don't fall for the BS about expensive alkaline (high pH) water being healthy for humans to drink. It does not have any proven health benefits[iv].

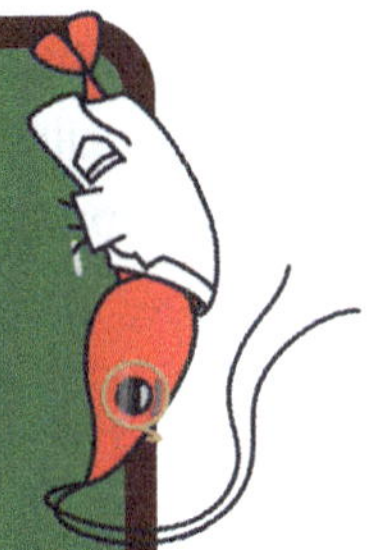

Temperature

Why Temperature Matters

Shrimp are ectothermic creatures, meaning they cannot regulate their body temperature. They haven't evolved the brains to make clothes (yet), so their bodies match the temperature of their environment. This helps shrimp conserve energy but means they are at the whim of their environment, so **tank temperature has a large effect on how their bodies function**.

To dig a little deeper – the rates of all chemical reactions are affected by temperature, with higher temperatures leading to faster reactions and vice versa for lower temperatures. Shrimp digestion, movement, growth rate, reproduction, and every other aspect of their life is at least somewhat dependent on temperature.

Traditionally, we typically hear that cooler temperatures lead to slower growth and breeding, along with a longer lifespan. Warmer temperatures have the opposite effect. According to the research, this is...

Let's take a look at some studies.

According to Tropea et al. 2015, the growth rate of *Neocaridina* kept at different temperatures (24°C [75°F], 28°C [82°F], 32°C [90°F]) varied significantly depending on both temperature and development stage (Tropea C, 2015). For example, the 28°C and 32°C groups both showed faster growth than the shrimp kept at 24°C for juveniles in the first 30 days of their life. That pattern reversed later on, with 30-90 day old juveniles growing faster at 24°C than those kept at higher temperatures. The study hypothesized that the change in growth rate may be because the shrimp kept at higher temperatures started redirecting energy from growth to the development of sexual organs (ovaries and gonads).

Other studies support this finding that the relationship between growth rate and temperature differ depending on the stage of development for multiple species of crustaceans (Maszczyk & Brzezinski, 2018). However, **lower temperatures do generally lead to slower growth once crustaceans reach adulthood.** Lower temperatures also cause slower reproduction (Tropea C, 2015).

That being said, slower growth as a result of temperature *does not lead to smaller body sizes overall.* Lower temperatures do reduce molting frequency but they also extend shrimp lifespans and let them molt more, leading to larger body size over a longer life. This follows the Temperature-Size Rule of ectothermic organisms, which states that "individuals reared at colder temperatures reach maturity at a larger size than when reared at warmer temperatures", a phenomenon observed in 83% of ectothermic species (Forster & Hirst, 2012). Interestingly, the difference in body size appears to be at least partially a result of larger cells at lower temperatures, rather than a greater quantity of cells (Maszczyk & Brzezinski, 2018).

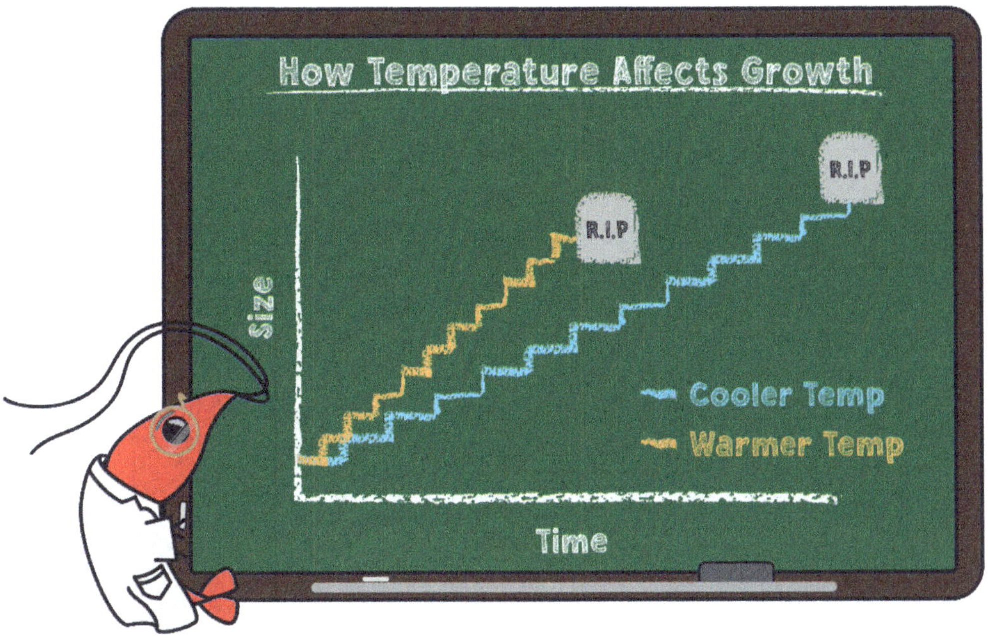

Graph showing how temperature affects growth over time. Horizontal changes indicate the time between molts and vertical changes indicate the size increase with a molt.

All this information suggests that the ideal temperature of a shrimp tank depends on our goals. If we want to increase reproduction rate to get more shrimp, then we may want to keep the temperature at the higher end of what's recommended for the species. On the other hand, if our goal is to have large shrimp for shrimp shows (p218, 248), or longer-lived shrimp for any reason, then keeping them at a lower temperature in their recommended range is ideal.

Concerns at Higher Temperatures

Increasing temperature may help breed more shrimp but it does come with some additional concerns. Shrimp metabolism is faster at higher temperatures, so they need more food and produce more waste. This, in turn, leads to more maintenance and a higher risk of overfeeding, both of which can destabilize an ecosystem if we're not careful.

Side note: Some hobbyists believe temperature affects the sex ratio of offspring, with more females at lower temperature and more males at higher temperature. Qualified research does not currently support this theory (Tropea C, 2015). A slightly higher female:male ratio was found at lower temperatures, but *the difference was not statistically significant.* Check out our video for a more detailed (and entertaining) explanation (Shrimply, Breed More Shrimp With Ice, 2024)[v].

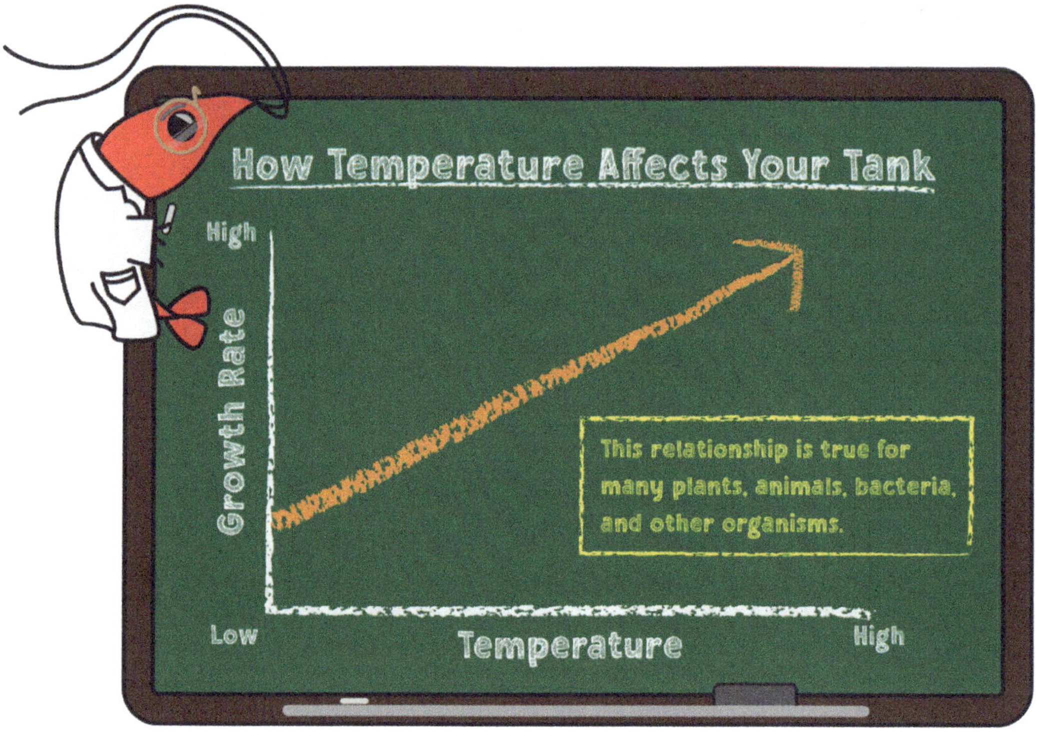

Additionally, temperature also impacts everything else in the ecosystem. Bacteria exposed to higher temperatures reproduce rapidly. A small bacterial infection at cooler temperatures may be fought off by a shrimp's immune system, whereas infections at higher temperatures have a larger bacterial load that is more difficult to fight off, resulting in additional stress or death (Shields J. , 2019).

Higher temperatures also increase the decomposition rate in aquatic ecosystems. Decomposition uses up oxygen and releases organic matter into the water, both of which can stress shrimp. When combined with the stress of increased bacterial growth rates, higher temperatures may lead to significantly more stress on shrimp. This is not to say they can't be kept at higher temperatures – potentially even above the ranges we recommend (especially Neocaridina) – but doing so requires more care and a good understanding of disease symptoms, as discussed in the Shrimp Diseases section (p277).

Additional considerations for tank temperature are discussed in the following sections:

1. **The Different Types of Shrimp Tanks** (p225)
2. **Breeding Bigger Shrimp** (p218)

Controlling Temperature

Many shrimp keepers keep their homes within the preferred temperature range of shrimp (65-75°F [18-24°C]). If your home is not within that range, or if you want more control over the tank temperature, then a heater or cooler/fan may be useful. This topic is discussed more in the Heater/Cooler (p73) section.

Testing Methods

1. Thermometer (Recommended method)
2. Temperature Sticker

pH

Why It Matters

pH tells us how acidic or basic the water is. It is measured on a scale of 0-14, with pH below 7 being **acidic**, pH above 7 being **basic**, and a pH of 7 being **neutral**. The pH affects chemical reactions in the water and in shrimp bodies, either stopping, slowing, or speeding them up. Shrimp can adapt to a range of pH to some degree, but anything outside of their preferred range may cause stress.

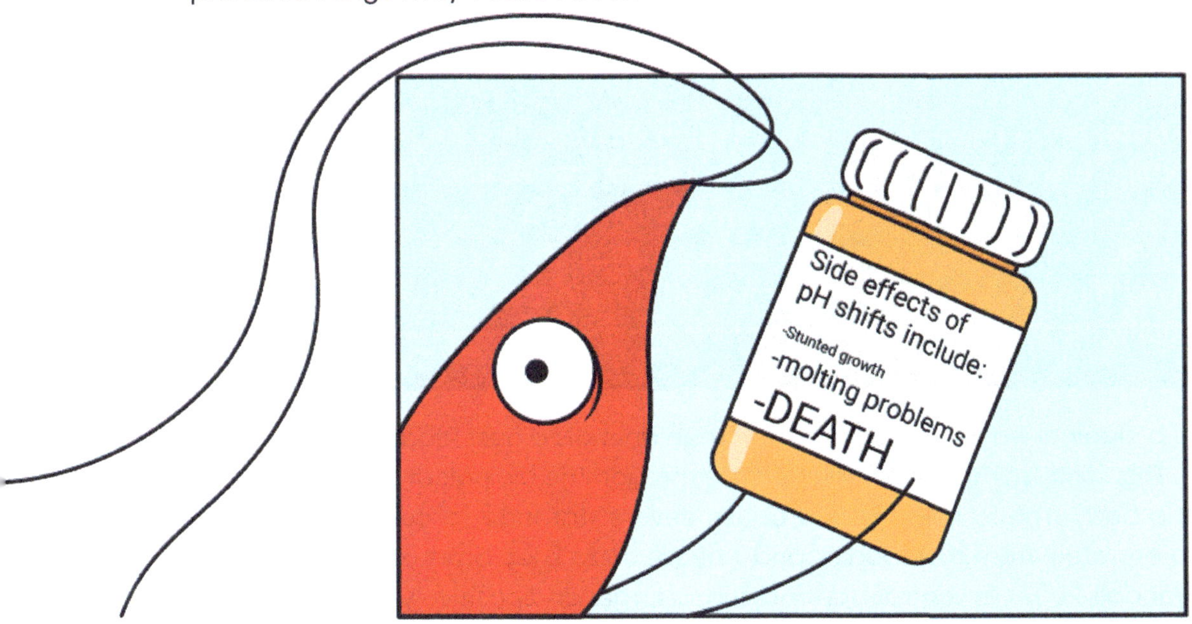

Frequent pH changes OR stable pH outside of the recommended range can cause problems, including but not limited to: stunted and slow growth, low reproduction rates, molting problems, diseases, loss of eggs, death, etc. As such, understanding why pH changes and how to control it are very important to your success as a shrimp keeper.

The Science of pH

pH quantifies the concentration of hydrogen ions (H^+ - AKA protons) in a liquid, with a lower pH indicating a higher concentration of H^+ and vice versa for higher pH. The scale is

logarithmic – just like the earthquake scale - meaning **a one-unit drop in pH represents a 10x increase in H^+**. To put it in other words, if one substrate drops the pH to 6.5 and a second drops the pH to 5.5, then the second substrate has a 10x stronger effect on pH.

pH is incredibly important for aquatic life because the concentration of H^+ affects how many H^+ ions bond to chemical compounds in the water. Even a single extra H^+ can have a large effect on the structure and function of a compound. For example, take a look at ammonia (NH_3), and ammonium (NH_4^+), the protonated form that 100% of ammonia is converted to by 6 pH:

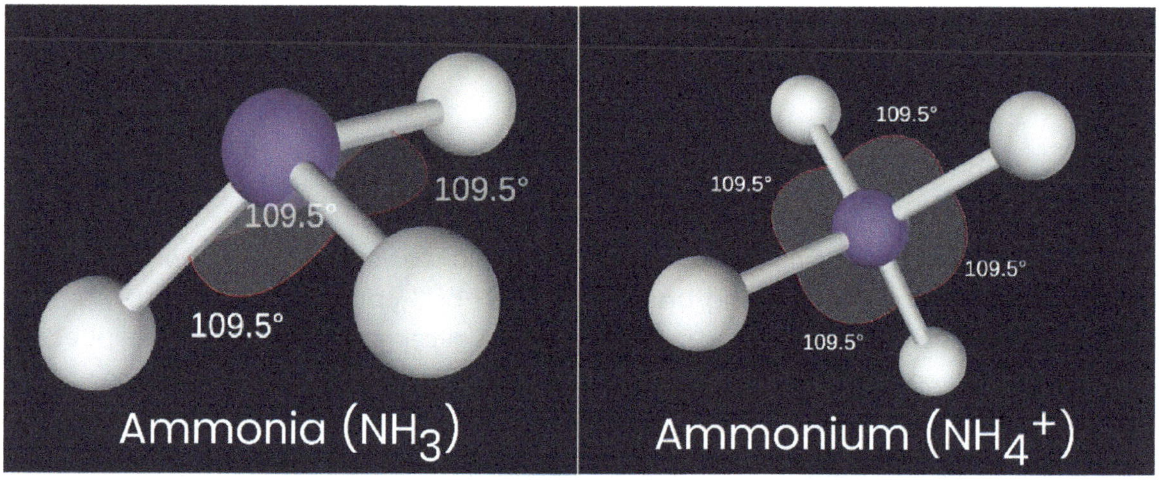

Side Note: pH-dependent compounds often aren't all in a single form. Instead, a higher concentration of H^+ *increases the likelihood that H^+ bonds at any given moment.* At a low enough pH, a compound like ammonia may be surrounded by enough H^+ to be 100% ammonium but, as pH increases, more individual compounds lose their H^+ and become ammonia, as shown in the graph on p19.

Ammonia is a biological waste product formed when animals digest food (more on it in a later section). This small compound is free to pass through the biological barriers like gills to re-enter the body from the water and cause severe damage to cells – that's why animals try to excrete it from their bodies and why we try to keep ammonia levels low in aquariums (Randall, 2002). In contrast, ammonium is larger and positively charged, which significantly lowers the chance of passing through an animal's gills at lower pH (Ip, 2010).

The addition or loss of H^+ also affects every other chemical compound in the water and in shrimp bodies, either stopping, slowing, or speeding up chemical reactions. Shrimp can adapt to a range of pH to some degree, but anything outside of their preferred range may cause stress because their body can't perform optimally.

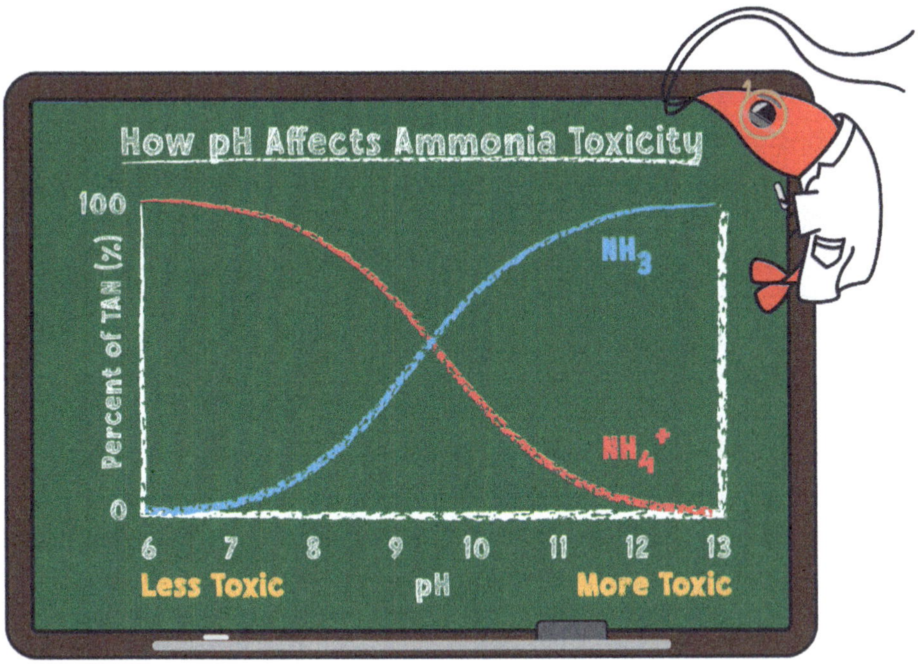

What Causes pH Change?

Three factors commonly cause pH change:

1. Excess waste
2. Water changes
3. Photosynthesis cycle

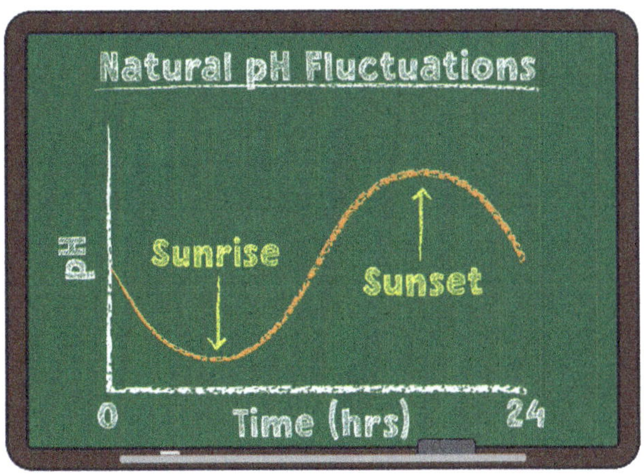

The first two factors are managed through proper tank maintenance, which we cover in a later section. The third factor, the photosynthesis cycle, is a natural process that turns light into stored energy (food) for plants during the day and removes CO_2 from the water in the process. CO_2 is acidic so its removal causes pH to increase. At night, this stored energy gets used up, releasing CO_2 into the water and decreasing pH. **In nature, photosynthesis causes *gradual* changes in pH throughout the day, so the change is not a problem unless it dips too low or high**. Many tanks are not set up naturally, however, which is why excess plants can actually *cause* pH instability in some setups. More on that in the Plants section (p50).

A Word on Water Changes

Water changes also typically cause pH changes, which leads to some anxiety when shrimp keepers consider doing large water changes to treat diseases or adjust tank parameters. While we should be cautious, **a large but temporary pH change that goes**

back to a healthy pH range in 30-60 minutes is not going to kill *healthy* shrimp. It may lead to a little bit of stress, but shrimp will be fine *if* the new water entering the tank is healthy and if the water change doesn't kick up a bunch of debris in the substrate.

This is demonstrated by numerous shrimp breeders/sellers who keep caridina shrimp (a more sensitive species) and regularly do 30-40% weekly water changes. There are some caveats though. These shrimp keepers always use exactly the same type of water (Remineralized RO/DI water - more on that later) that is guaranteed to be healthy, unlike tap water. They also do not pour the water into the tank all at once but typically let it flow in over 10-20 minutes. This leads to a less significant pH change while also preventing substrate disturbances.

Problems With Low pH

Planted tanks and Caridina shrimp tanks are often kept at low pH (<7 pH) by acidic soil, which is part of why Caridina can be more difficult to keep. Low pH has been found to decrease decomposition rates (Krachler R. F., 2009), leading to a buildup of debris in a tank. If pH increases due to chemicals or at the time the soil runs out of acid, then decomposition rates drastically increase, resulting in the release of WAY more nutrients than the tank ecosystem is used to (more on those in the Ammonia section [p30]). These nutrients destabilize the ecosystem and can significantly stress shrimp so it's especially important to monitor pH in low-pH tanks.

> **Decomposition**: The process in which bacteria, fungi, and archaea break down complex plant and animal matter into usable nutrients like ammonia, phosphates, etc.

Testing Methods

1. Test strips
2. Liquid Test Kit (Recommended)
3. Electronic Meter

[vi]*Links to recommended products available at www.ShrimplyExplained.com/book-links.*

Controlling pH

pH is closely tied to KH so control methods are combined in the pH/KH Control section (p261).

KH

What is KH?

KH stands for **carbonate hardness**. That is correct – There is not a K in the English spelling, but the acronym comes from the German spelling (karbonate). It may also be called **total alkalinity**. The units of measurement are **parts per million (ppm) when using test strips**, or **degrees KH (dKH) for liquid test kits**. 1 dKH is equivalent to 17.9 ppm (round up to 20 ppm for easy math). Shrimply Explained also has a handy unit calculator available online[vii].

Why Does KH Matter?

KH technically measures the concentration of carbonate (CO_3^{2-}) and bicarbonate (HCO_3^-) in water, but what's the point of measuring it? What does carbonate actually *do* for a tank?

More carbonate in a tank raises pH. It also protects against pH swings – an effect known as **buffering**. Having some buffering capacity in the tank is critical for a stable shrimp tank, especially for newer shrimp keepers who may accidentally overfeed or disrupt their tank in other ways. pH fluctuations (like the daily pH swing) do still happen, but they are generally less extreme at higher KH levels.

The only time KH is not needed is when we use a buffering substrate (AKA **active substrate** [p63]) that takes over the role of KH in controlling pH. We discuss KH more when covering Biological Processes (p38), along with the Substrate Types section (p62). While understanding the exact science of KH buffering is not required for shrimp keeping, we have an article available if you are interested[viii]!

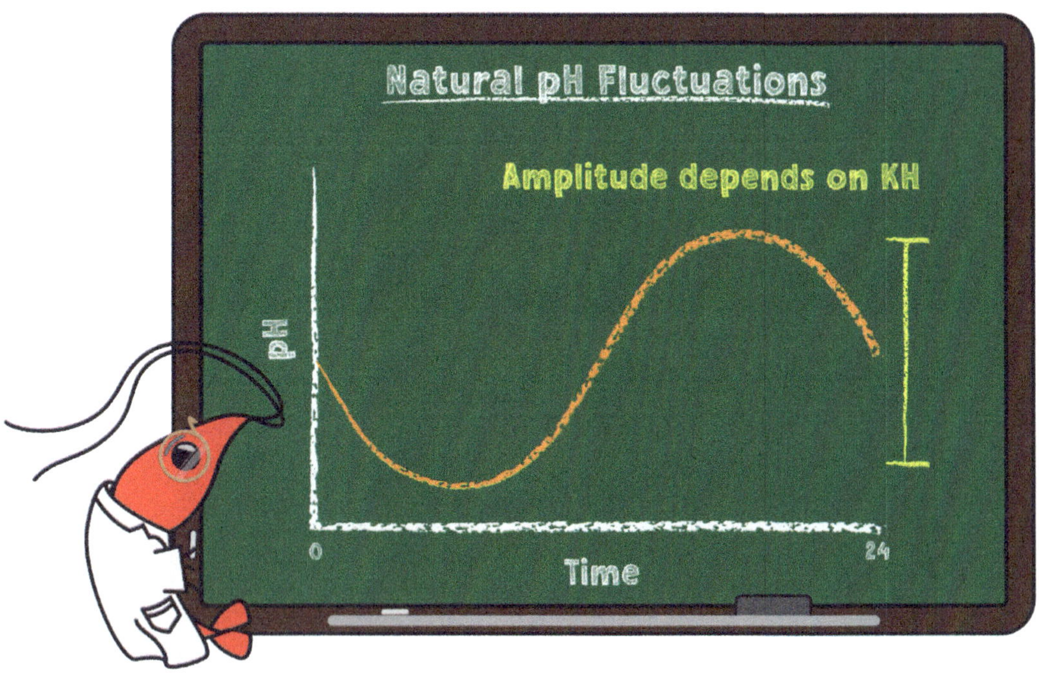

Side Note: Some people claim that Neocaridina need at least *some* KH to molt properly, which is not true. While carbonate is needed to build the shells of crustaceans, carbonate forms naturally inside a shrimp's body throughout the day. We know this because living animals produce CO_2 through aerobic respiration and shrimp body's internal pH (around 7.3 pH) converts some of that CO_2 to bicarbonate (HCO_3^-), which then gets converted to carbonate.

KH is depleted over time by processes that break down waste, so it must be replenished periodically through water changes. The more food that enters the system, the more frequently KH needs to be replenished.

A Metaphor For KH

Having some buffering capacity in a tank is like having an airbag in a car. We never know when we might be involved in a crash, but an airbag is ready to be deployed anytime the speed suddenly changes too much. It gives our bodies that bit of extra time to slow down so we aren't thrown against the steering wheel and killed instantly. That being said, if the crash is bad enough, even an airbag might not be enough to save us.

Buffering capacity is like a chemical airbag that slows down pH changes to make them more tolerable for animals in a tank. A large enough pH change can still overwhelm the buffer and lead to a harmful crash, but that's significantly less likely to happen in tanks with sufficient KH.

Testing Methods

1. Test Strips
2. Liquid Test Kit (Recommended)

[ix] *Links to recommended products available at www.ShrimplyExplained.com/book-links.*

Controlling KH

KH control and adjustment is discussed in the pH/KH Control section (p261).

GH

What is General Hardness?

~~General Hardness is Shrimply's stage name at his other job.~~

General hardness (GH) is a measure of how many minerals are in the water. It may also be called **water hardness** or **total hardness**. Just like KH, the units are in parts per million (ppm - for test strips) or degrees hardness (dGH - for liquid test kits) and the conversion is 1 dGH to 17.9 ppm.

Shrimp shells are made of complex structures of calcium carbonate, magnesium calcite, and calcium phosphate, along with chitin and protein. These compounds form structures that vary throughout the shell to control how stiff the shell is (This paper offers some incredible microscopic images of the different exoskeleton structures (Fabritius, Karsten, & et al., 2012)). If it's too stiff, then the shell can become brittle and break when hit, whereas too flexible of a shell may lead to insufficient protection or difficulty escaping it during molting.

Shrimp need access to minerals to form their shell every time they molt. They spend time between molts storing up minerals, which **shrimp get from food and by absorbing them from the water column**. There is some evidence that mineral supplementation in food is helpful but it is not enough to satisfy the requirements of molting in crustaceans (Ding, Yu, Peng, & et al., 2024). Ergo, we need to know whether there are healthy amounts of minerals in the water column, which brings us to GH.

Why Is GH Important?

To understand GH, we need to recognize that shrimp bodies are basically tiny bakeries that cover themselves with their baked good (the shell) for protection. Throughout their lifespan, they spend days collecting the perfect combination of ingredients (minerals), then shed the old shell to grow and make room for a new shell. If shrimp can't get the right ingredients, then their new shell may be too hard or soft, which impacts how well it works. Just like a baker needs the right ratio of dry and wet ingredients to get the right taste and texture, **a shrimp needs the right minerals to form a shell that is strong enough to protect them without being so brittle it cracks or so flexible that it is difficult to escape from**. The wrong ingredients lead to molting problems like the White Ring of Death (WRoD) (p267).

Research shows that crustacean shells are mainly composed of calcium carbonate, along with critical magnesium calcite structures to control how flexible the shell is (Fabritius, Karsten, & et al., 2012) (Raz, Hamilton, Wilt, & et al., 2003). These minerals are found in food

but not in sufficient quantities to satisfy the requirements of molting (Ding, Yu, Peng, & et al., 2024). Unlike human bakers, shrimp also can't just go and buy their mineral ingredients from a store. Instead, **shrimp mainly get minerals from the water column by absorbing them as water passes through their gills**. This makes it extremely important to know the mineral content of our water, which is where GH comes in.

We can easily measure GH in tap water using liquid test kits. By counting the number of drops, we find out the dGH value of our water – for example, 6 dGH, which falls in range for both Neocaridina and Caridina. Does that mean this water is good for shrimp?

It's impossible to know based on this information.

Technically, GH is based on the concentrations of calcium and magnesium in the water, as shown by the following formula[x]:

$$GH\ (ppm) = (2.497 * [Ca^{2+}\ (ppm)]) + (4.118 * [Mg^{2+}\ (ppm)])$$

You see, GH measures the *amount* of calcium and magnesium in the water**. It does not tell us what *ratio* of those minerals are in the water**. Therefore, 6 dGH *could* be the right ratio of calcium and magnesium to bake healthy shells, but it could also be entirely calcium, leading to molting problems. This is a common issue with tap water and is often frustrating to new shrimp keepers because "*the GH is in the right range*!"

Why The Ratio Matters

One big question that needs to be answered is how we know that shrimp need a specific ratio of minerals. What if each mineral just needs to be above a minimum level in the water – say, 10+ mg/L of magnesium and 20+ mg/L of calcium for healthy shrimp? Let's look at some research.

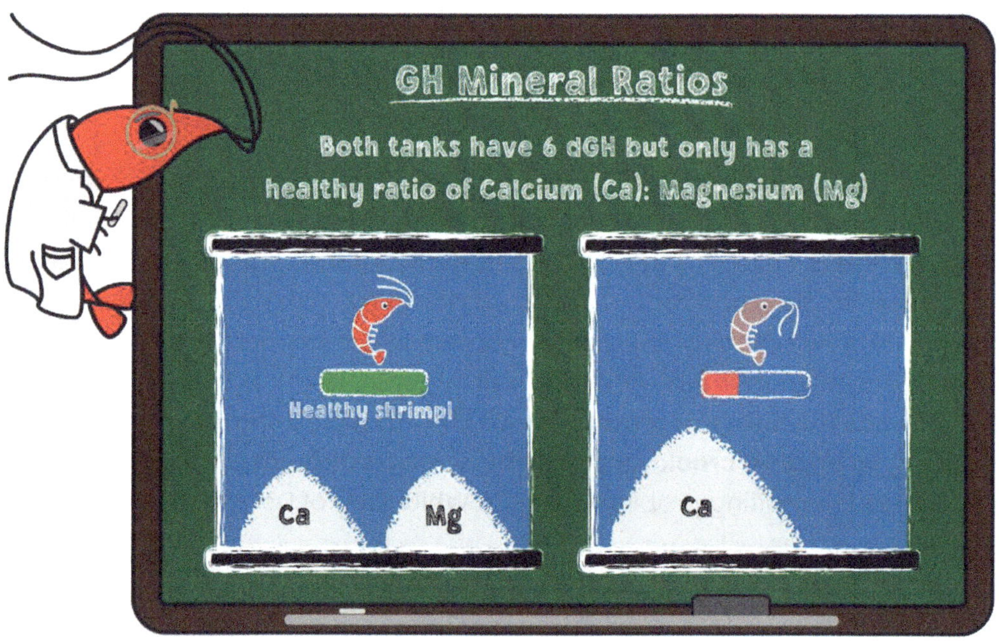

One study manipulated the Ca:Mg ratios in water to compare how different ratios affected the reproductive performance and offspring quality of *Macrobrachium rosenbergii* (Tavabe, Rafiee, Shoeiry, & et al., 2015). This species is frequently farmed for food so is one of the most well-researched freshwater shrimp species.

There were two treatment groups kept at either 150 ppm (8.4 dGH) (Group 1) or 190 ppm (10.6 dGH) (Group 2). This was done by adding calcium and/or magnesium to the control water that started at 110 ppm GH. Each group contained subgroups with varying Ca:Mg ratios (A-E), as shown in the table below:

Treatments	GH	Ca:Mg (% mols added)	Ca:Mg (weight)	Ca (ppm)	Mg (ppm)
Hatchery Water	110 ppm	Control treatment (1A)	3.8	30.9	8.1
	150 ppm (8.4 dGH)	0:100 (1B)	1.7	30.9	17.7
		20:80 (1C)	2.2	34.1	15.8
		50:50 (1D)	3.0	38.8	12.9
		80:20 (1E)	4.35	43.5	10
Hatchery Water	110 ppm	Control treatment (2A)	3.8	30.9	8.1
	190 ppm (10.6 dGH)	0:100 (2B)	1.1	30.9	27.5
		20:80 (2C)	1.6	37.3	23.6
		50:50 (2D)	2.6	46.8	17.8
		80:20 (2E)	4.7	56.5	11.9

Intermolt period (time between molts), egg hatchability, egg dry weight, and ESI (weight of eggs vs female weight) were all significantly affected by the Ca:Mg ratio and by water hardness. Treatment groups 1+2D (2.6-3:1 Ca:Mg by weight) showed the shortest intermolt period, along with the highest egg hatchability, dry weight, and ESI, indicating better reproduction than lower or higher Ca:Mg ratios. There also appears to be a sweet spot for GH, with the groups kept at 150 ppm showing better reproduction than those at either 110 or 190 ppm. While the experimental subjects are not Neocaridina or Caridina, it provides evidence that **at least one freshwater shrimp species does prefer a certain *ratio* and hardness**, as opposed to simply meeting minimum values of calcium and magnesium.

This points to an issue with common advice in the hobby that suggests adding calcium through water supplementation (crushed coral, wonder shells, etc.) or in the diet (GlasGarten Mineral Junkie[xi], etc.) whenever shrimp have molting problems. Doing so *might* help with the problem – but only if molting issues are caused by a low Ca:Mg ratio (insufficient calcium). Calcium supplementation may instead *increase* molting issues if there's already excess calcium, in which case magnesium supplementation is what's needed.

Experimenting With the Ca:Mg Ratio

Another study on *Macrobrachium rosenbergii* tested the effect of different concentrations of calcium and magnesium in the water column (Li, Lan, & Dai, 2024). The researchers found that increasing calcium concentration led to more frequent molts and changes in

hormone expression, but high calcium concentrations reduced survival rate. Increased levels of magnesium led to faster metabolism. Overall, too high or low levels of calcium and magnesium had varying negative effects, so a balance of both minerals leads to the healthiest shrimp.

A different paper gives us some additional insight into mineral ratios. In this study, researchers varied the calcium concentration between 25 and 205 ppm. They found the group at 145 ppm showed the highest molting rate of the groups, while the highest group (205 ppm) showed the lowest molting rate. The researchers conclude that moderate calcium supplementation can improve molting rate but excess calcium is detrimental. The magnesium concentration was not discussed.

This led to an experiment where we placed Neocaridina into three tanks at the same Ca:Mg ratio but drastically different GH levels. 10 shrimp were placed into each tank with water remineralized to 2 dGH, 8 dGH, or 30 dGH and observed for two months. Over this time, we found each group of shrimp successfully completed their lifecycle by molting, mating, and having babies that survived. 1-2 adult shrimp were lost in the 2 and 30 dGH tanks over the time of the experiment. The ages of our test shrimp were not known and there were not enough replicates to see a trend.

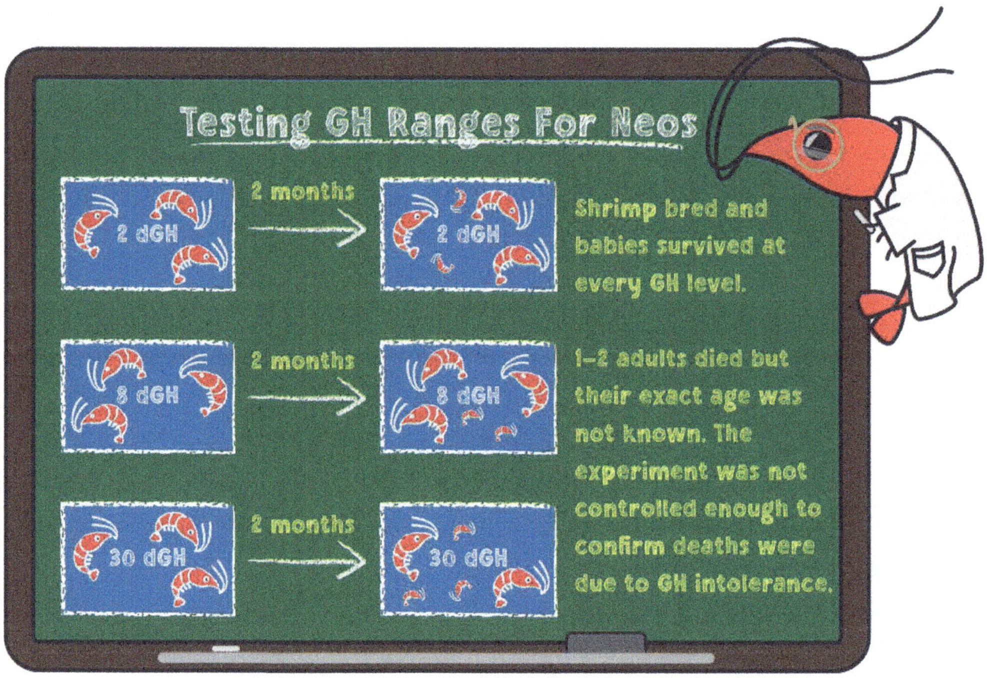

While there was only one replicate for each group, **this experiment demonstrates that it is possible for Neocaridina to survive and reproduce in a wide range of GH, at least if the Ca:Mg ratio is healthy** (guaranteed by using Salty Shrimp GH+ remineralizer). If Neocaridina are dying in a tank within the range of 2-30 dGH, then it is most likely an issue with Ca:Mg ratio or separate from GH entirely. This experiment also demonstrated that matching GH is not critical for Neocaridina, as the shrimp were originally raised at 7-8 dGH but they survived and reproduced at both 2 and 30 dGH.

Caridina, on the other hand, do seem to be more sensitive to water parameters than Neocaridina and there's not much research on their hardness preferences. As such, **we recommend trying to match the GH of the breeder Caridina are bought from**.

What Are the Right Ratios for Neocaridina and Caridina?

While there is not any research focused on the mineral ratio needs of Neocaridina and Caridina, the previous research - in combination with available product information – give us enough information to make a reasonable estimate. Doing so gives us the ability to modify tap water to make it suitable for shrimp. *Please note this mostly applies to Neocaridina since Caridina are typically kept in purified water.*

We've compiled Ca:Mg ratios of different remineralizers on the market based on their ingredient percentages or our testing. For reference, **remineralizers are liquid or powder products that contain necessary minerals (mainly calcium and magnesium) to raise the GH of water in a controlled manner**. Product ingredients – in combination with scientific research – provide the data we need to figure out the right mineral ratios. Doing so gives us the ability to modify tap water to make it suitable for Neocaridina. Caridina are a different story though – We'll get to them later.

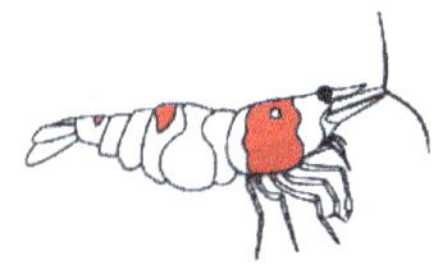

Remineralizer Ratios	Ca:Mg Ratio (by weight)
Seachem Equilibrium	3.3
Flip Aquatics Shrimple GH+	3.2
Salty Shrimp GH+*	2
Liquid Gardens GH+	3
*Experimentally determined estimate	

Side Note: While shrimp need a variety of minerals, it seems that calcium and magnesium are the most important to supply in the water column, based on the fact that some remineralizers (Like Flip Aquatics Shrimple GH+[xii]) *only* have calcium and magnesium and still work well.

Based on available product information, commercial remineralizers span 2-3.3 Ca:Mg, which matches the range of ratios that performed best in the Ca:Mg ratio experiment for *M. rosenbergii* (Li, Lan, & Dai, 2024). Tap water in the United States averages a ratio of 3.3 so many areas have tap water suitable for shrimp. That being said, the Ca:Mg ratio differs depending on a region's geology so some areas may have much higher or lower ratios that are not suitable for shrimp.

Shrimply Explained has developed a calculator to determine the Ca:Mg ratio of tap water and take the guesswork out of it for you[xiii]. It does require an additional $10-20 of testing equipment, however, so there may be some hobbyists who prefer to try adding shrimp to their tank if the GH is in the right range. Mineral additives, like CrystalPro Shrimp Minerals, are one cheap option to fix small mineral imbalances and may be the simplest fix. It also

helps to find local hobbyists who successfully raise shrimp in the area to confirm the local water is acceptable. The other option is to use remineralized RO/DI water, which is far easier than doing all the testing but is also more expensive. We discuss how to choose the best water for a tank in a later chapter (p88).

How GH Changes Over Time

GH changes over time for two reasons:

- **Water evaporation**
- **Plant growth**

Water Evaporation

H_2O naturally evaporates and leaves the tank while everything in the water (organic matter, minerals, bacteria, etc.) remains. This causes minerals to become more concentrated as the volume of water decreases, leading to an increase in GH. If 50% of the water evaporates, then GH doubles, which doesn't create stable conditions for our shrimp. **It's a good idea to top off the tank with new water whenever more than 10% of water evaporates to avoid large shifts in parameters**.

There's an additional problem with top offs known as **GH creep**. If we top off the tank with tap water or remineralized water, **then we're adding minerals to an already concentrated volume of water**. Over time, this causes an increase in GH that may eventually be harmful, depending on how often water changes are done. **To avoid GH creep, always replace evaporated water with an equal amount of *purified* water that doesn't contain any minerals**.

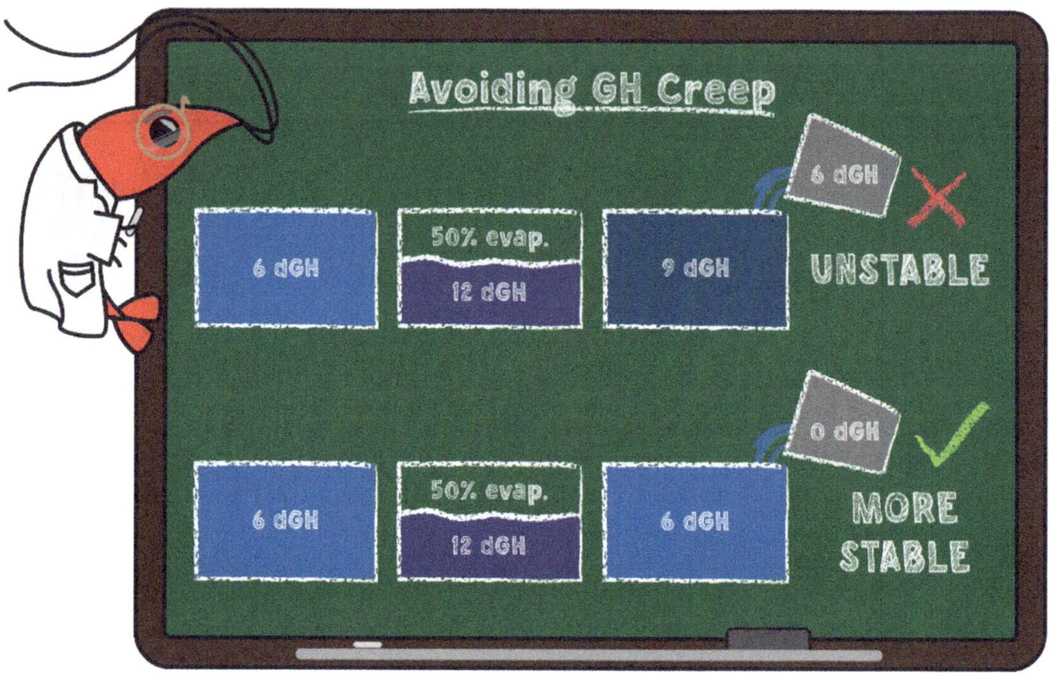

Plant Growth

This phenomenon isn't a common issue but it does explain why some tanks experience molting problems even though they've run without issue for months or even years.

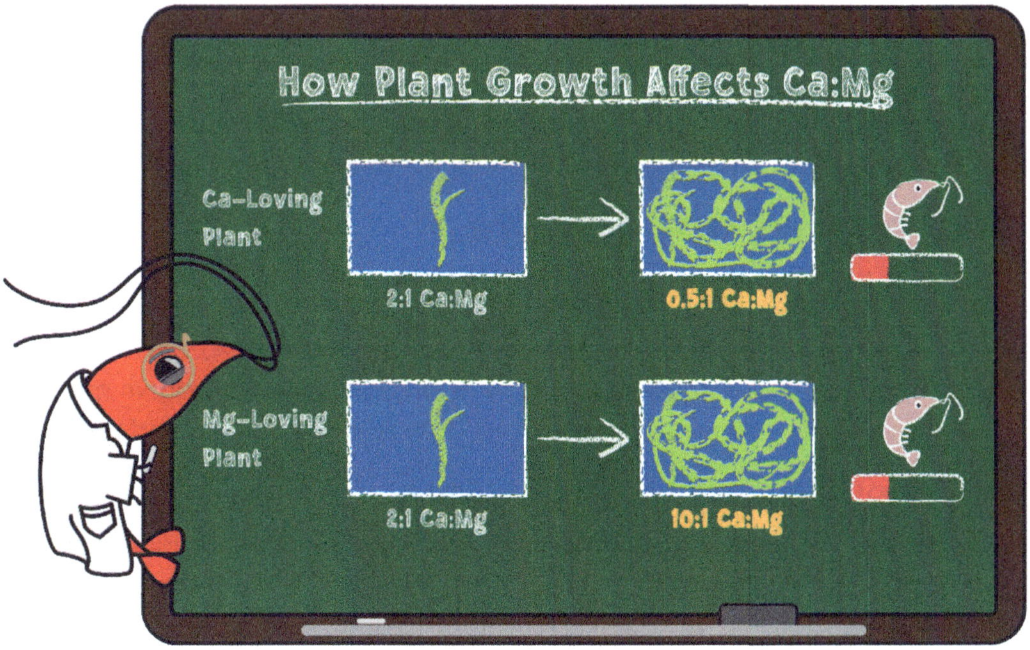

Somewhat weirdly, plant growth affects our tank's mineral ratios. As pointed out in a paper comparing the nutrient uptake of different aquatic plants (technically emersed plants), different plant species uptake nutrients at different rates (Tanner, 1995). Certain plants may uptake more magnesium than calcium or vice versa, potentially leading to unhealthy ratios over time. This is one explanation for why shrimp keepers suddenly experience molting problems in tanks that have been established for 6+ months. **There are three ways to reduce the risks of plant growth causing unhealthy mineral ratios:**

- **Consistent water changes**
  Water changes (using water with healthy Ca:Mg ratios) help reset the ratio back to healthy levels.
- **Diverse plant species**
  Increasing the number of plant species in a tank helps average out nutrient usage. One plant may uptake a lot of calcium but another may not, so they balance each other out. There's not much data on nutrient uptake for common plant species in the aquarium hobby so we can't choose specific plants to balance each other out, but simply having a diverse set of species reduces the risk of problems.
- **Slow-Growing Plants**
  Fast-growing plants take up more nutrients and are more likely to cause a mineral imbalance over time. Any mineral imbalance takes longer to occur when only slow-growing species are present.

Controlling GH

Right now, we're focused on understanding key water quality concepts. A supplemental section explains how best to control and adjust GH to get it in the right range (p264).

Testing Methods

1. Test Strips
2. Liquid Test Kit (Recommended)

[xiv]*Links to recommended products available at www.ShrimplyExplained.com/book-links.*

Ammonia, Nitrite, and Nitrate

What Are Ammonia, Nitrite, and Nitrate?

Ammonia (NH_3 or NH_4^+, depending on pH), **nitrite** (NO_2^-), and **nitrate** (NO_3) are nitrogen-based (nitrogenous) compounds released from animal waste and dead organic matter. They are measured in ppm or milligrams per liter (mg/L) by test strips and liquid test kits.

Why Nitrogenous Compounds Matter

Ammonia is a waste product released by animals, bacteria, and decaying organic matter. It is *extremely toxic* (in most circumstances - toxicity depends on pH). It gets broken down by nitrifying bacteria through the **nitrogen cycle**, turning into nitrite (less toxic) and eventually into nitrate (even less toxic but still a concern). More on that in the Cycling chapter (p99). Additionally, this Aquarium Co-Op video has a great visual explanation[xv].

Our goal is to have **no measurable amount of ammonia or nitrite at any time**, along with just a small amount of nitrate (15 ppm or less is ideal). These compounds can be removed through water changes, but they also get absorbed by plants for building material. This is just one of the many reasons plants are so important for shrimp tanks.

Controlling Nitrogenous Compounds

An important concept to understand is this: Ammonia, nitrite, and nitrate are always present in an *established* tank, even when they aren't present in measurable amounts. These compounds are constantly being released by animal waste, bacteria, and dying plants, but they get absorbed or converted to something else just as quickly as they get released so they often don't show up on tests. There is a whole biological production line for nitrogenous waste that turns it into something usable by plants and other organisms in a tank. **Our goal is to set up a strong production line and avoid disrupting it once it's established**. We'll explain more about how to set up this production line in the Cycling chapter (p99) and how to avoid disrupting the cycle later on (p129).

Testing Methods

1. Test Strips
2. Liquid Test Kit (Recommended)

[xvi]*Links to recommended products available at www.ShrimplyExplained.com/book-links.*

Dissolved Oxygen

What is Dissolved Oxygen?

Dissolved oxygen (DO) is a measure of oxygen gas (O_2) in the water. Oxygen gets dissolved into the water through water agitation and photosynthesis. It is measured in ppm, mg/L, or percent saturation. With proper setup, this parameter does not need to be measured because it is almost always at reasonable levels.

Why Is Dissolved Oxygen Important?

You know how we need oxygen to survive?

Well, it turns out shrimp do, as well, along with a lot of other animals and bacteria in your tank. When oxygen falls, it stresses or kills animals (Dean & Richardson, 1999) and promotes the growth of harmful bacteria (Liu, 2021). High DO levels are incredibly important for a healthy tank.

More DO must constantly be added to the water as organisms and decomposition processes use it up. In nature, oxygen enters through surface agitation caused by wind, waterfalls, river rapids, along with release by aquatic plants. Since our aquariums are little cubes of water without any natural water agitation, we have to add plants and/or artificial agitation (via filtration) to maintain healthy DO levels. We'll explain more about the different types of filters available in a future section so you can choose what's right for your tank.

Preferred DO Levels

Scientists recommend DO levels of 5 mg/L or higher for most freshwater fish and shrimp (Dean & Richardson, 1999).

Unfortunately, there is no economical way for a hobbyist to test DO levels but it is rarely a concern *if properly sized filters are used*. In addition, basic observation of the tank notifies us if something is wrong. When DO levels get too low, nearly all the shrimp will be near the top of the waterline trying to move water with their legs to agitate it. There is nothing to be concerned about if only a small percentage of your shrimp are at or near the waterline as they are likely just eating algae.

TDS - Honorable Mentions

TDS *can* be a very useful measurement in many shrimp keeping situations – it can also be entirely misleading and cause chaos for shrimp keepers. **Let's dive into the science behind TDS – what it's measuring, why certain ranges are recommended, and when TDS measurements are useful.** We'll also learn about the relationship between TDS and **conductivity** - a more common measurement type for European shrimp keepers.

What Does TDS Actually Measure?

Unsurprisingly, TDS measures all dissolved solids, including salts, nutrients, and toxins and gives us an estimated concentration in parts per million (**ppm**). Scientifically speaking, it measures the concentration of charged particles (ions like calcium [Ca^{2+}], magnesium [Mg^{2+}], sodium [Na^{+}], sulfur [S^{2-}], chloride [Cl^{-}], carbonate [CO_3^{2-}], etc.) in the water.

What Affects TDS In a Shrimp Tank?

Increases TDS	Decreases TDS
• **Fertilizer** • **Minerals** (remineralizer, leaching from rocks) • **Pollutants** (chemicals from non-aquarium-safe décor) • **Evaporation** • **Dechlorinator** • **Decomposing Food**	• **Water Changes** (with water that has lower TDS) • **Plant Growth** (nutrient absorption) • **Active Substrates** (depletes KH)

When Is TDS Useful?

TDS is useful in two situations: When we have multiple measurements over time AND/OR when we know a lot about our water.

How To Use Multiple TDS Measurements

A single TDS measurement is typically useless. That being said, *it becomes useful with multiple measurements over time* because it gives us a snapshot of how our water quality is changing. For example, if we see a sudden increase in TDS compared to a previous measurement, then we know that something has entered the water recently. Maybe minerals are leaching from a rock, or maybe it's from the fertilizer we added the other day. Likewise, if we know the average TDS of our tap water and we suddenly see a large increase in TDS one day, then it indicates the water has changed in some way. This might be a seasonal change at the water treatment plant during a rainy season that is harmless, or it could be as bad as a chemical spill leaching into the pipes. The more data points we have, the better we can understand when to worry about elevated TDS.

How To Incorporate TDS With Other Water Parameters

Now, let's examine how more information helps us analyze TDS. That starts by understanding how other parameters like GH and KH affect it. As mentioned previously, GH and KH may be measured in either drops of hardness (dGH/dKH) or ppm (1 dH = 17.9 ppm). In ppm, they can be used to identify a large portion of the TDS value of our tank. For example, let's say we have a tank with 107 ppm of GH (6 dGH) and 54 ppm of KH (3 dKH), along with a TDS of 200 ppm. By adding up the GH and KH values, we see that 161 ppm (107 + 54 ppm) of the TDS comes from our GH and KH. If we also know that we have 15 ppm of nitrates in the shrimp tank, then we account for 176 ppm and only have 24 ppm coming from unknown sources. Considering the sources are often harmless or beneficial nutrients for plants (sodium, phosphate, etc.), this would be considered pretty healthy water for a shrimp tank. If we are running a high-tech planted tank and measuring nutrient levels like phosphate, then we may be able to reduce this unknown amount of TDS even further.

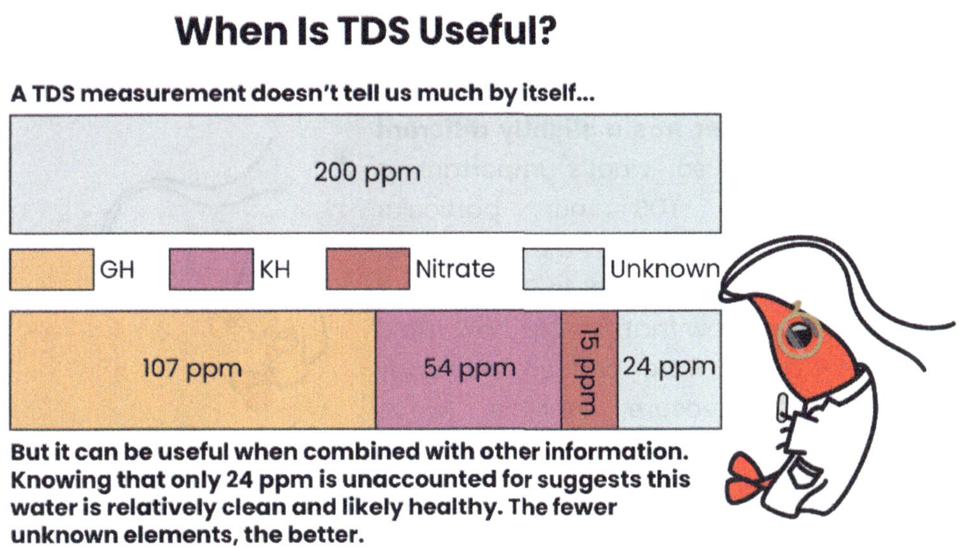

The most controlled setting possible for shrimp keepers is to use purified water with a remineralizer (a product that increases GH and sometimes KH). RO/DI water may start with 5 ppm of TDS – a very small amount of unknown substance that is typically harmless - which we then increase by adding enough remineralizer to get the desired GH/KH range. In this example, let's say we remineralize to 71 ppm of GH (4 dGH) and 0 KH for a Blue bolt Caridina shrimp tank. This means we would expect a TDS of around 76 ppm after remineralization, including the initial TDS of the RO/DI water. We will never quite see the TDS calculated from adding together GH/KH because remineralizers always have additional salts to balance them out, which increases the TDS.

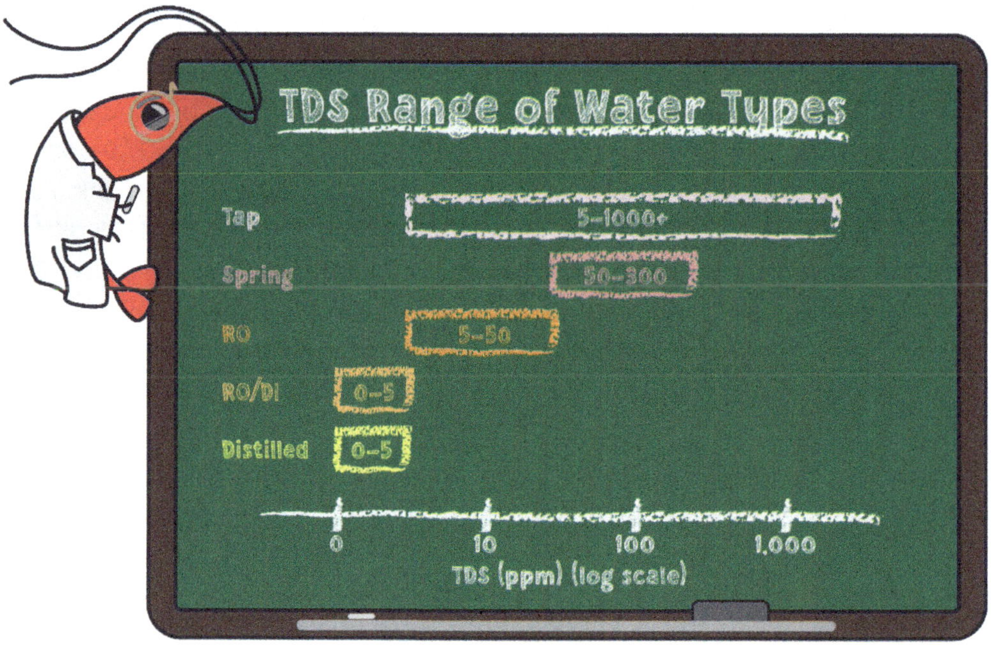

Each remineralizer has a slightly different blend of salts, so what's important is knowing what TDS our particular remineralizer reaches. In this case, our remineralizer raises the TDS to 90 ppm at 4 dGH. Since we know that, we can now use a TDS meter during water changes to quickly and easily measure whether we've remineralized new water to the right GH, rather than using the slightly less convenient liquid test kits. **The first time using a new remineralizer does require testing GH/KH, but a TDS meter may be used every time after that**. The benefit of TDS is the speed and ease with which we can get information.

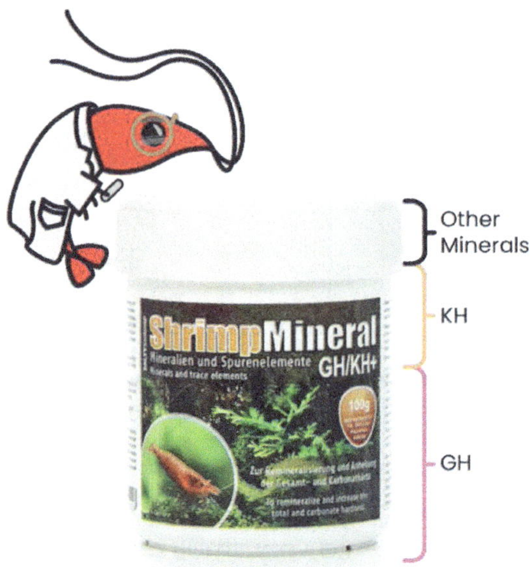

This remineralizer increase both GH and KH by a 2:1 ratio. It also includes other minerals to balance the water.

Now, let's compare that experience to measuring TDS in a shrimp tank with tap water. Out of this hypothetical tap, the water has 71 ppm of GH and 0 ppm KH - just like the remineralized water from the previous example. The difference is that the total TDS is 450 ppm, meaning there's around 375 ppm of TDS that's unaccounted for. That could consist entirely of toxins like copper or nitrates, or it could be all sodium and relatively harmless. That's why *a single TDS measurement* does not tell us any useful information about how healthy tap water is but the more we know about our water, the better we can analyze its quality.

Common Uses for TDS Measurements

- **Monitor water quality changes over time**
- **Detect leaching**
- **Identify changes in tap water** (seasonal variations, contamination)
- **Verifying remineralizer levels**
- **Assessing water purity in combination with other parameters**
- **Fertilization monitoring**
- **Equipment troubleshooting** (verifying RO filters work)
- **Evaporation compensation** (calculating amount of pure water vs remineralized water to top off tank with)

Why Are Certain TDS Ranges Recommended?

Using the relationship between GH, KH, and TDS, we can also understand how the recommended TDS ranges for different shrimp species are calculated. Let's take the recommended parameters for Neocaridina (p46) as an example:

Neocaridina	**ppm**	**dH**
Recommended GH	107-179	6-10
Recommended KH	36-143	2-8

Commonly, the recommended TDS range for Neocaridina is 200-300 ppm. The combined minimum values of GH and KH give us 143 ppm, while the combined maximum values give us 322 ppm. As we mentioned before, the salts that increase GH and KH always involve other solids that increase TDS, so even the lowest levels of the recommended GH/KH ranges will be higher than 143 ppm and typically close to the 200 ppm of TDS, which explains the low end of the recommended range.

As for the high end, 10 dGH and 8 dKH (322 ppm total) are relatively rare because most natural mineral sources for tap water and most commercial products increase GH/KH by a ratio of 2:1. This means that the 10 dGH upper limit tends to limit KH to around 5 dKH, giving us a combined 269 ppm. When accounting for the additional TDS of the salts required to achieve these GH/KH values, we arrive close to 300 ppm, which explains the upper limit of the recommended range.

These same calculations can be done with other shrimp species to understand their recommended TDS ranges.

Again, **surpassing this upper TDS limit is not *necessarily* a bad thing** but the risk depends on the shrimp species. Hardier species like Neocaridina are not very sensitive to TDS, so they are often kept at 400-500 ppm without issue – assuming there are no dangerous levels of pollutants in the water, which is difficult to determine with tap water. With remineralized RO/DI water guaranteed to have no pollutants, we've successfully kept Neocaridina at 600+ ppm (30+ dGH) without issue because the water is healthy and the mineral ratios are balanced[xvii].

On the other hand, more sensitive species and breeds of Caridina shrimp are affected by TDS outside of the range they are bred in. They can adapt to higher ranges assuming the

water is healthy (right mineral ratios, low nitrates, etc.) but it does seem to be less likely[xviii]. Newer breeds of Caridina like Blue BOAs are especially sensitive to TDS. Our hypothesis is that this sensitivity is a result of inbreeding depression, which should improve over time as breeders work to reduce inbreeding.

How To Measure TDS At Home

TDS is relatively easy to measure in shrimp tanks at home – you just need a TDS meter. The technology is simple so there are many mass-produced models for $10-15 that work well for hobbyist purposes[xix]. These meters are very easy to maintain. Unlike pH meters, there aren't disposable electrodes that go bad over time. Instead, we just need to brush the metal electrode tips every so often to clean off any debris that dries on them. If we don't clean off the debris, then we occasionally may see artificially low TDS values because the debris blocks the signal.

How TDS Meters Work (Advanced)

TDS meters are essentially voltmeters, if you're familiar with electronics. They compare two points to measure how well a material carries electricity – AKA **conductivity**. How well water **conducts** electricity is determined by how many charged particles (ions like Ca^{2+}) are in the water. Freshwater like what we have in our shrimp tank is relatively low TDS (0-1,000 ppm) because it does not have many ions, at least compared to something like saltwater (>35,000 ppm). Perfectly pure water has a TDS of 0 ppm, while most RO/DI (filtered) water is around 5-10 ppm due to imperfect filtration.

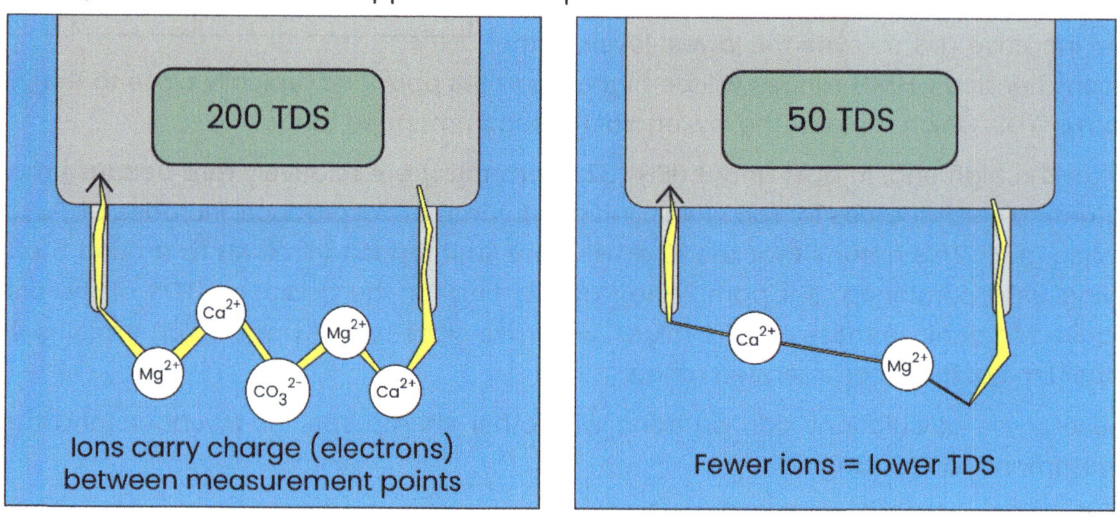

How TDS and Conductivity Differ (Advanced)

TDS is derived from conductivity values. Conductivity is typically measured in milliSiemens or microSiemens per centimeter (mS/cm or uS/cm), which then gets converted electronically to a TDS value in ppm (equation: *conductivity value * X conv. factor = TDS*). A problem often arises because *there is no set conversion factor between conductivity and TDS*. Specifically, an accurate conversion factor varies depending on the types of dissolved solids in the water, as each ion (Ca^{2+}, Mg^{2+}, etc.) changes the conductivity in different ways. For example, water with only calcium carbonate dissolved in it will have a

different conductivity value than water with the same amount of magnesium carbonate dissolved in it. If we search "conductivity to TDS conversion factor", we get a variety of different answers ranging from 0.55 to 0.8[xx] – a huge range! The standard conversion factor for freshwater is generally around 0.64-0.67[xxi] but **we don't know what conversion most manufacturers use for their TDS meters, which may account for large variations in TDS readings between different meters.**

Conclusion

From monitoring changes in water quality to quickly checking if we've added enough remineralizer, TDS can be a very useful tool to ensure the health of our shrimp tanks. The key is understanding that *TDS is only as meaningful as the context we provide it*. A reading of 350 ppm could indicate perfectly healthy water in a remineralized RO/DI setup, or it could be a warning sign in tap water with unknown dissolved substances.

Remember the two golden rules: TDS is most valuable when tracked over time, and it becomes exponentially more useful the more we know about our water's composition. Whether we're troubleshooting sudden changes, streamlining the remineralization process, or simply wanting greater insight into our shrimp's environment, a $15 TDS meter paired with knowledge of GH and KH values puts us ahead of most shrimp keepers.

The biggest mistake we see is shrimp keepers panicking over TDS readings without considering what's actually dissolved in their water. Don't let a number dictate actions – always look at a tank holistically. Our shrimp are happier for it.

*Red Cherry (*Neocaridina davidi *juvenile) riding a rabbit snail.*

Biological Processes

Now, let's apply all this knowledge about water parameters to understand how biological processes like the breakdown of organic matter (OM – includes food, excrement, dead plant/animal/cell tissue) affect our tank. This section also gives us a quick look into some of the fascinating complexities of biofilm.

OM is full of useful nutrients for different organisms. It starts out as large chunks that get eaten by macrofauna like shrimp and snails. Digestion breaks down OM into simpler organic compounds that are more readily available for smaller organisms like copepods and detritus worms (p112). These organisms break down OM into even smaller pieces that are available for various heterotrophic bacteria to eat in a process known as **decomposition**.

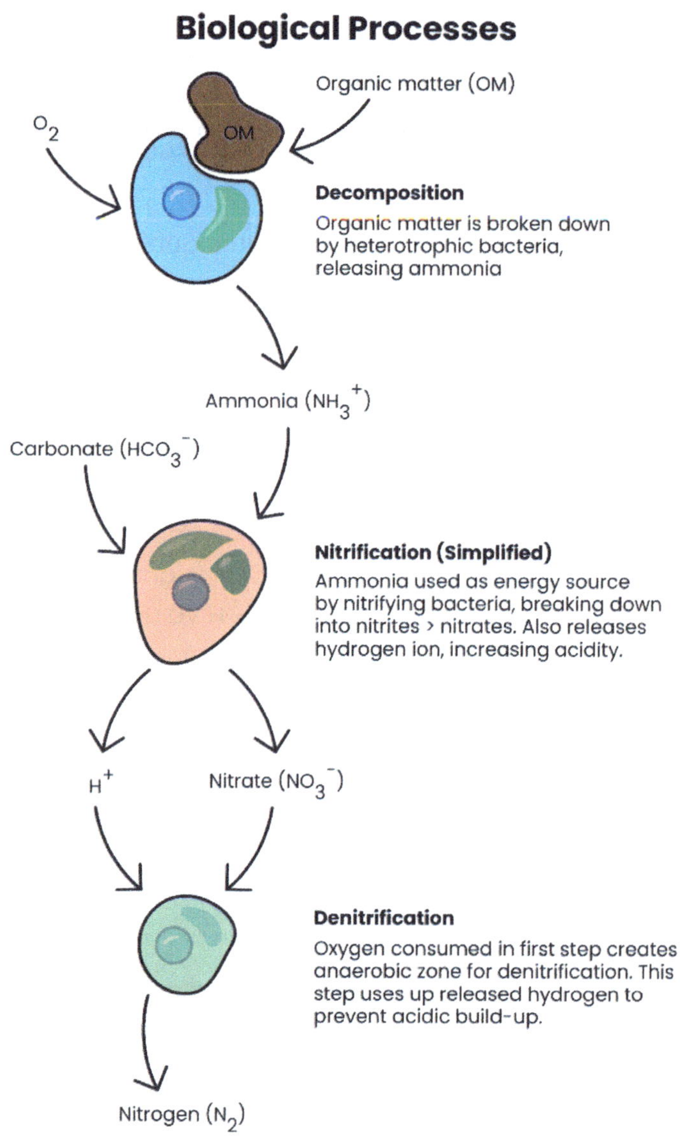

Cellular Processes - Simplified breakdown of inputs and outputs involved in decomposition, nitrification, and denitrification that occurs in biofilm

Decomposition

Decomposition requires two inputs – OM and oxygen. Heterotrophic bacteria ingest small pieces of OM and break them down using aerobic processes to produce ammonia and other byproducts. This is an incredibly necessary and beneficial process to prevent the build-up of waste in an aquatic environment, but too much decomposition can be harmful. When a shrimp keeper accidentally feeds too much, the heterotrophic bacteria reproduce rapidly in response to food availability. They may break down the excess waste quickly but, in doing so, may use up too much oxygen AND release a lot of ammonia. Low oxygen levels combined with ammonia stress can be very dangerous to animals, which is why overfeeding is such a big concern for shrimp keepers

(or any aquarist). That means we must be very cautious when adding food to a filterless tank. Even in a tank with typically adequate filtration, overfeeding or an excessive bioload causes more decomposition and can still create dangerously low dissolved oxygen (DO) levels. **If 75% or more of our animals are near the top of a tank, then that's a good sign DO levels may be low.**

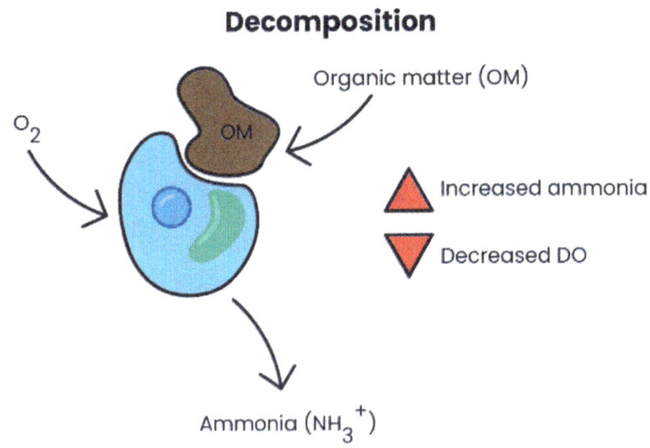

Controlled feeding and sufficient oxygenation via filtration are critical for a healthy tank. Even low levels of ammonia from decomposition can be harmful, however, which is why additional steps are needed to break down waste in a tank.

Side Note: Decomposition technically does not require oxygen and can be done anaerobically, but the process is much less energy efficient than aerobic decomposition. Aerobic decomposition *should* be the main form of decomposition in a healthy tank.

Nitrification

The next process, nitrification, also has a huge impact on tank stability. As ammonia gets broken down into nitrites, followed by nitrates, the bacteria and archaea responsible for nitrification use up carbonate (HCO_3^-) while releasing hydrogen ions (H^+). As we know from the Water Parameters chapter, carbonate is responsible for keeping pH stable and hydrogen ions lower pH. This means **nitrification has an outsized effect on pH since it removes the stabilizing agent while also encouraging a change, potentially leading to large pH drops when excess nitrification occurs**. Overfeeding or inadequate maintenance are common reasons why waste builds up and why nitrification may occur in excess.

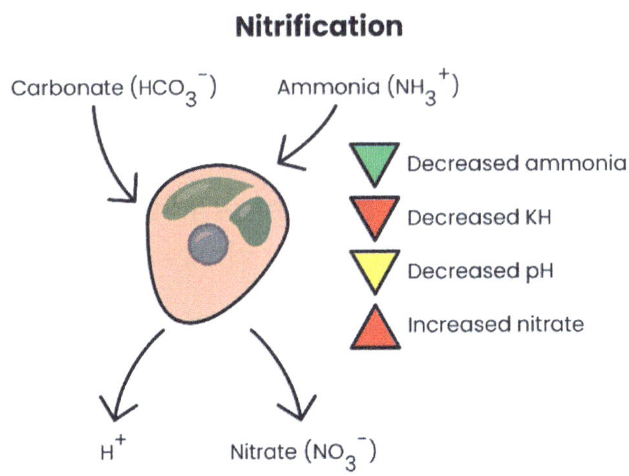

Side Note: There are a few different types of bacteria involved in nitrification, some of which only process ammonia (ammonia-oxidizing bacteria [AOB] – *Nitrosomonas* spp., *Nitrosococcus* spp., *Nitrosospira* spp., and others), others which process nitrites (nitrite-oxidizing bacteria – [NOB] – *Nitrobacter* spp., *Nitrococcus* spp., *Nitrospira* spp., and others), and still others that do both steps of the process (comammox – *Nitrospira* spp.) (van Kessel, Speth, Albertsen, & et al., 2015) (Samocha & Prangnell, 2019).

Typically, carbonate (KH) stays at a consistent level because it is replaced by CO_2 from the air that gets converted to bicarbonate or carbonate in water, even as KH is used up in nitrification. **If we notice KH dropping (and know it's not simply from an active substrate), then this may be a sign maintenance is required to handle a buildup of organic matter**. If our goal is a low-maintenance shrimp tank, then this is also a good sign to reduce feeding, since there's plenty of organic matter available for shrimp to feast on.

If, however, we're actively feeding a lot to encourage growth, then a drop in KH may be normal and simply requires consistent observation and water changes to maintain levels. Check out the Controlling KH section (p261) to learn more about adjusting KH.

Complicated But Potentially Useful Science

The optimal pH for nitrification is 7.2-8.0, while rates drop significantly below 6.7 pH (ECOS Environmental Consultants Limited, n.d.). In Caridina tanks, the pH generally needs to be below 6.7, which increases the likelihood of ammonia buildup. As we've mentioned before, ammonia becomes the non-toxic ammonium at low pH levels, but a sudden pH increase (like from a water change) may increase toxicity to harmful levels and potentially kill our shrimp.

Okay, but we can measure ammonia and prevent this problem, right?

Problems With Low pH Caridina Tanks

6.0 pH
-Normal conditions
-OM buildup (overfeeding)

7.0 pH ▲
-Depleted substrate
-Rapid decomposition
-Increased ammonia

6.0 pH ▼
-Rapid nitrification
-Increased nitrates
-Reduced pH
-Dead shrimp

To some degree, but OM may act as a ticking time bomb in low pH environments. You see, low pH reduces decomposition rates and leads to OM buildup (Krachler R. F., 2009) (Boyd, 2017). When pH increases from a water change or depletion of active substrate, excess OM breaks down rapidly, leading to a sudden increase in ammonia and subsequent toxicity.

Understanding this science helps explain why low-pH Caridina tanks may be more difficult to maintain than Neocaridina. The tanks don't necessarily show the same warning signs of ammonia/nitrite/nitrate buildup but that can change quickly. By the time we see our shrimp acting stressed, the ammonia may already be converted to nitrates and, in doing so, drop the pH back to normal levels for Caridina (since nitrification releases H^+). Understanding this chemistry helps explain why Caridina tanks may be more sensitive, requiring more caution during maintenance and feeding. This is discussed more in The Dangers of Active Substrate (p64).

Denitrification

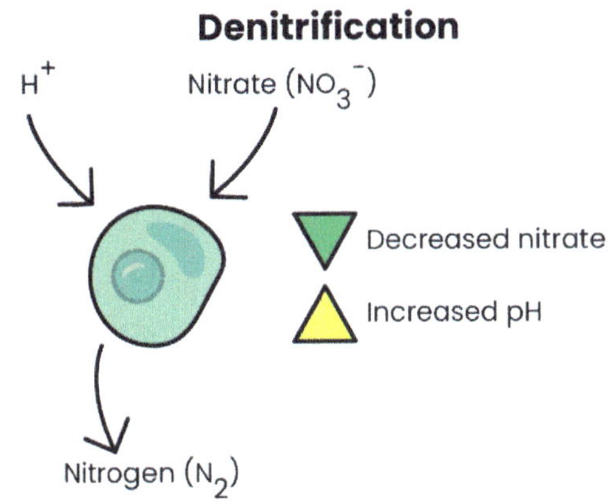

Denitrification is a natural process that prevents toxic buildup of products from nitrification. This process converts nitrate into gaseous nitrogen (N_2), which then gets released into the water, subsequently floating up and out of the tank. It does so through the following chemical reaction:

$$2NO^{3-} + 10e^- + 12H^+ \rightarrow N_2 + 6H_2O$$

From the above equation, we see that denitrification uses up the products created through nitrification (H^+ and NO^{3-}) (LibreTexts - Biology, N.D.). Again, this prevents toxic nitrate buildup while also preventing pH changes by removing H^+. It's a beautiful synergy created through billions of years of evolution.

Many of the organisms that perform denitrification do so in **anoxic** conditions (no dissolved oxygen) (Hong, Wu, & Shu, 2019). In a tank with plenty of water flow and aeration, there aren't many anoxic spots for this process to occur. We can make anoxic spaces with proper tank setup.

One option is a 2-3" thick substrate with a couple of inches of dirt capped with sand (exact depth for anoxic conditions is still debated and depends on the substrate structure). Another option is an anoxic filter like a plenum filter recommended by Dr. Kevin Novak (@anoxicfiltrationplenums - Dr. Kevin Novak, 2023). A third option is to use aquasoil which have anoxic spaces in the center of aquasoil pellets where water flow is low. Lastly, **some anoxic zones are always present in a tank to promote low levels of denitrification, and these zones give us insight into what biofilm is and how it functions** (FISHTORY, 2024).

Please note that denitrification does happen in aquariums but generally at very low levels due to the lack of large anoxic spaces (like deep substrate). It does still happen in biofilm to a limited degree in all tanks. There's also some evidence that it may occur in aerobic conditions in tanks when certain bacterial species are present (Ponpornpisit, 2024).

The Role of Biofilm

Anoxic zones form in biofilm naturally because they are carefully organized communities of cells – basically microscopic cities with highways for nutrient transport and waste disposal. It all starts with heterotrophic bacteria.

These bacteria decompose OM and release ammonia, which is toxic to them if it builds up. Fortunately, nitrifying bacteria/archaea use ammonia as a food source and gather around heterotrophic bacteria. This leads to a mutually beneficial relationship where the heterotrophic bacteria get protection from elevated ammonia levels while the nitrifying bacteria get food.

Both the decomposition and nitrification processes use oxygen to make nitrite and nitrate. This creates an anoxic zone full of nitrates, leading to the perfect conditions for denitrifying bacteria to grow, take in nitrate, and produce nitrogen gas (N_2).

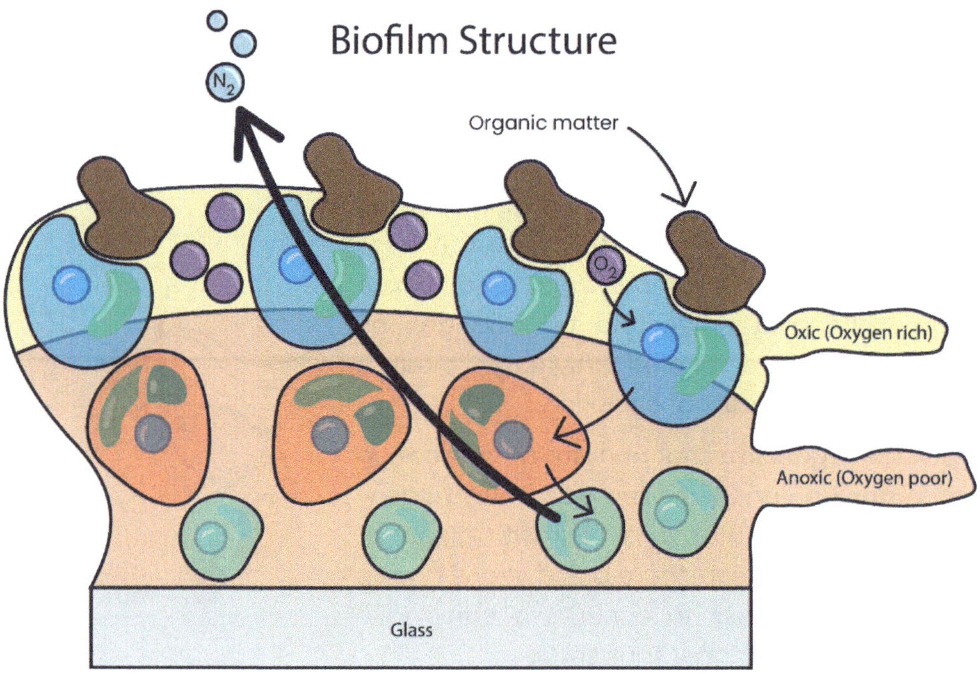

The microorganisms involved in decomposition, nitrification, and denitrification are held together by a substance they excrete known as extracellular polymeric substance, or EPS (Donlan, 2002). EPS is a mix of sugars, proteins, and other organic compounds that hold cells in place and facilitate the transfer of compounds between cells.

When these bacteria are together in biofilm, they work to transport compounds like ammonia efficiently between cells and create micro-environments for each species to thrive. Of course, biofilm structure is more complex than what we're describing, likely

containing hundreds of different microorganism species in any area of biofilm growth, but the relationship between heterotrophic bacteria, nitrifying bacteria, and denitrifying bacteria gives some insight into the fascinating way microorganisms work together. This also helps us understand why biofilm matters and why promoting healthy biofilm growth with diverse surfaces (wood, substrate, plants, etc.) and bacterial powders can be so advantageous for a healthy shrimp tank.

Types of Water Tests

Before getting into the details of different water parameters, we want to provide a brief overview of the various ways to measure them. There are four types of testing products available, including alert devices, test strips, liquid test kits, and electronic meters. Here are the pros and cons of each product.

Alert Devices

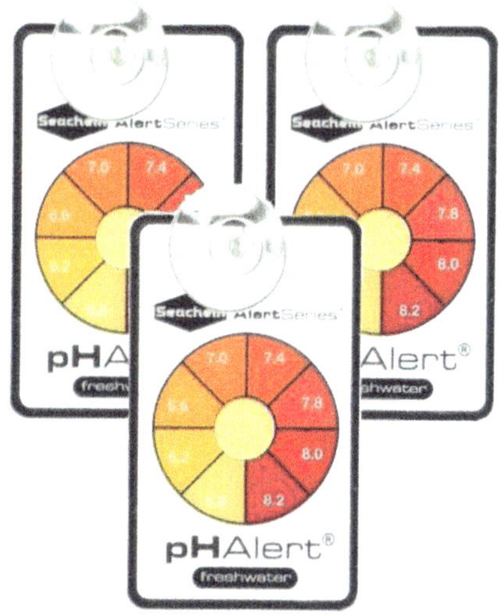

Alert devices are plastic cards coated with a chemical that changes color depending on a parameter level. A suction cup holds them to the wall of a tank so we can quickly glance at them to check that conditions are good. **They are somewhat reliable but are not a replacement for regular water testing**. Each alert card requires replacement every 3-6 months, making this more expensive than other testing methods ($7-8 per card). They also must be placed in a highly visible area in the tank, so they often end up blocking viewing areas.

Pros

- Potentially alert us of sudden changes in our tank

Cons

- Not cheap
- Unreliable
- Limited to pH and ammonia
- Not aesthetically pleasing

These devices are a cool idea but, in general, we don't recommend them due to their expense and lack of reliability.

Test Strips

Test strips are pieces of paper with colored squares attached to them. These squares are made of materials that change color based on water parameter levels. Simply dip one into your tank, then pull it out and wait a minute for the squares to change color. They are the cheapest and fastest way to test every important water parameter (excluding ammonia) at once. Each square can be difficult to interpret because the squares may be more than one color, or the colors may run. They are also less accurate than other methods, especially for GH and KH.

Pros

- Cheap ($0.01-0.02 per test per parameter)
- Fast
- Pretty easy to use
- Measures most parameters

Cons

- Can be difficult to interpret
- Need to buy separate ammonia test kit ($15)

Liquid Test Kits

Liquid test kits come with chemicals that change color based on a given water parameter. To use them, take a tank water sample, drip some of the chemicals into it, then wait up to 5 minutes for a color change. After that, compare the color to the reference guide included with each kit. Interpreting them isn't always easy without bright light and good eyesight, but they are very consistent, relatively fast, and fairly easy to use. The kits are somewhat more expensive upfront than test strips because you need one kit ($8-10) for each parameter, totaling around $50 for 120 tests of each parameter.

Liquid test kits are the most widely used testing method in the hobby because they are a great compromise between accuracy, reliability, speed, and price.

Pros

- Reliable
- Accurate
- Relatively easy to interpret
- Measures every important parameter

Cons

- More expensive than test strips ($0.06-0.07 per test per parameter)
- Slow (up to 5-10 mins for preparation/waiting depending on test)

We highly recommend liquid test kits to measure pH, GH, KH, ammonia, nitrites, and nitrates.

Electronic Meters

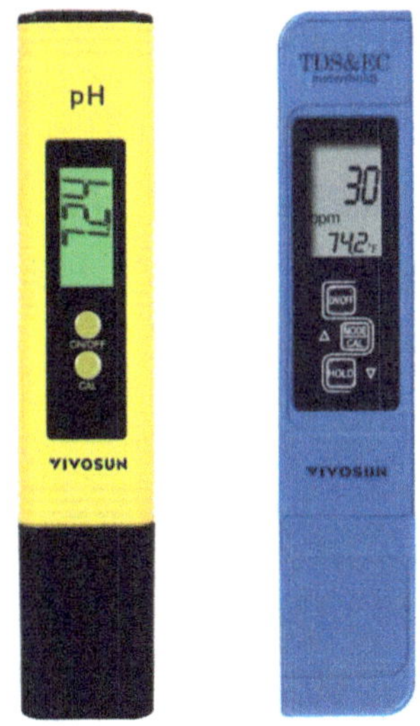

Electronic meters use chemical-sensitive electrodes to provide a precise digital output when placed in water. They are very fast and accurate, so meters are great for frequent checks. Unfortunately, they require some maintenance to keep them reliable. In addition, lower-cost meters ($10-15) can only measure pH and TDS. There are meters for other important parameters, but they are too expensive for most hobbyists ($100+).

Overall, we recommend them if you need to check the pH or TDS in multiple tanks frequently as they'll save you quite a bit of time.

Links to recommended products available at www.ShrimplyExplained.com/book-links.

Pros

- Fast
- Accurate (when calibrated)
- Reliable (when calibrated)

Cons

- Expensive
- Limited parameters
- Maintenance required

Electronic meters are best for measuring TDS – they're also the only way to do it...

Summary and Recommendations

In general, **we recommend liquid test kits because accuracy and reliability are especially important for shrimp keeping**. A TDS meter is also extremely helpful for quickly estimating water quality when you have a good understanding of how it plays into water quality.

Recommended Water Parameters

Detailed information about testing methods is included in the next chapter.

Parameters	Neocaridina	Crystal	Tiger	Taiwan Bee
pH	6.8-7.5	6.0-6.7	6.0-6.7	5.5-6.5
GH (dGH [ppm])	6-10 [108-179]	4-6 [72-108]	4-6 [72-108]	3-5 [54-90]
KH (dKH [ppm])	3-8 [54-143]	0-1 [0-18]	0-1 [0-36]	0-1 [0-18]
Ammonia (ppm)	0	0	0	0
Nitrites (ppm)	0	0	0	0
Nitrates (ppm)	0-15	0-15	0-15	0-10
Temperature (°F [°C])	65-78 [18-26]	65-73 [18-23]	65-75 [18-23]	65-73 [18-23]
TDS (ppm)	200-300	100-180	100-180	80-150

Importance	Description
High	Shrimp do best at or near this range and there isn't much flexibility outside of the range. Failure to maintain parameters at or near this range may result in death. For example, Taiwan Bee shrimp may do fine at 6.6 pH but are unlikely to survive and reproduce successfully above 7 pH.
Medium	Shrimp do best at or near the listed range with some flexibility outside of the range. For example, many Tiger shrimp can survive at 7.5 pH but may not breed very well.
Low	Shrimp do best around this range but there is a lot of flexibility. The listed TDS ranges are for optimal water conditions (remineralized RO/DI) but there can be many harmless salts that increase TDS outside of these ranges without impacting shrimp health.
Depends on setup	This is a category exclusively for KH. Neocaridina and certain Tiger shrimp may be kept on either active or inert substrate (p62). When kept on active substrate, no KH is needed because the substrate buffers pH. Tanks with inert substrate benefit immensely from a moderate level of KH (3+ dKH).

What Water Parameters Don't Tell Us

Testing our water parameters gives us specific values *for a specific point in time*. That can certainly be useful, but we also need to understand the limits of what water parameters can tell us. For example, pH is heavily affected by photosynthesis. We can measure pH anytime throughout the day and get a consistent value without realizing how low or how quickly pH dips when the lights turn out and photosynthesis stops.

Water parameters are useful but they have their limits, which is why we also need to use senses like sight to identify issues. One example is with molting problems. As we mentioned, GH is a measurement of calcium, magnesium, and a few other elements that react with the test chemicals. A GH test gives us a *quantity* of minerals but does not tell us whether those minerals are available *in the right ratios* for healthy molting. **When using water with unknown mineral values, we need to be aware of what molting problems look like and how to fix them when they come up**, which is why there's a whole chapter dedicated to molting problems (p267).

Along the same topic - we can overfeed our tank and put it in danger, even if ammonia, nitrite, and nitrate levels are all healthy. When food sits around, it may pollute the water, or it may instead be eaten by microorganisms that store those nutrients. Microorganisms use those nutrients to reproduce but, if they run out of food or get killed by a pH swing, then nutrients get released back into the tank and can cause a sudden ammonia spike that kills our shrimp. Again, **tests are useful and might catch the problem *as it's happening*, but our eyes are much better at judging whether leftover food is building up in the tank before a test ever does**. Even if we can't see the actual food particles, we can look out for sudden increases in microorganism population that signal an excess of food.

Another thing we should be using - our nose. Does the tank smell like clean water? It should. If we notice our room suddenly smell like a stagnant pond, it might be a sign a filter isn't working properly.

Does the tank smell like rotten eggs? Then something is decomposing and producing excess sulfur dioxide in the tank.

As for hearing – Does the room sound different? Is the filter gurgling correctly? We've walked into our tank room and immediately known that a filter wasn't working properly just because it didn't sound "normal". It might be time for cleaning or a sign of an airline malfunction.

The point is that water parameters are useful, but we can't solely rely on them to know if a tank is healthy. We have the best chance of success with shrimp tanks when we carefully observe them using both tests and our senses. That being said, success is heavily influenced by tank setup, which is covered in the next chapter.

Red cherry shrimp (*Neocaridina davidi* female) on Java moss (*Taxiphyllum barbieri*).

Building An Ecosystem

In this chapter, we cover everything there is to know about tanks, substrates, plants, décor, filters, and heaters. Each of these subjects affects the overall health of an ecosystem and our chance of success as a shrimp keeper, so shrimp will thank us for taking the time to understand these options.

Recommended equipment for a low maintenance setup is provided on page 96.

Tanks

Tanks are pretty simple things – just a few panes of glass or acrylic glued together. The main choice is how large of a tank to get.

We recommend at least a 5-gallon (19L) tank for Neocaridina or a 10-gallon (38L) tank for Caridina, especially if you're just getting into the shrimp keeping hobby. Caridina tend to be more sensitive so benefit from a larger and more stable tank. It is certainly possible to keep shrimp in something smaller, but doing so reduces the chance of success. **Every additional gallon of water helps to stabilize parameters, which in turn leads to a healthier ecosystem and happier shrimp**.

Think about it this way – What's better to pee in: the ocean or a bathtub that you're sitting in? Most people will pick the ocean because it's so large that we barely even come into contact with any excretions. On the other hand, peeing in a bathtub leaves us bathing in waste.

A larger tank allows for better dispersal of ammonia/nitrites/nitrates and leaves time for bacteria and plants to soak up that waste before it reaches harmful levels. More space also lets more plants and bacteria grow to process waste more effectively.

There are a few other factors when considering tank size though, which we've included below:

Cost

The smaller the tank, the less we need to spend on substrate, filters, lights, décor, etc. This lets us save money to get more tanks!

On the other hand, a larger tank allows for a larger colony of shrimp to sell, which is something worth considering if we're trying to convince our parents or significant other to let us get a large tank. The conversation might go something like this:

> **Mean Person (MP):** "You don't need a larger tank."
>
> **Reasonable Person (RP):** "But [mom/SO], a larger tank will let me learn marketing and sales by selling shrimp to strangers on Craigslist!"
>
> **MP:** "Oh, okay honey. That's fine then."

Problem solved – You're welcome.

Parameter Stability

As mentioned previously, smaller tanks experience faster parameter changes. A 1-gallon (4L) jar in front of a window will experience much greater temperature swings throughout the day than a 50-gallon (190L) tank. With smaller tanks, try placing them in insulated areas where temperature doesn't change drastically. Aside from temperature, other parameters also change quickly so smaller tanks must be tested more frequently to ensure they remain stable.

Visibility and Accessibility

Visibility and accessibility are very important to consider when breeding high quality shrimp. A larger tank makes it harder to see and catch shrimp so it can be more difficult to select the highest quality shrimp for breeding projects. When breeders are focused on breeding for shrimp shows, their tanks rarely exceed 20 gallons (76 L).

Other

Larger tanks are more difficult to move, fill, and clean. They also spill more water on the floor if they break.

As a side note, we highly recommend checking with your home or renter's insurance provider to see if they cover property damage from a broken tank.

Plants

Plants are a natural part of almost every ecosystem on the planet and that's for a good reason. Their ability to photosynthesize turns them into a key energy source for other organisms up the food chain. Their roots, leaves, and stems all support different microbial communities to increase biodiversity while improving water quality and substrate health (Srivastava, Chandra, & Kalra, 2017). Plants are incredibly beneficial for shrimp, in particular, since they offer hiding spots, help break down waste, and provide surface area for **biofilm** growth. Having a lot of plants in a tank is great, although doing so may block visibility and accessibility of shrimp, which is why breeders typically have fewer plants in their tanks.

Biofilm - A community of bacteria, fungi, archaea, and algae that form naturally on a surface. One example of biofilm is plaque on our teeth. In tanks, biofilm accumulates nutrients and turns them into food for shrimp, which is why shrimp are constantly scraping their claws against surfaces in a tank.

Now, it may seem daunting to try caring for *another* living thing in the tank—especially if you don't have much luck with terrestrial plants—but don't worry. There are plants of all different care requirements out there, some of which require very little effort to maintain.

We guarantee you can find something that both you and the shrimp will be happy about. Let's start by diving into how plants affect our tank, then cover care requirements and plant recommendations.

How Plants Impact Water Parameters

We've already mentioned that plants improve water quality by absorbing nutrients, but they also have an impact on pH and dissolved oxygen (DO). Plants perform two important processes that have enormous effects on these water parameters – **photosynthesis** and **cellular respiration**.

Photosynthesis is the process of converting light into stored energy (sugar). In doing so, plants release oxygen and absorb CO_2 (an acid). The removal of CO_2 leads to an increase in pH, while the production of oxygen increases DO levels. *These effects only happen when light is available.*

Cellular respiration is a chemical reaction in all plants and animals that converts *stored energy* (fats, sugars, etc.) into usable energy for cells. This process consumes oxygen and sugar to produce energy and CO_2, leading to a decrease in DO and pH. This process happens regardless of light level (Crash Course, 2024).

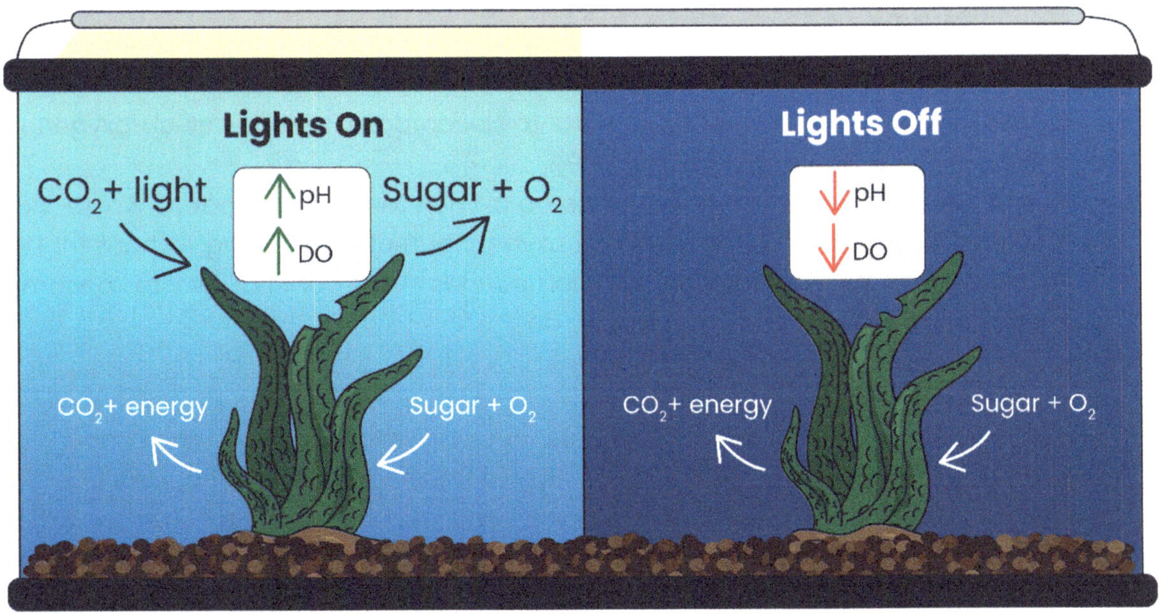

Photosynthesis absorbs more CO_2 than cellular respiration produces, so pH and DO increase during the day. At night, only cellular respiration occurs so pH and DO levels drop. Overall, tanks with plants in them have larger pH swings throughout a 24-hour period because of these chemical reactions. Fast-growing plants have a larger impact on parameters than slow-growing plants. **This pH change is gradual enough that it doesn't stress out shrimp**, but the DO drop at night is a serious concern.

Filters and adequate water flow ensure a healthy DO level. Shrimp in **filterless tanks are much more likely to experience stress from low DO levels**, which is part of why we always recommend a filtration system.

How Plants Impact Dissolved Oxygen Levels

Another way photosynthesis and cellular respiration impacts our tank is by changing dissolved oxygen (DO) levels throughout the day. As noted in the photosynthesis overview, oxygen gets created during photosynthesis and used up during cellular respiration. This causes DO levels to increase during the day and decrease at night, reaching the lowest point just before sunrise – the same time as the lowest pH. The more plants that are in the tank, the more extreme the drop in DO.

Animals need oxygen to survive, so low levels lead to unhealthy conditions in most species (Dean & Richardson, 1999). Shrimp tend to require less oxygen than fish because they have less muscle, but still show low survival, worse immune function, and lower growth when kept at DO levels of less than 2 mg/L (Nonwachai, Purivirojku, Chuchird, & et al., 2011). Shrimp raised at levels above 4 mg/L showed significantly higher growth, so our goal is to maintain DO levels above 4 mg/L.

The best ways to avoid low DO levels in our tanks are to:

- **Use properly sized or oversized filters for our tanks.** Many breeders use two double-sponge filters (p78) to ensure sufficient oxygenation and flow while adding surface area for shrimp to graze on. It is possible to have too much oxygen in a tank but it is very rare. If it happens, a *large majority* of the shrimp population will attempt to crawl out of the tank or be at the surface of the water.
- **Avoid overfeeding.** Leftover food starts to decompose, which uses up oxygen in the tank (Texas AgriLife Extension, 2009).
- **Avoid overstocking.** More animals use up oxygen faster and produce waste that decomposes and uses up additional oxygen, potentially leading to harmful DO levels. Often, excess nitrogenous compounds from animal waste become problematic before DO concentrations do.

The Benefits of Plant Diversity

Every plant needs the same *types* of nutrients, but each plant varies in the *amount* of nutrients they use. This is demonstrated in a study that tested the nutrient uptake of eight different **emergent plants** (aquatic plants that grow up out of the water) (Tanner, 1995). They analyzed a total of twelve different nutrients to compare how each plant incorporated them into above- or below-water growth. Some nutrients were incorporated at similar concentrations while others varied significantly between plants. For example, the submersed growth of *Baumea articulata* (jointed rush) absorbed at least six times the manganese (Mn) as the submersed growth of *Bolboschoenus fluviatilis* (river tuber-bulrush). *Cyperus involucratus* (umbrella plant) absorbed over four times the amount of magnesium (Mg) as *Phragmites australis* (common reed). While none of the plants included in the study are typically found in home aquaria, this study helps us understand why certain nutrients may build up in our tank over time.

As a hypothetical example to demonstrate the importance of this study, let's say we just have *Bolboschoenus fluviatilis* in our tank. It doesn't take in much Mn, so an excess of Mn may build up, and water changes would be needed to prevent Mn from reaching harmful levels. Knowing that, we could add *Baumea articulata* into our tank since it absorbs more Mn than the other plant species. Doing so would balance the Mn uptake, reducing the risk of buildup and the need for water changes.

Baumea articulata - *Harry Rose from Dungog, Australia, CC BY 2.0 <https://creativecommons.org/licenses/by/2.0>, via Wikimedia Commons*

The problem is that we don't have good data on all aquatic plants in our tank and, even if we did, the amount of work going into picking *just the right plants* to even out nutrient usage and prevent buildup of any single nutrient is more than most people want to deal with. **Don't worry too much about it – Just understand that having a variety of plants reduces the chance of any single nutrient building up to unhealthy levels in the tank**. This, in turn, reduces the need for water changes and creates a more stable ecosystem.

Bolboschoenus fluviatilis - *Harry Rose from Dungog, Australia, CC BY 2.0 <https://creativecommons.org/licenses/by/2.0>, via Wikimedia Commons*

This may go against the setups that many shrimp breeders have, in which they only have one clump of moss or just a few red root floaters. Keep in mind that breeders have different goals than many hobbyists. The goal of a breeder is to maintain the best water quality for shrimp health, along with the best visibility for selection and observation, even at the expense of spending more time on maintenance like water changes. A hobbyist's goal often differs and may be more focused on a low maintenance tank, in which case **a variety of plants reduces the chance of a given nutrient building up in the water and reduces the need for water changes**.

How Plants Impact Microbial Diversity

As we've already discussed, biofilm is made of a complex web of organism interactions and is one of the main food sources for shrimp. **The diversity of organisms in biofilm influences our shrimps' diet variety, with greater diversity leading to a more well-rounded diet**. Too much of a single food type is never good for shrimp, humans, or any other animal.

Plants, as it turns out, play a huge role in promoting diverse biofilm and microbial communities (Srivastava J. , Chandra, Kalra, & et al., 2017). They do so because plants offer unique surfaces and release byproducts that specialized microbes benefit from. For example, microbes in the rhizosphere (around the roots) receive nutrients, carbon, and oxygen from the root surface.

The microbes aren't just taking advantage of the plants though - Plants also benefit from a close association with microbes around their leaves and roots. In return for nutrients, microbes help plants by sequestering heavy metals and making minerals more available for absorption (Srivastava J. , Chandra, Kalra, & et al., 2017).

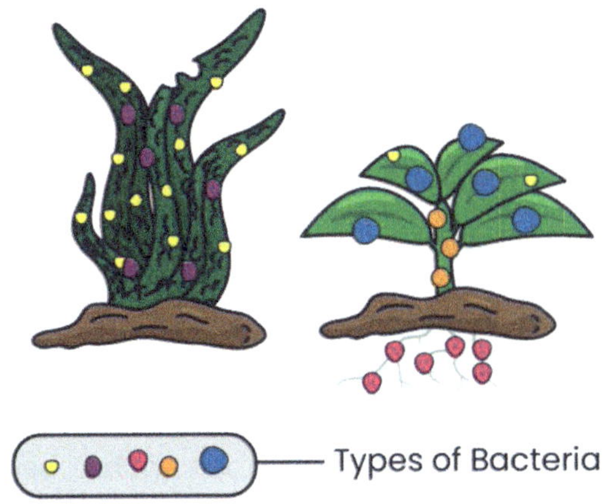

According to Crump and Koch (2008), different plant species also host unique microbial species; another reason why having a variety of plants benefits our ecosystems. One of the ways they do this is by releasing antimicrobial agents like zosteric acid, which "limit bacterial and fungal colonization of plant surfaces and allow only certain microbes to become established" (Crump and Koch, 2008). Plants effectively farm microbes on their surfaces to encourage the growth of microbes and biofilm that are beneficial for them. The diversity of microbes even varies between a plant's leaves and roots. Overall, greater plant diversity leads to more diverse biofilm growth and a more diverse diet for shrimp.

Summary of Microbial Impact

There's a lot of interesting research around microbe-plant interaction but we're not going into it further. The takeaways for shrimp keepers are two-fold:

1. **Plants offer unique surfaces that encourage diverse microbial communities, leading to a more diverse diet for our shrimp.**
2. **Each plant species encourages the growth of unique microbial communities, so a greater variety of plants increases diet diversity for shrimp.**

The Basics of Plant Care

All plants need *light* and *nutrients* to survive. The light can be natural or come from an artificial **grow light designed for planted tanks**. We specify grow lights because they have the red and blue spectrum that plants need, whereas extremely cheap lights generally have suboptimal spectra for plant growth and may increase the growth of undesirable algae.

There are also different sources for plant nutrients. A natural option is animal waste or leftover food that breaks down in the tank. With enough animals, even the most nutrient-hungry plant can get most of their required nutrients. We say "most" because some plants require more micronutrients than others, so the correct balance may not be found in animal waste. We recommend doing research on nutrient needs of plant species to decide what is best for the setup. Options include nutrient-rich substrate, liquid fertilizer, or solid fertilizer (root tabs) placed into the substrate (Bhuij, 2016).

In summary, each plant needs certain methods of planting, nutrient delivery, or lighting, so we recommend carefully researching each species you want before buying everything for your tank. A couple of basic plant categories are defined below:

Lighting Needs

Light brightness is measured in PAR (photosynthetic active radiation). Higher PAR indicates brighter light. (PAR Range source - (Aquarium Co-Op, N.D.)

Low-Light: Plants that only require indirect light (10-20 PAR) to grow. Best for low-tech tanks.

Medium-Light: Plants that require slightly more light (20-35 PAR) to grow.

High-Light: Plants that require lots of light (40+ PAR) to grow. They also benefit from CO_2 and fertilizer regimens to support their growth.

Nutrient Needs

Root Feeders: Plants that absorb most of their nutrients from the substrate. Root feeders often benefit from a nutrient-rich substrate or root tabs.

Stem Feeders: Plants that absorb most nutrients from the water column. Stem feeders often benefit from the addition of liquid fertilizer.

One major concern around plant fertilization is whether the product is shrimp-safe. Many fertilizers (root tabs or liquid) contain some amount of copper, which is poisonous to shrimp and other invertebrates at even relatively low doses. We see a lot of questions about which brands are shrimp-safe, but the truth is that **nearly all commercial fertilizers are safe**. We've never heard of one that isn't. That being said, it's always a good idea to be cautious and do a quick search for "[fertilizer] shrimp safe?" to double-check.

Our Plant Recommendations

We recommend at least three types of plants in a low-maintenance shrimp tank:

1. Mosses

Plants in the *Vesicularia* genus (Christmas moss, java moss, weeping moss, etc.) have been shown to increase Neocaridina survival rates (Vazquez, Sganga, & López Greco, 2022). *Vesicularia spp.* survive in a wide range of conditions, are easy to keep, and are widely available. Please note that mosses do best in water with lower pH, GH, and KH, so they may struggle in harder water conditions.

2. Slow Growing Rooted Plants

Plants like *Cryptocoryne* species have roots that dig into the substrate to aerate it and improve nutrient cycling. This reduces the need for substrate cleaning and improves overall water quality. They are also very easy to care for with just a little light.

3. Floating Plants

Floating plants like Red Root Floaters have access to air, which lets them soak up more nitrates than other types of plants. This significantly reduces the need for water changes while also protecting shrimp from harmful nitrate spikes. We recommend Red Root Floaters because they are easier to control than some of the smaller, more invasive floaters like duckweed that eventually end up in every aquarium in the house. **One concern with floating plants is that they can block out light from other plants and algae below them, so we recommend removing extra floating plants periodically to keep around 30-50% of the surface clear**. *Please note that floating plants do best in lidless tanks because lids create too humid of an atmosphere.*

> **Note**: If you have harder water, a limited budget, or a complete lack of a green thumb, then **guppy grass** is a fantastic option that offers the surface area of mosses with the nutrient absorption of floating plants.

We always recommend quarantining plants before putting them into a tank, unless they are from tissue culture. (Quarantine procedure[xxii])

A diverse tank at Shrimply Explained HQ – The more plants, the better!

Substrate

Substrate refers to anything that covers the bottom of your tank, whether that's sand, dirt, gravel, or aquasoil (p63). There are tons of different methods of adding substrate to our tank depending on our goals with appearance, plant health, etc., but going into all the science behind substrates is a book on its own. Instead, we're focusing on how substrate affects shrimp, so we have a way to evaluate which substrate is best for each type of shrimp. Here's how the substrate affects a shrimp tank:

1. Substrate increases the surface area in a tank
2. Substrate color affects shrimp coloration
3. *Some* substrate changes pH/GH/KH

Impact On Surface Area

The type of substrate we have affects how much surface area (SA) is in a tank and, therefore, how much biofilm is available. As an example, let's start by comparing the SA in a bare-bottom 5-gallon (19L) tank versus a tank with aquasoil in it.

Bare-Bottom Tank Calculations

	Dimensions (inches)
Length (L)	16.5
Width (W)	8.75
Height (H)	10.75

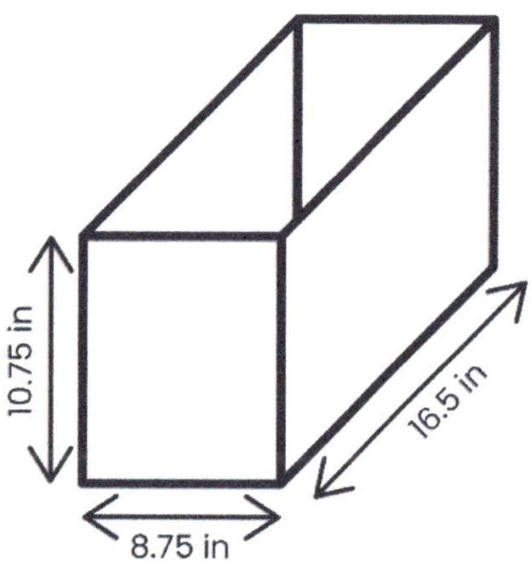

$$SA = LW+2LH+2WH$$

$$SA = 687.25\ in^2$$

The surface area of the sides and bottom of a 5g tank are 687.25 in^2. Now, let's look at the surface area of a single ball of aquasoil with a radius of 0.2 inches:

$$\text{SA of sphere} = 4\pi r^2$$

$$SA = 4\pi * 0.2^2$$

$$SA = 0.50\ in^2$$

Amazonia ADA V2 Substrate closeup - Source: Amazon

The SA of one ball of our theoretical aquasoil is approximately 0.50 in^2, which doesn't seem like much. But we're adding *2000 of them* to cover the bottom of the tank with an inch of them, bringing the total additional SA to 1005 in^2 and increasing the surface area of the otherwise bare-bottom tank by approximately 140%. In theory, that means 140% more biofilm is available for our shrimp to eat! In reality, much

of that SA may not be within reach for our shrimp, but real aquasoil is also not a perfect sphere – it's rough and semi-porous, so SA is significantly higher than that of a smooth sphere.

Additional surface area is still good even if shrimp can't reach it though. Biofilm also harbors beneficial bacteria and other microorganisms that help purify the water, so having **additional surface area makes the ecosystem healthier *in addition* to providing more food**. It's a win-win for our shrimp.

Admittedly, a bare-bottom tank is easier to clean and that's why we may see some hobbyists keep shrimp without any substrate - especially with Asian-Style UGF setups. A bare-bottom setup also makes sense for shrimp importers/sellers who want to sterilize the tank between groups of shrimp to prevent disease transfer. For regular hobbyists who want a nice-looking low-maintenance tank, having some substrate is a really good idea.

Impact On Shrimp Coloration

Having vibrant shrimp is a huge part of the hobby for a lot of people, whether for the purposes of selling, winning shows, or as accent pieces for aquascapes. Studies have shown that substrate has a huge impact on the coloration of our dwarf shrimp and that's what we're going to talk about here. Let's start with the basics.

Shrimp shells are covered by chromatophores - basically little pockets to store color pigments. These pigments are the same compounds that give other organisms their color, like orange carotenoids in carrots, yellow astaxanthin in lemons, or chlorophyll in leaves. The major pigments in shrimp tend to be carotenoids, most of which is stored in the shell to block UV radiation (Vazquez, Sganga, & Lopez-Greco, 2022), much like melanin in human skin. One key difference between humans and shrimp is that humans can make melanin, whereas shrimp cannot. **They need to get carotenoids from their diet**. In addition to UV protection, pigments like carotenoids also act as antioxidants in the bloodstream to protect cells from damage (Palozza & Krinsky, 1992).

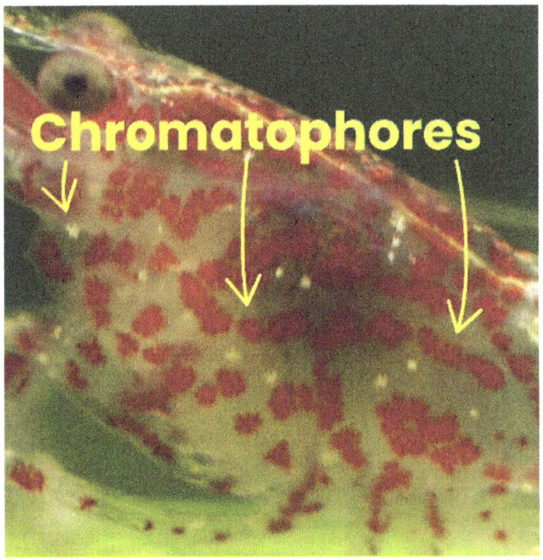

Cherry Shrimp with distinct chromatophore spots - Emilia Murray from Edinburgh, Scotland, CC BY-SA 2.0 via Wikimedia Commons - Edited

Most shrimp keepers aren't keeping our shrimp out in the sun where UV damage comes into play, so what we care about more is what causes the chromatophores and pigment amount to change. Chromatophores can shrink or expand in response to factors like stress from disease, water quality, or *background color*. Take a look at the difference in pigment between cherry shrimp kept on a white, red, or black background by viewing the images on page 780-781 (Tomas, Sganga, & Lopez-

Greco, Effect of background color and shelters on female pigmentation in the ornamental red cherry shrimp Neocaridina davidi (Caridea, Atyidae), 2020).

It's very clear that substrate color influences shell color, but why?

The study hypothesized the color change increased survival rate of wild Neocaridina (i.e. ones with normal brown coloration) by letting them camouflage better with their background. On lighter backgrounds, the darker pigments (red – carotenoids) contracted while the lighter pigments (white – purines) expanded to help the shrimp become lighter-colored and blend in more effectively, and vice versa on black backgrounds.

There were two other questions this study also asked:

- **Was the color change in response to specific substrate colors or just dark or light the substrate is?**
- **Was the color change a result of physically changing the shape of chromatophores or did carotenoid concentration change within their bodies?**

Do shrimp care about substrate color or brightness?

Knowing whether shrimp change color based on the specific color of the substrate or if they're just reacting to brightness (the amount of light reflected) can be useful to shrimp keepers. For example, if we want to have the reddest shrimp and find out that they try blending into red substrate by expanding the red pigments on their body, then we want red substrate. If all that matters is how light or dark the substrate is, then substrate color doesn't matter.

To test whether the color change was in response to substrate color or brightness, the study tested how shrimp reacted when placed on a red background. If shrimp were reacting to the substrate color, then red chromatophores should expand on a red background. Instead, researchers found that neither type of chromatophore expanded, suggesting the red background was sensed as neither bright nor dark.

This discovery indicates that **shrimp react to substrate brightness (the amount of light reflected), NOT the specific color.** Darker colored substrates lead to larger chromatophores that shrimp naturally consider "dark", like red chromatophores, while lighter ones do the opposite.

Are visual differences due to changes in chromatophore structure or carotenoid concentration?

Understanding how the color change happens is also useful for shrimp keepers. It gives an indication of how easily reversible color changes may be. Carotenoid concentration takes time to build up, whereas change in chromatophore structure may be relatively fast and easily manipulated for improved color. This question was tested in the following experiment from the same study:

Shrimp were kept on specific backgrounds for 90 days and then tested for the concentration of carotenoid in their bodies. The goal was to see whether the visual

difference was due at least partially to a change in carotenoid concentration or just a change in the contraction of the pigment. They found that **shrimp kept on the dark background had *two times the carotenoid content* in their body as compared to those kept on white or red background**.

Based on the results, it appears that shrimp accumulate carotenoids at different rates depending on substrate color, meaning **long-term adaptations occur when shrimp are moved from one substrate color to another**. That could be useful if we want our shrimp to look best at a specific time, like a contest. For example, if we want to emphasize the white on a shrimp, then it's worth keeping the shrimp on a lighter substrate. Even if the contest uses a black substrate, the shrimp will still have lasting chromatophore changes from their previous substrate.

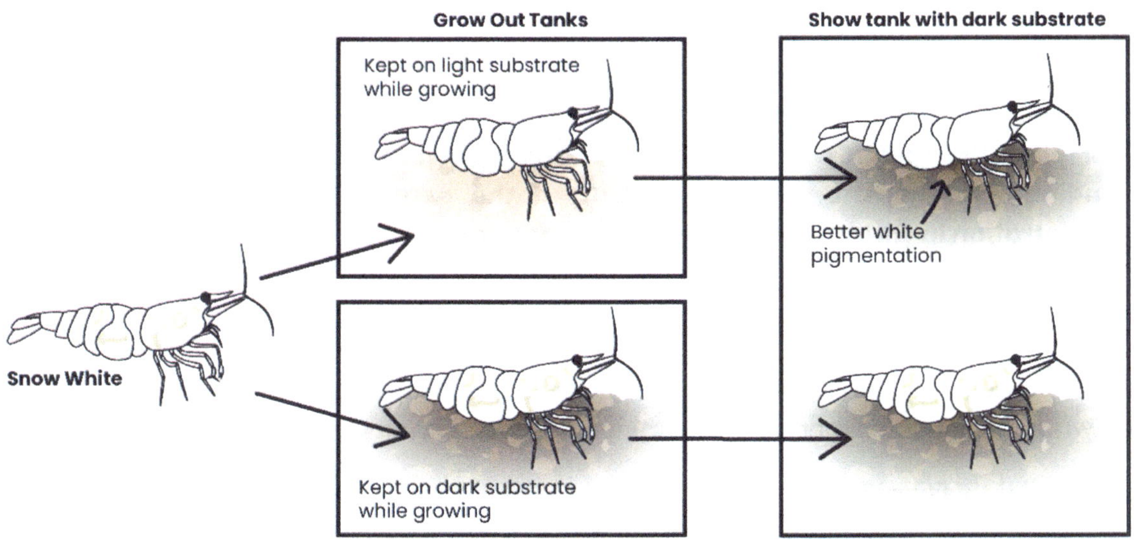

That being said, **it's also likely that chromatophore structure changes rapidly with the environment**. When we first receive shrimp in the bags they're shipped in, we often find shrimp look much paler than they do in pictures on the website. This can frustrate some hobbyists because they think they've been scammed and given low quality shrimp, when, in reality, the shrimp are just stressed and being kept in the dark, so their chromatophores don't know how to react. They typically recover their color after living in a healthy ecosystem for a few hours or days. If they still aren't vibrant after a few days, then the seller may have sent low grade shrimp.

Pure-Line Crystal Black Caridina with slight translucency on it's carapace that may be improved if raised on a lighter substrate.

Knowing that, **the contest example from the second to last paragraph may be a bad thing to do, depending on if a shrimp relaxes and regains their color in time for judging**.

If the contest uses dark substrate, then raising shrimp on a light substrate makes it difficult to know what they may fade to during the contest. We might choose the wrong shrimp and end up wasting valuable time and money.

Takeaways from the effect of substrate on shrimp coloration

1. Shrimp pigments change based on substrate color.
2. This color change is believed to be an adaptive trait to help shrimp hide from predators.
3. Shrimp react to how bright or dark the substrate is, as opposed to specific colors.
4. Shrimp can change their pigmentation relatively quickly (hours to days) but substrate color also has long term effects on the amount of pigment that shrimp store. Darker substrates cause more pigment to be stored.

For most colors (red, green, yellow, orange), a darker substrate is recommended to obtain a darker, more intense coloration. Darker colored shrimp (blue, black, dark green) tend to blend into dark substrates though, so the best brightness depends on your goal. We recommend lighter substrates if there's any concern about shrimp blending into them, or darker substrate if the goal is to increase the pigment concentration in shrimp before, say, a show or a sale.

Substrate Types

There are two main types of substrates for shrimp tanks: **inert** or **active**. Understanding the difference between these two is critical for shrimp keeping success because choosing the wrong substrate either creates a minor headache or, at worst, kills our shrimp.

Inert Substrates

Inert substrates do nothing – They don't provide nutrients or affect water quality in any way. They just hold down plants and offer a variety of different options to customize the appearance of a tank. They also are typically the cheapest options.

Please note, a variety of plants can still be supported by using liquid fertilizers for stem-feeders or root tabs for root-feeders, even with inert substrate.

If the goal is to have a low-maintenance Neocaridina tank, then the type of substrate doesn't matter very much - or if there's any substrate at all. The only things they care about are biofilm availability and healthy plants, which are not problems if the tank is properly set up. While substrate provides more room for biofilm, a **filter with good surface area for biofilm is sufficient for small shrimp populations**. The substrate (or lack thereof) is mainly based on aesthetic and plant preferences in Neocaridina tanks. As an aside, bare-bottom tanks are easier to clean, which is beneficial for breeding tanks with large populations that need frequent feeding (discussed in Breeding-For-Profit Tank Setup section [p226]).

In contrast, **most Caridina require a specific substrate to even have a chance at surviving long-term**. They need stable low pH (and low KH) conditions and **the only way to achieve that is with active substrates.**

Active Substrates

Active substrates are substrates that *do something* to water chemistry when they enter your tank, whether that's changing pH, KH, GH, or nutrient levels like ammonia. They are typically made by compacting and cooking volcanic clay into 1-5mm shapes that hold together even after long periods of submersion. Some active substrates are intended for plants, so the volcanic clay is mixed in with nutrient-rich soil or soaked in nutrients during the manufacturing process. These nutrient-rich substrates are called **aquasoils** and are typically best for moderately- to heavily planted tanks. They may also be referred to as **buffering substrates** because they help maintain a specific pH.

Here are some key points every hobbyist should know about active substrate:

1. They lower pH and KH (excluding crushed coral).
2. They are the only way to maintain stable and low pH levels over an extended period of time.
    a. They are sometimes called **buffering substrates** because they control pH.
3. **The buffering capacity (and any nutrients) of active substrates eventually becomes depleted and may require replacement**. Their lifespans are typically 8-24 months.
4. Active substrates last the longest when used with GH-remineralized RO/DI water. KH shortens the lifespan of the substrate and purified water reduces the amount of KH (carbonates) going into the system.
5. Active substrates may be used with Neocaridina if they do not drop the pH too low (<6.0 or so).
6. **Active substrates *must* be used for Caridina that need low pH. They are the only way to maintain stable and**
7. **low pH.**
8. Do not rinse active substrates even if the manufacturer says to rinse them.

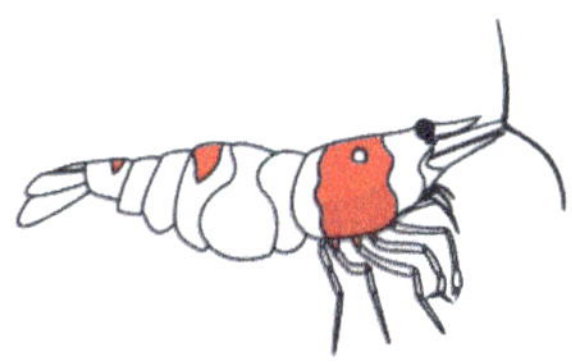

Now, let's go into the chemistry of active substrates.

Based on our definition of active substrate, crushed coral substrate also falls into the category. Crushed coral has a very different effect on water chemistry (raises pH/GH/KH), so the following chemistry does not apply to it. We do not consider crushed coral a good substrate for shrimp.

How Active Substrates Change pH and KH

The clay used in active substrates is mixed in with tons of organic matter (OM) – basically decomposed plants, animals, bacteria, fungi, and archaea. OM has humic acid in it that gets released over time from the substrate. As it enters the water, it lowers pH and KH, which is why many of these substrates advertise that they keep pH below 7.

This isn't quite true.

Active substrates release acid at a certain rate and only have a finite supply to begin with. This leads to two key points:

- **Active substrates lower pH and KH but can only lower them so much.** If we start with water at 8 pH and a high KH (7-10 dKH), then we may see a small drop to, say, 7.8 pH and 6 dKH, but won't see our pH drop into acidic ranges (<7 pH) like advertised. Active substrates continually decrease pH and KH but it may take a long time to reach acidic conditions – especially if water changes add KH in. *The pH range achieved by active substrates all depends on the water we start with.* The best way to achieve acidic pH is to start with purified water that is already neutral or acidic and has a low KH. This is why many shrimp keepers who use active substrates add purified water (RO/DI water) that they know is low in KH and pH.
- **Active substrates have a finite supply of acid so only lower pH and KH for a certain amount of time**. After they run out, they are **depleted** and may need to be replaced if the goal is to maintain stable acidic conditions like those required for soft water caridina shrimp. We can get the longest lifespan from active substrates by adding as little KH as possible to the tank, which is often achieved by using purified water.

Please note that purified water by itself is not suitable for life. Minerals need to be added to it so plants and animals can survive. This is typically done through powdered minerals or concentrated mineral solutions called remineralizers, which are discussed in-depth in the Remineralizer section (p93).

The Dangers of Active Substrate

Since active substrate changes the chemistry of a tank, it's important to consider how that change affects biological processes – specifically decomposition and nitrification.

Decomposition and nitrification (collectively called "decomposition" from now on) are heavily influenced by pH and the presence of KH. At low pH and KH, both of these processes occur much slower, which leads to a buildup of waste – especially when overfeeding (Evans & Sober, N.D.) (Sundberg & Jönsson, 2008). **This waste buildup turns into a ticking bomb for a tank once the substrate starts to lose its buffering capacity**.

As active substrates lose their buffering capacity over 9+ months, the pH rises and decomposition rates increase. The initial small pH increase is hardly noticeable since decomposition releases hydrogen ions, causing pH to decrease in turn. **Once the substrate loses more buffering capacity, pH rises further, leading to even faster decomposition and possible ammonia spikes**.

It gets worse though.

While decomposition decreases pH and fights against the lost buffering capacity of the substrate, photosynthesis actively works to increase pH during the day. Tanks that have been around long enough for active substrate to lose buffering capacity often have good plant growth, which increases the amount of photosynthesis and leads to stronger pH increases. With strong plant growth and a failing substrate, the natural pH decrease from

decomposition gets overridden, leading to a steady increase in pH throughout the day and significantly higher decomposition rates than before. In turn, this leads to ammonia spikes and a sudden drop in water quality that quickly kills sensitive shrimp.

This all ties back into the danger of overfeeding. With the lowest pH and decomposition rates when a tank is first set up, we can overfeed without seeing elevated ammonia or nitrate. The only sign of waste buildup may be an increase in microorganism populations. By continuing to overfeed, we set the stage for an eventual collapse once the substrate loses its buffering capacity. This is part of the reason why many shrimp keepers reset Caridina tanks every 12-24 months, depending on substrate lifespan. It is not an issue if we follow the rules set out in the Food chapter (p137) to prevent overfeeding, but a tank reset is still required eventually.

How Active Substrates MAY Change GH

This aspect of active substrates isn't often discussed because it only happens with some substrates and some batches of substrates. As we mentioned before, active substrates are clay-based, and clay has an affinity for nutrients. Some of these nutrients include calcium and magnesium, which determine GH of water. When these minerals get pulled from the water, they decrease GH, potentially leading to levels outside the recommended range for our shrimp. This happens until the substrate is fully saturated with nutrients. **How much a substrate attracts nutrients like calcium is known as the cation exchange capacity, or CEC**.

Many active substrates are already saturated with nutrients, so they have no impact on GH. Some batches may have fewer nutrients than others though, so the effect on GH can vary. When using active substrates, just be aware of how much GH changes in the first couple of weeks and be ready to do water changes until GH stabilizes.

To be clear, most active substrates do not have a significant effect on GH, but it's good to be aware of the possibility. If there is an initial drop in GH, perform a few large water changes (30-50%) to bring the GH back up. These water changes may need to be done during the cycling period (p99) if GH goes below 2 dGH.

How Active Substrates Impact Different Shrimp Species

Active substrates have a different impact depending on the type of shrimp.

Neocaridina

Some sources online say that Neocaridina cannot handle low pH and KH conditions so active substrates are bad for them. This is... somewhat true. Let's address the rationale behind this claim.

Myth: Neocaridina Need Carbonate in The Water

There are some hobbyists who think shrimp need some KH because their shells are partially made of carbonate. If there's no carbonate (KH) in the water, then how can shrimp get it?

While shrimp do need carbonate for their shells, that carbonate *doesn't need to come from the water.* Carbonate is created from CO_2. Shrimp, just like humans, take in oxygen, use it up, and turn it into CO_2 inside their bodies. Some of this CO_2 gets converted into carbonic acid, bicarbonate, and then carbonate, which gives them enough material to make their shells. We have proven this by keeping Neocaridina in 0 dKH water, during which they successfully molted repeatedly. In addition, some shrimp farms in Asia exclusively keep Neocaridina in 0 dKH water.

Instead, the main problem with active substrates for Neocaridina is pH stress. The recommended pH range for Neocaridina is around 6.5-8.0 pH. The exact upper and lower limits aren't set in stone though. For example, at Shrimply Explained we've kept Neocaridina with Caridina in 6.0 pH water and they've survived for 8-12 months. We didn't see breeding but some hobbyists report Neocaridina breeding and thriving at that low of a pH. It's not that they can't survive and even thrive, but the further outside the recommended pH range, the more stressed they get.

Some active substrates (typically the cheaper ones) only bring pH down around 6.5-7.0 pH, so these are fine for Neocaridina. Again, even if a tank is at a pH of 6.2, Neocaridina will still likely do well. They're very adaptable. The problem happens with more acidic active substrates that achieve pH below 6.0, at which point success with Neocaridina becomes much less likely.

What makes this difficult is that **the acidity of some substrates depends on the batch**. Each batch is made from clay/soil collected from *slightly* different areas, some of which may be much more acidic than others. This creates very different experiences for hobbyists depending on the batch they get, leading to conflicting reviews like:

> "WAH, [X SUBSTRATE] DROPPED MY PH TO 5 AND NOW I CAN'T ADD SHRIMP!!! IT ALSO LEACHED A TON OF AMMONIA! WHAT DO I DO???"
>
> Or "[X substrate] is a fantastic substrate that works well in Neocaridina tanks and doesn't leach ammonia at all. 10/10!"

Preparing For Different Batches

Since the main problem for Neocaridina is pH stress from exceedingly low pH, all we need to do is pay attention to the pH and raise it if it approaches 6.0 pH. This is easily done by adding baking soda (sodium carbonate) to the tank. Any baking soda in your kitchen is food-grade, so it is perfectly safe to add. Doing so raises the KH to combat acid being released from the substrate. Just keep in mind that adding KH does deplete the substrate's buffering capacity.

Note: Only add baking soda directly to a tank if shrimp are not yet introduced. If shrimp are already in the tank, then adding carbonate changes pH quickly and may stress out shrimp. Instead, add baking soda to new water outside of the tank and do a slow water change (30-120 mins to add in all the water).

Another option for maintaining a healthy pH level is to add some crushed coral to the tank. Crushed coral dissolves over time to release minerals and carbonate into the water, which raises pH and GH. We don't want to raise GH excessively, so always start with a conservative amount of crushed coral (1 tsp per gallon). More information about adding baking soda and crushed coral may be found in the pH/KH Control section (p261).

Active Substrates for Caridina

Unlike Neocaridina, soft water Caridina want and, in fact, *need* a low pH around 5.5-6.7 (depending on the breed and breeder). It's possible to add acid buffering products to water to get that level but doing so doesn't keep pH low or stable in the long run. **Active substrates are the only way to achieve stable and low pH, so all soft-water Caridina tanks must have an active substrate**. That being said, we've already discussed how active substrates get depleted over time and do need to be replaced eventually (approximately every 8-24 months). We have control over some factors that affect how quickly active substrates deplete and those are discussed in the next section.

How to Extend the Buffering Capacity of Active Substrate

Limit KH

The most important thing anyone can do to extend substrate buffering capacity is to limit the amount of KH (carbonate) that enters a tank.

KH is a basic compound (>7 pH) that increases pH *and* resists drops in pH, so it quickly neutralizes acidic compounds in active substrates. With enough KH, an active substrate can lose its buffering capacity in a month, no matter the quality of the substrate. The main source of KH for many tanks is tap water, which is why shrimp keepers who use active substrates typically use remineralized RO/DI water (p93) without any KH. That also means using the right remineralizer that *only* affects GH and not KH.

Regardless of what type of water we use, limiting the amount of water changes helps extend the lifespan of active substrates. Here's a metaphor about a roller coaster to help explain what happens:

> *There are X number of people waiting to ride a roller coaster. Each group of riders must wait until the previous group gets off the ride before they can get on. As each group of riders get on and off the ride, the line depletes until there's no one left. If the ride is longer or if there are technical issues, then it takes longer to clear the line.*

That's how acids in active substrates work. When a substrate gets added to the tank, the first group of acid leaves the substrate to ride around in the water column. The remaining acid waits in the substrate until the acid in water gets removed via a water change, at which point the next group of acid can take their place in the water. Each water change reduces the line of acid waiting to enter the water, until eventually there's no acid left. Decreasing the frequency between water changes keeps more acid in the substrate for a longer time.

Side Note: There are times when water changes are necessary (high nitrates/ammonia, etc.) vs unnecessary routine maintenance. Necessary water changes are still, well... necessary, but we recommend reducing the amount of unnecessary water changes in tanks with active substrate if the goal is to extend substrate lifespan as much as possible. **Shrimp colonies may do better or worse with fewer water changes depending on the setup, so we need to balance their health with the water change frequency and substrate lifespan.** Some breeders change their substrate every 9-12 months after doing regular water changes while others do so every 3-4 years by only topping off their tanks. This is discussed more when we compare two different Caridina breeding setups on page 235.

Does this mean a substrate can last forever if we don't do water changes? Not quite. Going back to the roller coaster metaphor - what happens when a roller coaster breaks down or takes too long to get people on?

Some of the people standing in line will leave and the line gets shorter.

The same is true for active substrates even when we don't do water changes. Acids stuck in line in the substrate can be removed through biological processes from plants and animals and bacteria so the substrate still depletes over time, although at a slower rate than with regular water changes.

Buy High Quality Substrate

Every substrate has a different **acid capacity (AC)** and **acid release rate (ARR)**. For example, ADA Amazonia V1 tends to keep the pH lower and maintains that pH for a longer time than Fluval Stratum, indicating that Amazonia V1 typically holds more acid (higher AC) and takes longer to release it (lower ARR). Because of this higher performance, Amazonia V1 tends to be 20-30% more expensive than Fluval Stratum.

ADA Amazonia V1 was the original go-to substrate for most shrimp breeders and still is for Singaporean breeders/sellers like Kelvin Wong (KWshrimps.com) and Lawrence Lee (ShrimpHub.com.sg). These two breeders liked the substrate so much they managed to stockpile hundreds of bags of V1 before it was discontinued. For everyone else in the world, active substrate choice is more difficult now because Amazonia V2 is made of volcanic soil from a different area than V1. As a result, V2 doesn't have quite the same longevity or nutrients in it so other substrates like akadama have become competitive. Based on polls

and discussions with other shrimp breeders, we've compiled a short list of the best-performing active substrates:

Best Active Substrates for Caridina (in no specific order)

- Brightwell Rio Escuro
- ADA Amazonia V1 (hard to find)
- Akadama Double/Triple Red Line
- SL Aqua Soil
- We keep an updated list at shrimplyexplained.com/book-links/[xxiii]

Please note that all these substrates are made of volcanic clay but some of them have additional nutrients (Like ADA V1). As a result, **they may leach ammonia for 2-3 months after starting a tank, whereas Akadama substrate does not leach ammonia at all**. Not having ammonia sounds like a great feature since we don't need to worry about ammonia spikes when replacing substrate, but it comes with some challenges. Be sure to do research before purchasing a substrate to understand its nutrient contents.

As we've discussed in other sections, ammonia and its byproducts are critical nutrients for plants and algae. It forms the base of a strong ecosystem once beneficial bacteria and plants are established enough to prevent ammonia from reaching harmful levels in a tank. **Active substrates with ammonia may be easier to set up and cycle because they already have a consistent ammonia source**. There's also some evidence that ammonia is required for green walls in a tank. This makes substrates like Amazonia excellent for starting a tank and growing healthy plants but more time-consuming to replace. If you choose to use an active substrate that doesn't leach ammonia, then fertilizers may be required to obtain the healthiest ecosystem. This entirely depends on what types of plants are in the tank.

Don't Rinse Active Substrate

An active substrate is active because of all the compounds inside of it. Rinsing the substrate removes some of those critical compounds we paid for, so do not rinse it even if the instructions say to rinse!

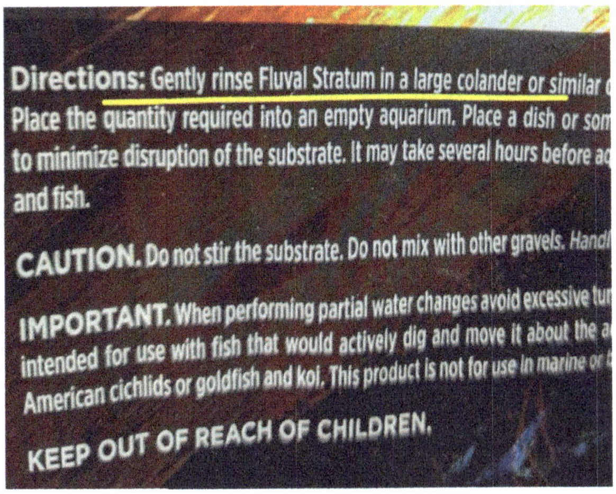

Yes, this is a direct callout of Fluval Stratum - WHY DO YOU INSTRUCT PEOPLE TO DO THE WRONG THING, FLUVAL???

Add More Substrate

A simple way to increase the lifespan of active substrate is to add more of it to an established tank every 6-12 months. We can do this by dumping loose substrate in, adding a new undergravel filter (UGF) filled with substrate, or by adding a mesh bag full of substrate for better control. **Active substrates that don't have ammonia may be added in without any preparation**. On the other hand, active substrates with ammonia should

be soaked in a bucket until they are done leaching ammonia. Admittedly, even a little ammonia leaching shouldn't be an issue in most Caridina tanks because they're at a low enough pH that ammonia isn't toxic.

Summary – How to Extend Active Substrate Lifespan

Do the following to extend the life of active substrates:

1. Use purified water that doesn't have KH in it (RO/DI)
2. Use remineralizers that don't have KH
3. Do fewer water changes (unless necessary)
4. Buy high quality substrate
5. Don't rinse substrate
6. Add more substrate periodically

Substrate Additives

There are many powders and other products that shrimp breeders use to varying degrees of success. Most of these have no evidence to support their use. The one additive we found with scientific evidence to support it is montmorillonite clay (MC), AKA calcium bentonite clay.

If you have proof of other additives being beneficial for a tank, share it with us at contact@ShrimplyExplained.com so we can add it to the book!

What Is Montmorillonite Clay?

MC is a silica-based clay with a strong negative charge (Yaghmaeiyan, Mirzaei, & Delghavi, 2022). This negative charge causes anything in the water with a strong positive charge to bind to it, including dissolved organic matter, heavy metals, and minerals like calcium (Jha, Mishra, Kumari, & et al., 2011). **The clay comes in two forms: sodium- or calcium-based MC (Na-MC or Ca-MC)** and the Na or Ca are released when more positively charged compounds bind. After binding with a compound, MC sinks down into the substrate, which makes bound nutrients more accessible to plant roots, bacteria, and other organisms in the substrate (Tietjen, Vähätalo, & Wetzel, 2005). MC is commonly used in ponds and wastewater plants to bind organic matter in the water column to improve water clarity. If a tank frequently has cloudy water from organic matter or a bacterial bloom, then MC is a good option to solve those issues.

Interestingly, there have been a number of experiments testing the effect of MC as a dietary supplement in shrimp food. In one experiment, researchers fed diets with varying levels of Ca-MC (0%, 0.25%, 0.5%) to different groups of shrimp (Ng, Mong, & Abdul-Hamid, 2023). After two weeks, they exposed the shrimp to a bacterial disease and measured their survival over time and found that the experimental groups fed MC had at least twice the

survival rate compared to the control group without any dietary MC (80% or more vs 40% survival, respectively). This study found no change in growth over the trial period but a separate study found an increase in growth along with improved immune function in shrimp (Pourjam, Sotoudeh, Bagheri, & et al., 2024).

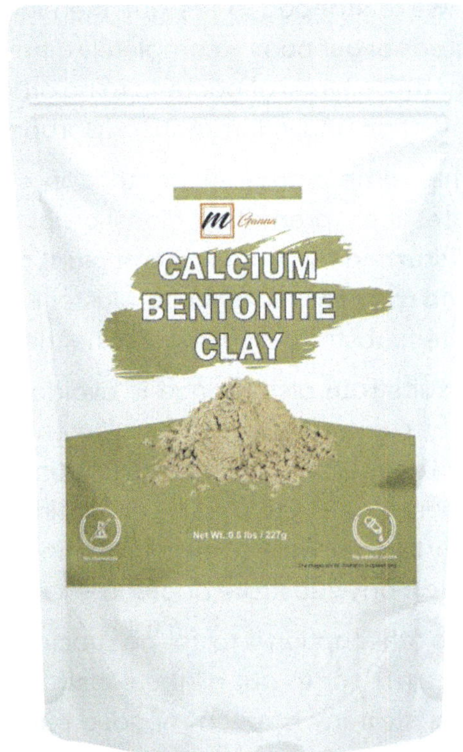

One problem with MC is that it binds minerals like calcium (Ca) and magnesium (Mg) so can decrease GH. How much it does so depends on the source, as MC from different areas have different amounts of Ca and Mg bound to them before entering the water. When it does enter water, MC can actually *raise* GH because it releases bound minerals in exchange for more positively charged organic matter and heavy metals. An increase in GH is more often seen with Ca-MCs. As such, only small amounts are recommended or needed.

Summary

Overall, research supports the addition of Ca-MC to promote cleaner water and better immune function in shrimp. It may be added to the tank by mixing a sprinkle (1-2g per 100L [~40g]) of MC into the substrate when setting up the tank. Some shrimp keepers also add a very small pinch of MC to the water during water changes or anytime the water isn't crystal clear. It can float so some stirring may be required. To be safe, we recommend checking GH levels periodically when first using it to get a better understanding of how much to add.

Check out shrimplyexplained.com/book-links for our recommended food-grade MC[xxiv]. Many sellers advertise it for healthy skin and questionable detox purposes, but it would be hilarious to suddenly have a bunch of reviews talking about how good it is for shrimp skin care...

The Disease Risk of Thick Substrates

As we've mentioned, substrates add surface area to promote additional bacterial growth. The bacteria that grow in deeper substrates prefer reduced water flow and oxygen concentration found there. They perform important roles in cycling nutrients throughout the ecosystem.

That all changes if substrate bacteria are disturbed. Outside of their normal area, **deep-substrate bacteria act differently and may cause disease**.

Think about it like bacteria in our intestines. Intestinal bacteria are necessary to help break down food efficiently, giving us the most resources from our burgers and fries.

If we're stabbed in the gut, then intestinal bacteria leave the intestines and enter other areas of our body – completely different microbial environments. Our body is not prepared for those types of bacteria to be other areas, so intestinal bacteria are very likely to cause infection and inflammation (Gorbach, 1996).

The same is true for bacteria in substrate. The substrate effectively acts like a tank's intestines, promoting digestion of waste from animals and plants. **When substrate is disturbed by vacuuming or plant removal, substrate bacteria fly into the water column and may spread disease.** Master Breeder Kerric Peyton says that he often noticed deaths after substrate cleanings, so he tries to avoid substrate disturbances at all costs.

If substrate disturbance is avoided, thick substrates create a low-oxygen environment that supports a variety of different bacteria which can help break down waste, improve water quality, and reduce maintenance frequency. Disease risk is very low IF the substrate doesn't get disturbed. If you do find that you overfed or need to disturb the substrate for any reason, we recommend dosing hydrogen peroxide (p295) soon after a disturbance to kill off any substrate bacteria left in the water column.

The other option is to use no substrate (bare-bottom), a thin layer of substrate (<0.5 inches [1.2 cm]), or to put all the substrate inside an Asian-style undergravel (UGF) filter (p79). This strategy is recommended if the goal is to breed a lot of shrimp, since that typically requires a large quantity of food and therefore a large amount of waste that needs to be removed.

Tank Equipment

Heater/Cooler/Cooling Fans

These pieces of equipment are self-explanatory – A heater raises the tank temperature and keeps it at a certain point, while a cooler does the opposite. There are three reasons to use one of these in a tank:

- If the tank reaches temperatures outside of a species preferred range, then temperature control is recommended.
- If the tank is in an area that experiences fast temperature changes (i.e. next to a window), then temperature control prevents stressful fluctuations. Slow changes throughout the day are not a problem, unless they dip outside of the recommended range for a shrimp species.
- If the goal is to breed or sell shrimp, then maintaining a tank at the upper-middle range of their preferred temperature increases growth and breeding rates while limiting disease risk. **With higher temperatures or denser populations, bacterial infections are more likely to spread and cause problems so regular maintenance is important.**
    - Going much higher than the recommended temperature causes females to lose their eggs, defeating the purpose of raising the temperature in the first place. Research shows that somewhere between 82°F (28°C) and 90°F (32°C) is when Neocaridina become too stressed to carry their offspring to term (Tropea, Stumpf, & Lopez Greco, 2015). There has been no scientific research on Caridina but breeders often start having problems above 75°F (24°C) so temperature control is even more important for them.

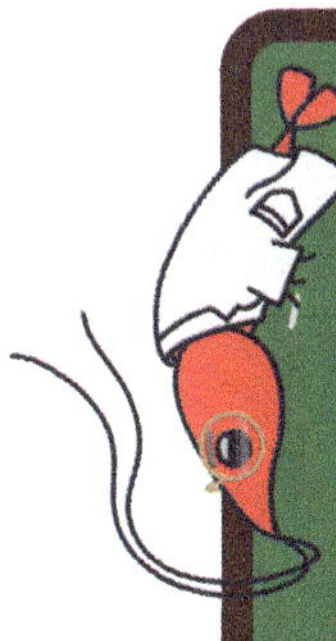

Purchasing Tip: Make sure the heater/cooler is rated for your tank size. If it isn't, then you may not get the control you're looking for and shrimp may get stressed. There's also a danger that a heater designed for smaller volumes of water may overheat and break, causing electrocution. For this same reason, **always install heaters somewhere they stay fully submerged and unplug them while performing large water changes.**

Lighting

Many sources say that lighting matters for plants and the overall appearance of our tank, but that it doesn't affect shrimp. It depends on our goal though. **If we just want to have a healthy shrimp tank, then the only concern with lighting is whether it is strong enough to promote healthy plant growth**. The photoperiod (24-hour light:dark cycle [i.e. 12:12]) also only matters as a means to control plant and algae growth.

If we want to optimize breeding and coloration though, then there's good reason to believe that lighting matters a great deal. First off, lighting influences the amount and type of biofilm that develops. Tasty biofilm and algae grows faster at higher light intensities (at least up to 200 μmol/m/s) and longer photoperiods (Wang, Jiang, Lai, & et al., 2021). The type of lighting also has an influence on the type of microorganisms that grow, as demonstrated in an experiment where more filamentous algae grew in response to fluorescent lighting (Coyle, Bright, Wood, & et al., 2011). If we want to copy the setup of a successful breeder, there are good reasons to pay attention to the types of lights they use. **In short, there's convincing evidence that lighting intensity, photoperiod, and programming does matter if we want to optimize shrimp breeding. Let's take a look at some research.**

Lighting Intensity

Numerous experiments have shown that higher lighting intensity increases shrimp growth rate. For example, one experiment showed an approximately 10% increase in final weight for shrimp kept with 24-hour lighting compared to those kept in the dark (Baloi, Arantes, Schveitzer, & et al., 2013). This may be partly due to an increase in food due to additional biofilm growth, but another experiment cleaned tanks twice a day to eliminate this confounding factor and *still* showed a 28% increase in weight gain of female shrimp in high-light vs low light conditions (211% growth in high-light vs 168% growth in low-light compared to starting weight) (Hoang, Barchiesis, Lee, Kennan, & Marsden, 2003). Researchers hypothesize this difference is due to activity level, as shrimp kept at higher light intensities show lower activity levels and may therefore have more energy put toward growth.

One study on *Macrobrachium tenellum* did not find that light intensity differences lead to significant differences in growth but did find it affected pigmentation (Vega-Villasante, Martínez-Ochoa, García-Guerrero, & Arrona-Ortiz, 2015). The researchers photographed multiple areas of each shrimp to count chromatophores (pockets that store color in the shell) before and after the experiment. They found shrimp exposed to 1500 lx had approximately 50% more chromatophores than those exposed to 30 lx and 25% more than those exposed to 600 lx. The theory behind this biological response is that more chromatophores protect shrimp from radiation, just like how humans develop more melanin in our skin (we tan more) with greater sun exposure.

Overall, while these studies are not specifically on Neocaridina or Caridina, they suggest that higher light intensity may lead to stronger coloration and faster growth rates.

Light Photoperiod

There is some research suggesting that photoperiods play a role in molting but results are inconclusive. One study showed a significant increase in molting rate as the day's photoperiod got longer (1.68 daily molting rate (DMR) for 13:11 photoperiod vs 0.32 DMR for 0:24 photoperiod for juvenile *Palaemonetes argentinus* shrimp) (Diaz, Sousa, Cuartas, & Petriella, 2003).

Another study tested a 12-hour photoperiod cycle (7:5 – 14 hours total light exposure per day) and found significantly higher growth when compared to a 12:12 photoperiod (Hoang, Barchiesis, Lee, Kennan, & Marsden, 2003). The molting rate remained the same, indicating the shrimp exposed to a 12-hour photoperiod had significantly higher weight gain with each molt. The increased growth remained consistently higher at the 7:5 photoperiod for the entire 12 weeks of the study.

These results suggest that a longer photoperiod of 13-14 hours may increase growth and/or molting rate. It is still unclear whether it is better to have the light on consistently for that time or to break up the lighting into two separate periods. In practice, constantly lighting for 14 hours may lead to unwanted algae growth, so breaking up the lighting periods may improve tank appearances.

How Lighting May Cause Stress

Many cheaper lights turn on and off immediately, leading to two potential problems. For one, plants change rapidly from energy storage to energy usage, leading to quick pH changes as plants shift their consumption of CO_2. This effect will be small in aquariums with adequate buffering from KH or substrate but is worth considering.

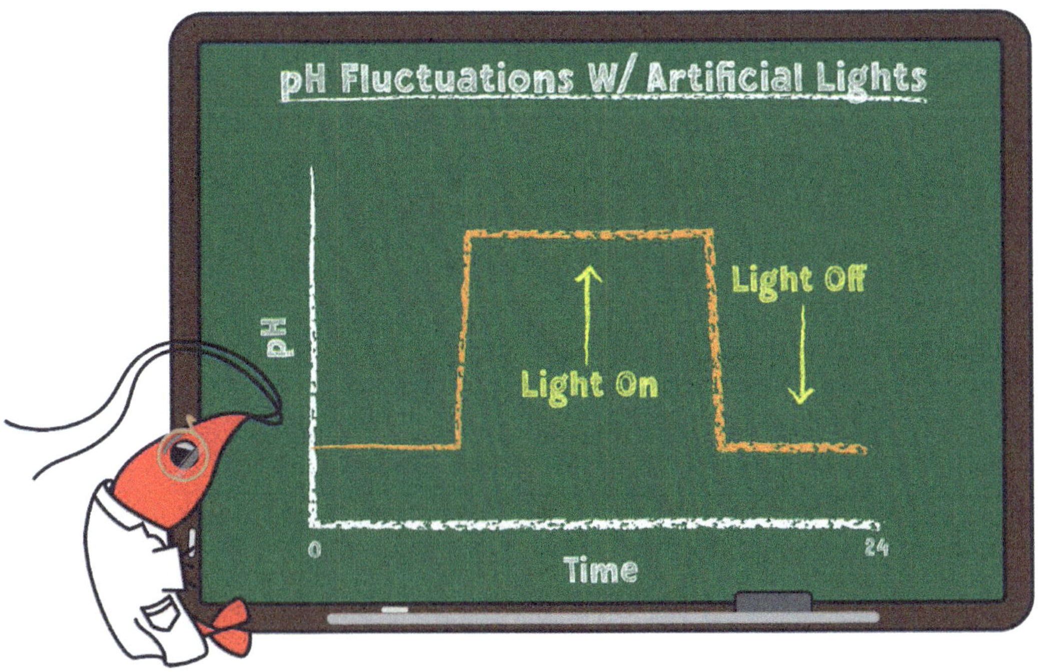

In addition, how would you feel if lights suddenly turned on or off outside of your control? We've noticed some shrimp appear startled when light changes suddenly so cheap lights may stress shrimp out more.

Overall, both concerns are unlikely to have a large effect on a tank. It is absolutely possible to have a healthy shrimp tank even with cheap lights, but better lighting is worth considering when breeding or keeping more expensive shrimp. **It may be worth buying a light with a sunrise/sunset mode that dims or brightens over time (instead of immediately turning off or on) to avoid sudden pH changes and limit stress.**

How Lighting Affects Our Perception of a Tank

Additional concerns of lighting include how the light spectrum and intensity affect our viewing experience. Different spectrums affect the contrast and brightness of different colors in the tank. A full spectrum light designed for planted tanks generally leads to better color contrast and a more striking tank appearance – and that applies to our shrimp, too! In addition, these same lights typically have higher intensity than the cheaper aquarium lights, making it easier to take good photos of our shrimp.

Tank lit by cheap LED shop light that doesn't display color well or encourage healthy plant growth.

Unfortunately, we could not find much information on how the light *spectrum* influenced shell pigmentation or shrimp growth but suspect it does play a role. It's also likely that shrimp kept at higher light intensities develop stronger colorations as a means of protecting themselves from UV rays. This leads to a few questions that we'd like to experiment with in the future:

- Does light spectrum affect shrimp coloration?
- Does light spectrum affect the tank's microbiome?
- Do light sources with more UV lead to better coloration in our shrimp?

Takeaways

There is still a lot to learn about how lighting impacts shrimp, but **the general points are that stronger lighting with longer daily exposure seems to have a positive impact on shrimp growth and coloration, whether due to behavior or microbiome changes**. In addition, opting for a light with sunrise/sunset mode may reduce stress. The research suggests that paying a bit extra for a planted tank light with more power and a better spectrum for plant growth is worthwhile for any shrimp tank, although budget lights work perfectly well for simple, low-maintenance setups.

To be clear, the light doesn't have to be super expensive. Some caridina breeders like Kerric Peyton breed award-winning shrimp with budget planted tank lights like the NICREW SkyLED Plus[xxv].

Filtration

Waste from animals or dying plants is constantly being produced in any tank. Without filtration, this waste floats around in your tank, reducing water quality and looking horrible. Dead zones (areas without water flow) may also develop due to a lack of water flow, increasing the chance of inconsistent pH, ammonia spikes and oxygen deficiency. Filters solve these problems by moving water around a tank and cleaning it through mechanical, biological, and sometimes chemical methods. A brief explanation of each of those methods is below, followed by sections about types of filters we can use in a tank.

Mechanical Filtration

Every filter has a porous layer (sponge, filter floss, etc.) that water flows through. This porous layer catches any large particles to keep water looking crystal-clear.

Biological Filtration

As we've learned previously, beneficial bacteria (BB) grow on all surfaces in a tank, including the sponge and other filter media. These BB eat ammonia released from the trapped waste in the sponge, which avoids toxic build-ups and promotes a healthy ecosystem.

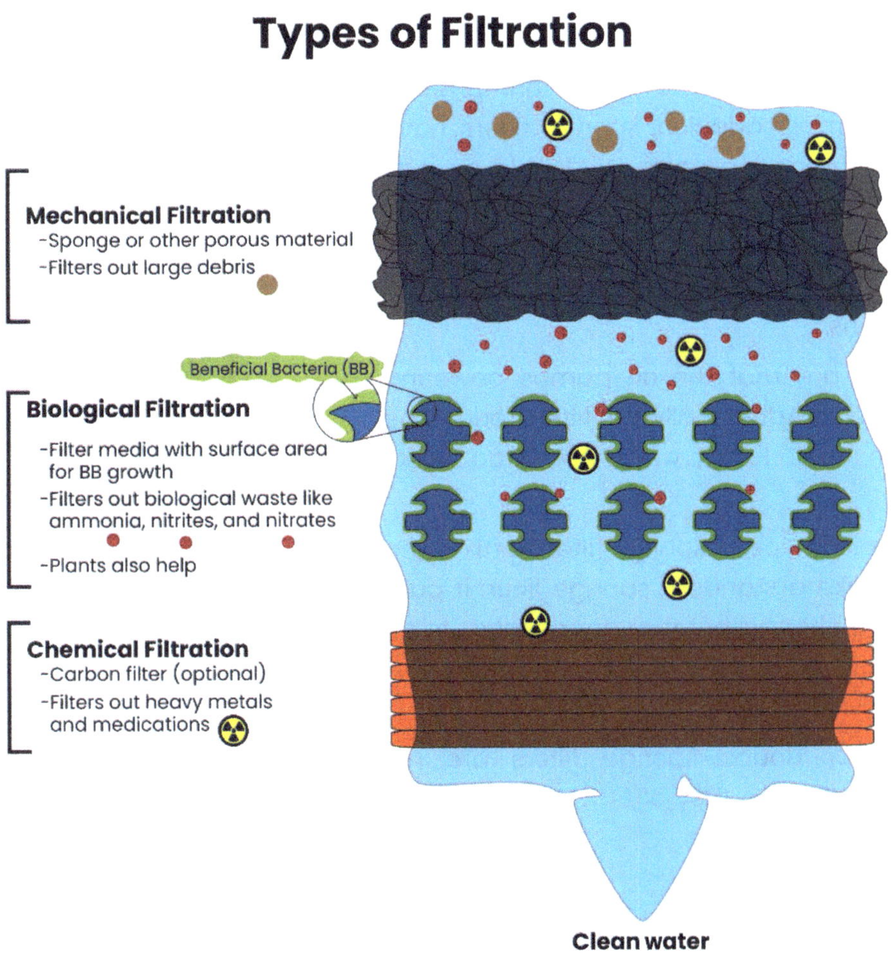

Chemical Filtration

Certain types of filters come with a layer of carbon. This carbon reacts with medications, metals, and some nutrients from the water to sequester them inside the carbon filter. We generally recommend *against* using carbon in shrimp tanks because plants need some of the nutrients that carbon takes away. The exception to this is when treating the tank with certain medications, as carbon may be the only way to remove those from the water column. Do research on specific medications to determine whether carbon filters are needed.

Types of Filters

There really are no bad filter options for shrimp, except for ones that suck them up and trap them inside --Those aren't great. Luckily, nearly all filters can be adapted to prevent that issue. The way different filters work, along with their advantages/disadvantages, are discussed below:

Sponge Filters

Sponge filters work by pumping water through a sponge, filtering out particulates (mechanical filtration) and pushing the water past BB (biological filtration). They move water through the tank very slowly but oxygenate it well.

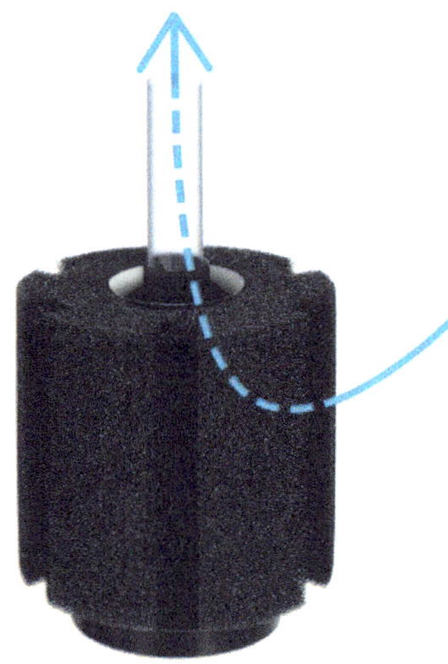

Sponge filters are most often used in shrimp tanks because there is absolutely no risk of shrimp being sucked into them. In addition, sponge filters have high surface areas (especially double-sponge filters[xxvi]) due to pores in the sponges, which helps grow more delicious biofilm and beneficial bacteria. Another benefit is that sponge filters tend to be the cheapest options.

The downsides are that the air pumps powering them can be noisy. Additionally, the filters tend to be very visible in most tanks, which some aquarists dislike.

A **Matten filter** is a type of sponge filter worth being aware of. Just like a standard sponge filter, it pulls water through a sponge, but the sponge tends to cover an entire side of a tank instead of just the area around the central tube. This provides even more surface area but tends to be more expensive.

Matten filters or double-sponge filters are recommended for Caridina tanks because they offer more surface area and filtration than standard sponge filters.

Asian-Style Undergravel Filter Boxes

Asian-style undergravel filters (UGFs) are another common type of filter mainly used in Caridina tanks with active substrate. They are basically plastic boxes that hold substrate and use airflow to pull water down into the box. This design pumps water through the substrate, which helps the substrate release acid consistently to keep the pH low. Another benefit of this design is that all the substrate is held in the box, making it easy to replace when the substrate loses buffering capacity.

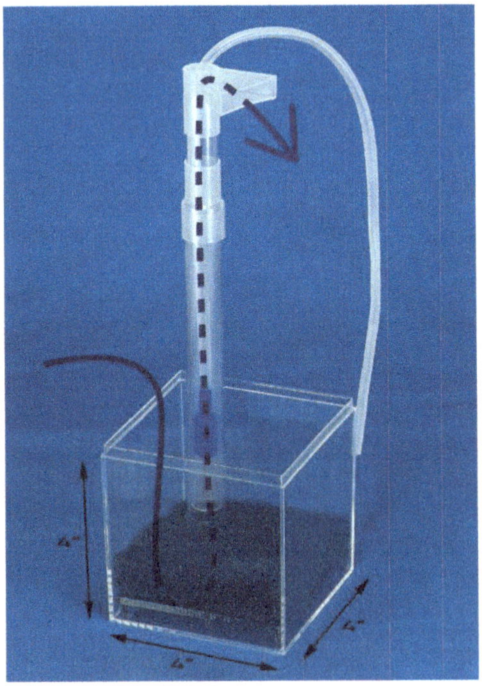

Asian-Style UGF from Plamski Shrimp Farms

One downside to these filters is the fact that they do not add as much surface area as sponge filters. Another is that the UGFs designed for shrimp tanks are often manufactured in Asia and may not be easily available in other parts of the world. Lastly, UGFs can clog and become ineffective quickly. To reduce how quickly UGFs clog, many breeders add a layer of coarse sponge and then place substrate on top of it.

To add surface area and ensure sufficient filtration even when UGFs clog, many shrimp breeders use 1-2 double-sponge filters in tandem with UGFs.

Hang-On-Back Filters

Hang-on-back (HOB) filters have an inlet tube that sucks up water into a chamber outside of the tank. This chamber is filled with various layers of materials that perform mechanical, biological, and chemical filtration. The layers are often fully customizable.

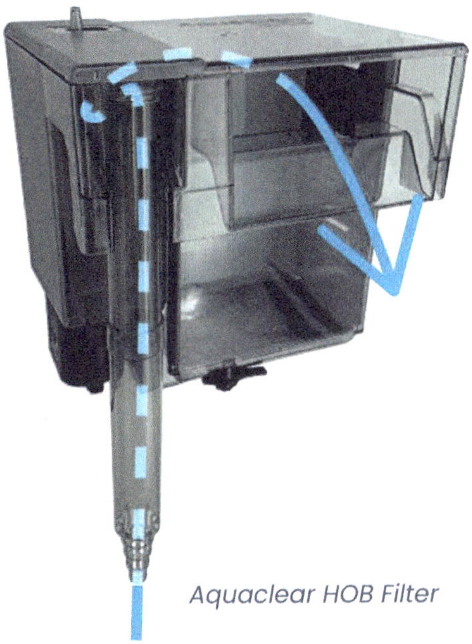

Aquaclear HOB Filter

HOBs often come with disposable cartridges that the manufacturer recommends replacing periodically. Feel free to ignore this advice - it's just a money grab. We can use the cartridges for much longer than manufacturers claim but they also may not be the best option for the filter. Instead, we can throw the cartridge in the trash and put in some **filter media** (p80). More on that after we discuss canister filters.

Due to the inlet tube, HOBs are normally not shrimp-safe initially, so they require a **pre-filter sponge** placed on the inlet. Once that modification is made, then HOBs become a great combination of safe, quiet (aside from a few gurgles), powerful, and relatively cheap. If you are on a budget and the sound of an air pump bugs you, then HOBs are an excellent choice.

> **TIP**: Simply place a pre-filter sponge or mesh fabric over the intake to prevent shrimp from being sucked up while also providing additional surface area for beneficial bacteria and biofilm to grow.

Canister Filters

Canister filters operate using a similar concept to HOBs. The only difference is that canister filters tend to be more powerful and better at cleaning water due to more layers of filtration material. They are also more expensive and can be tougher to shrimp-proof. This is because a pre-filter sponge may not fit on the intake and any mesh quickly clogs, thereby straining the expensive pump system. Frequent maintenance is often required to keep the filter running well. That being said, canister filters do provide the best possible filtration.

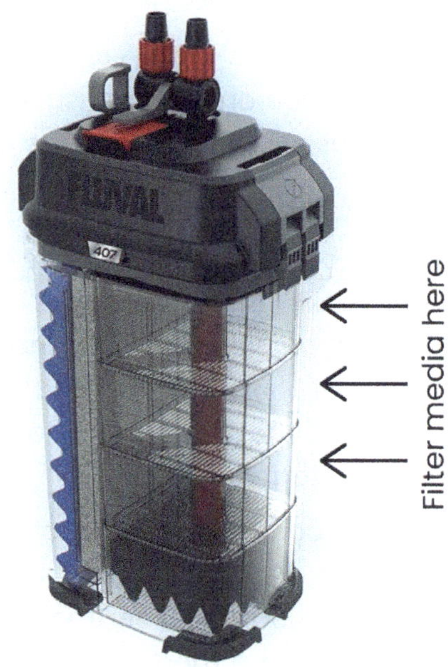

A Brief Word on Filter Media

We recommend filtered media because, as we've discussed, shrimp are easily stressed. Any signs of global conflict or adult situations may cause emotional damage, wreaking havoc on the gentle ecosystem in a tank. As such, we always recommend children's television shows or censored media.

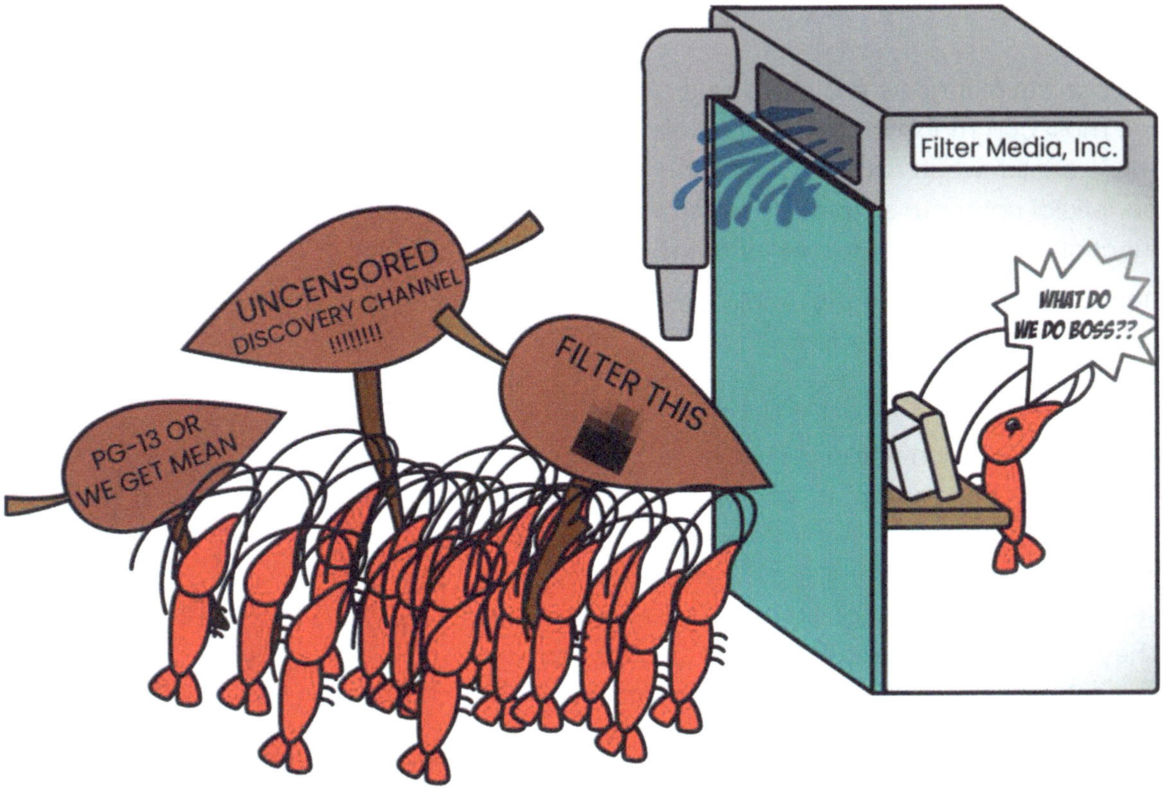

To be clear, that was a joke. Here's what's actually important about filter media:

Canister and HOB filters have compartments where various materials may be placed to increase surface area. These materials are called **filter media**, which may refer to additional layers of sponge, porous plastic balls, or a variety of other hole-y materials with lots of surface area. Any of these are great in a HOB or canister filter.

Botanicals

The word "botanicals" is incredibly broad – basically describing anything that **comes from plants AND is believed to have medicinal or health benefits**. Some botanical tinctures, infusions, decoctions, and other magical-sounding mixtures have good science behind them, while others are completely untested mumbo-jumbo meant to trick us out of our money.

We can't evaluate every claim about different types of botanicals here, but we can cover the most commonly recommended ones for shrimp, which include Indian almond leaves (IAL), cholla wood, alder cones, and other plant materials that supposedly release beneficial compounds into the water. These compounds claim to boost immune system function, improve growth rate, and have antioxidant properties.

But how beneficial are they really? Are botanicals worth going out of our way to order? Let's take a look at the science.

Why Botanicals Are Beneficial

The biggest reason botanicals have so many positive effects for our shrimp is because they introduce **tannins** into our tank. Tannins are acidic compounds found in "tree bark, wood, leaves, buds, stems, fruits, seeds, roots, and plant galls" (U.S. Forest Service, n.d.). They're everywhere inside of plants because tannins protect plants from many different attackers. For example, tannins in tree bark prevent bacteria and fungi from growing on the surface. Tannins in unripe fruits impart that bitter flavor that prevents animals from eating them until the seeds are fully developed and ready to be spread (U.S. Forest Service, n.d.).

They have similar protective properties for shrimp and other animals, including humans. One study showed that tannins in green tea prevent the hepatitis C virus from entering human cells (Calland, AlbeckaA., Belouzard, & et al., 2012). Similar effects were seen for human cells exposed to norovirus, HIV, and hepatitis B (Tong, He, Fan, & Guo, 2022)

That's cool, but let's focus on studies involving shrimp now.

Tannins Improve Shrimp Growth Rate

One research team studied the effect of Indian almond leaf extract on the survival and growth performance of black tiger shrimp (Penaeus monodon) (Ikhwanuddin, Moh, Hidayah, & et al., 2014), one of the most widely cultivated saltwater shrimp in the world.

They had a control group of juvenile shrimp, along with test groups exposed to different concentrations of tannins. As concentration increased, the researchers found a decrease in pH, which is to be expected since tannins are acidic. The team also found an increase in ammonia due to increasing amounts of plant matter that decayed while the experiment was being run. Again, nothing terribly surprising. The cool part – and what may be most useful for shrimp breeders – is that **the team found a full 24% increase in growth between the control (no tannins) and any experimental group (various tannin concentrations)**.

To summarize, having some level of tannins in the water improved the health of the shrimp, regardless of the tannin concentration. Tannins also appear to have a positive impact on growth.

Concerns With the Study

Before moving forward, we need to address some issues with this study. For one, the study only lasted 20 days so may not have identified long term health issues associated with tannin exposure. Another problem was that the lowest concentration of tannins used for the growth data was 1 mg/mL, which seems like a lot. A source online said the entire tank would need to be filled with leaves and look completely black to achieve that concentration.

Let's see if that's true by looking at some data from other sources:

- 1kg of chopped-up IAL soaked for 72 hours led to tannin concentrations of 12 mg/mL in 20L (~5g) of distilled water (Chansue & Assawawongkasem, 2011)
- 15 IALs (7″ or 18cm in diameter) weigh about 30g
    - One leaf weighs ~2g

With this information, we can estimate that we would need approximately 0.08 kg of IALs to achieve a concentration of 1 mg/mL (the lowest concentration used that found an increased growth rate) after 72 hours of soaking. That would require 40 IALs in a 5g tank.

Definitely possible, but most hobbyists aren't getting anywhere near those concentrations since most (understandably) do not want 40 IALs taking up so much space in their tank. This suggests that some growth improvement may be achieved by adding a few leaves, but doing so probably won't lead to 24% larger shrimp unless we STUFF our tank with IALs. There are numerous other benefits to tannins though, which is why even having a few leaves in your tank isn't a bad idea.

To be clear, **we do not recommend adding 40 IALs to your tank** because tannins can build up and reach toxic levels. The first study found that concentrations above 4.5 mg/mL killed about 50% of shrimp within 4 days. While we might only achieve 1 mg/mL after 3 days of soaking 40 IALs, they will continue to release tannins and toxic levels may eventually be reached without sufficient water changes. Additionally, they can lower pH below recommended levels for our shrimp.

Luckily, it only takes a small amount of tannins in the diet to see beneficial results. Let's take a look at some other studies.

Tannins Improve Disease Resistance

Shrimp can absorb some tannins from the water, but another means of getting tannins is by ingesting them. This next study tested the effect of adding different concentrations of purified tannins into the diet of pacific white leg shrimp (*Litopenaeus vannamei*) (Zhu, Guo, Li, & Zhu, 2021). **By adding 0.15% tannins (by weight) to food pellets, they found the diet "remarkably improve[d] the growth performance, antioxidant capacity, intestinal microflora", along with improved resistance against common bacteria**.

The way researchers test an animal's resistance to disease is pretty interesting. They start with healthy animals from each control and experimental group, then inject them with concentrated samples of the bacteria or virus in question (*Vibrio parahaemolyticus*) in this study) and see how many of them survive after X amount of time. This is called a "challenge test".

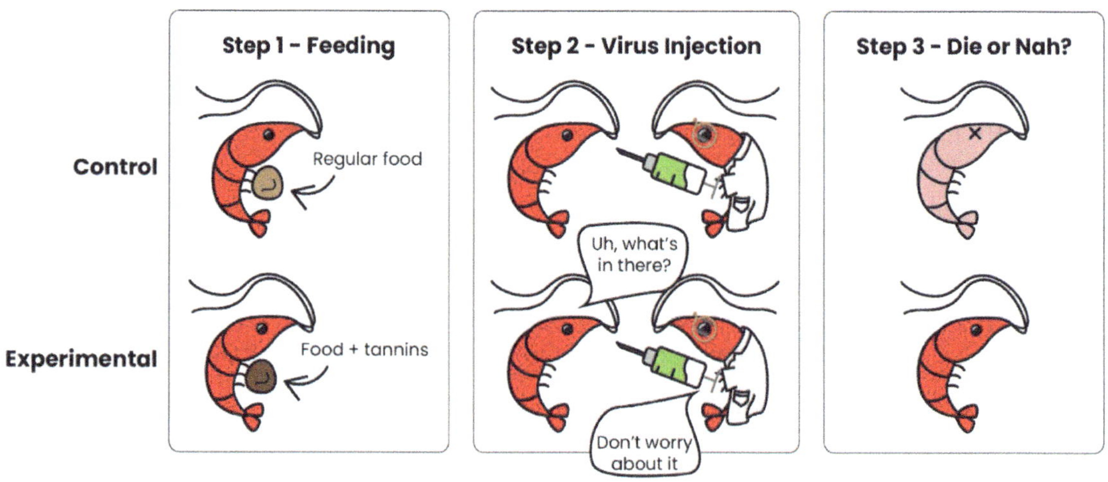

This study injected the shrimp with *V. parahaemolyticus*, waited 24 hours, and then counted how many survived. **They found shrimp fed the control diet (no tannins) had around a 40% survival rate, whereas animals fed with diets of 0.15% tannins had 80% survival**.

This disease resistance has been shown across numerous studies and species. For example, a similar challenge test was done on vundu catfish. After injecting the fish with a disease and observing for 14 days, they found 20% of the control group (no tannins) survived, whereas fish fed diets of 0.5-1.0% IAL aqueous extract had 80-90% survival rates.

There are clear benefits to adding tannins to the diets of our shrimp and many other animals.

How To Get Tannins in Shrimp Diets

We couldn't find any tannin dietary supplements for shrimp (let us know if you have a dealer) so the easiest way to add tannins to a shrimp's diet is by adding botanicals like IALs into their tank. These botanicals start decomposing relatively quickly and become an excellent source of tannins as shrimp graze on them. Here are some recommendations to add botanicals to a tank safely:

Master Breeder Kerric Peyton puts multiple cholla wood segments in his shrimp tanks. Photo credit: Kerric Peyton

- **Choose the right botanicals**
  As we mentioned earlier, there are tons of different plants with tannins in them - IALs, alder cones, and oak leaves are just a few. We've got a list below showing a few of the most common botanicals that are safe to add to a tank but the list is by no means comprehensive. You may have some around where you live that you'd like to try out in your tank, and that could be a great way to save some money**. Some plants are not recommended though, so be sure to do research**. It's best to avoid conifers and anything with a strong scent, like eucalyptus, as these tend to have naturally occurring pesticides. It's also best to only put plant matter into your tank that has been fully dried (before the boiling process).
- **Make sure botanicals are pesticide-free**
  Whether we like it or not, pesticides are sprayed all over our streets and on many trees. There are a lot of tannin-filled leaves on the ground but it can be tough to know whether they have harmful chemicals from agriculture, roads, etc. If we're ever unsure about an object, DO NOT put it in a tank. Some pesticides can cling to objects even after rinsing or boiling. When in doubt, buy from a well-reviewed source to ensure uncontaminated botanicals.

- **Boil botanicals for 3-5 minutes**

  If we do know the botanicals are pesticide-free, then it's still a good idea to wash and boil them. Doing so kills off bacteria and microorganisms that may be harmful to your tank and reduces the risk of harmful chemicals entering your tank – just in case. The water will turn brown and tea-like after boiling, which means some tannins were released during the process. There are still plenty of tannins in the leaves so losing some is not a serious concern. Losing the extra tannins also means our tank water won't be dyed as deep of a brown.

  In addition, boiling them helps some plant matter sink, as they otherwise take a few days to take on water and may float during that time. It also breaks down tissue to make it more digestible for organisms in our tanks.

List of Tannin-Packed Botanicals

(Not comprehensive)

- Oak leaves and acorn caps
- Sycamores
- Liquidambar (leaves and spiked pods)
- Magnolias (leaves and pods
- Maples – leaves and pods
- Alders
- Indian almond leaves (catappa leaves)
- Birch
- Cholla wood
- Cariniana pods
- Ramshorn pod
- Khaya pods
- Banana leaves
- Monkey pot and jacaranda pods (seed pods)

Jacaranda pods - Source: Yercaud-elango, CC BY-SA 4.0 via Wikimedia Commons

One Issue with Tannins

Be aware that a lot of tannins turn the water brown, creating what is known as a **blackwater environment**. This is not a bad thing but it is not what everyone wants their tank to look like.

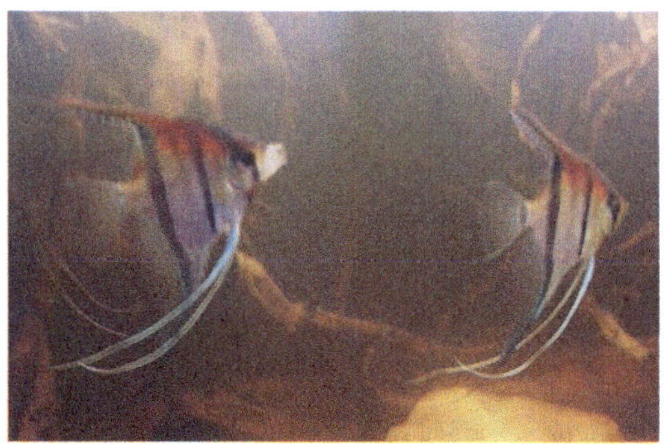

The brown tint to the water is caused by tannins in a blackwater tank - Source: Ewkaa, CC BY-SA 4.0 via Wikimedia Commons

Decor

Rocks

Rocks are common in shrimp tanks but they are also responsible for many problems. Certain types of rocks leach chemicals into the water that can increase pH, GH, and KH to harmful levels, along with releasing poisonous minerals. Some rocks may be perfectly fine in fish tanks but may be deadly in shrimp tanks because shrimp are more sensitive to mineral imbalances. Please take a look at this article on www.ShrimplyExplained.com for an updated list of shrimp-safe rocks[xxvii].

Plastic Decor

Some Facebook or Reddit group may rail against plastic decorations because they're "not natural" and don't belong in a shrimp tank. This includes things like plastic plants, the classic SpongeBob house[xxviii], and other artificial decorations.

Sure, plastic plants aren't as beneficial as real plants are to an ecosystem, but they do increase surface area. We always recommend some live plants to improve water quality and biological diversity in a tank but it's perfectly fine to mix real and fake plants if it helps achieve the aesthetic we want. That being said, there are some poorly made products that may leach chemicals into the water that are harmful to our shrimp, which is why we always recommend doing a thorough web search to look for reviews of a product.

Other Products

There are so many different products claiming to be good for shrimp - tourmaline balls, kallax balls, Dead Shrimp Powder (DSP), montmorillonite clay, calcium balls, etc. Some of these have obvious problems, like calcium balls leaching excess calcium and offsetting the mineral balance. Others may have an effect but have not been proven, like kallax balls or DSP. It seems odd that shrimp breeders push these products with claims of how they help shrimp breed, all without doing controlled side-by-side tests comparing tanks with the product compared to tanks without the product. It's reasonable to suspect that a bacterial powder is useful based on the ingredients list (Shrimply, 2022), but we've tried a few different ones and haven't seen any notable differences. Again, the lack of side-by-side comparison experiments from sellers speaks volumes. Montmorillonite clay is one additive that has some scientific backing, which we cover in the Substrates section (p70).

The Best Water for Shrimp

Consistently good water quality is key to successful shrimp keeping. Just as air tainted by chemical fumes or strawberry-unicorn's-blood-flavored vape smoke can harm humans, poor water can reduce the health of our beautiful shrimp. As such, our first big decision when setting up a tank is deciding what kind of water to use. This chapter provides a detailed look at that decision to help choose whether **tap water** or **purified water** is best for us and our shrimp. We'll start by covering the different types of water available. After that, we'll explain how to measure tap water and decide whether it's a good fit for shrimp.

Purified Water

Purified water is the gold standard – the holy grail - of shrimp keeping. It includes any water that has gone through a purification process to remove contaminants (chlorine, chloramines, minerals, bacteria, etc.) to produce near-pure H_2O. Purified water should have less than 20ppm of TDS but, ideally, less than 10 ppm. Effective purification processes include **reverse osmosis and deionization (RO/DI)** or **distillation**.

Reverse Osmosis

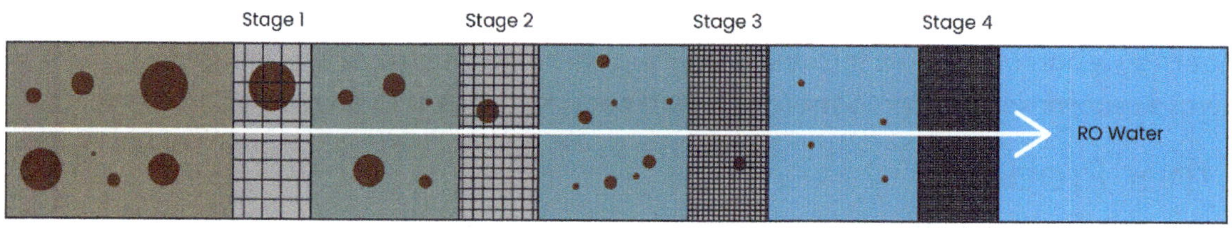

The RO process uses pressure to force water through two or more "stages" of filters with progressively smaller holes to catch contaminants. A DI filter may or may not be used at the end of the process to remove leftover ions from the RO water, leading to further purification. Many hobbyists call RO water "RO/DI" even when a DI filter isn't used, but even RO water is relatively pure.

Distillation

Water vapor cools off and recondenses as it travels through the coils

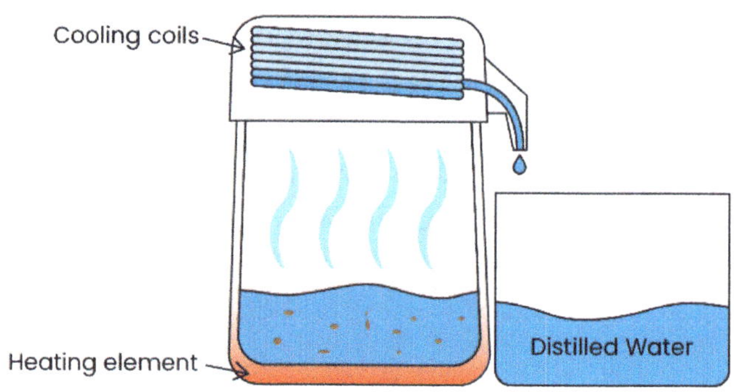

In contrast, distillation uses heat to purify water. When water evaporates it leaves behind contaminants. Distillation uses this phenomenon to force water to evaporate by heating it, then recapturing it by cooling the water enough to recondense. The recondensed water drips into a container while contaminants stay in the original chamber where the water evaporated from.

Spring water also falls into the category of purified water, although it is naturally pure instead of going through a purification process. Spring water varies in purity so we always

recommend checking the TDS to verify the quality. Spring water is not as pure as RO/DI or distilled water so it may have around 50 ppm TDS.

From now on, we'll be using RO/DI water to refer to all types of purified water since that's the common convention in the hobby.

Preparing RO/DI Water for Shrimp Tanks

By itself, RO/DI water kills shrimp and nearly every other living thing by making their cells explode. This happens because there's a huge difference in **osmotic pressure** between RO/DI water and the inside of a cell. This difference causes water to rush into the cell, making it expand until the cell membrane eventually breaks. That being said, our shrimp aren't going to blow apart if exposed to RO/DI water – they'll just die a slow and painful death over several hours to days. That's why we need to **remineralize** RO/DI water by adding healthy salts/minerals to it.

Osmosis – According to Merriam-Webster, osmosis is the "movement of a solvent (such as water) through a semipermeable membrane[xxix] (what living cell have) into a solution of higher solute[xxx] concentration that tends to equalize the concentrations of solute on the two sides of the membrane". Here's a classic example:

J Gen Physiol. 2023;155(10). doi:10.1085/jgp.202313332

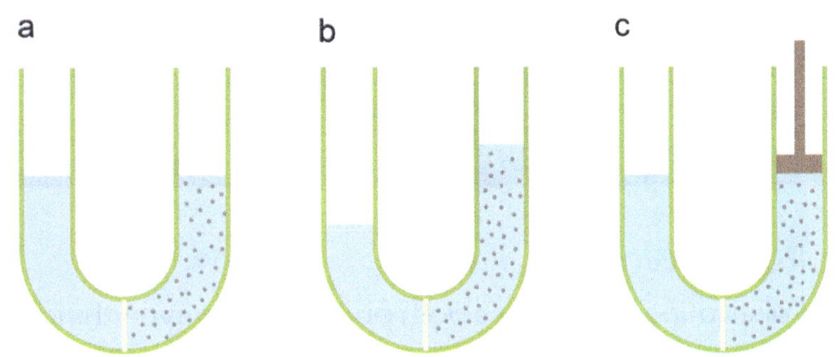

Figure Legend:

Classical demonstration of osmosis. (a) A U-tube with a semipermeable membrane separating pure water on the left from an aqueous solution with an impermeant solute of concentration c_s on the right. (b) With time, water will move from left to right, elevating the column of solution on the right, until its gravitational weight stops the flow. (c) Alternatively, the flow of water can be prevented if a piston applies a pressure equal to RTc_s (in the dilute regime).

Osmotic pressure forms across the semi-permeable membrane due to the difference in solvent concentration. Since cells actively hold in various soluble compounds like salts, there's guaranteed to be osmotic pressure between RO/DI water and the inside of a cell, resulting in water flowing into the cell and eventually ripping it apart like a balloon that's filled with too much air.

How To Remineralize RO/DI Water

Remineralization is done by adding commercial **remineralizers** – mineral powders/liquids – to RO/DI water. These products have the right ratio of calcium, magnesium, and other minerals to promote the right osmotic pressure on shrimp so they aren't stressed. **Using a remineralizer gives complete control over what is in our water to get the perfect pH/GH/KH and never need to worry about an imbalance of minerals or pollutants**. This reduces problems and simplifies troubleshooting, which is why remineralized water is *always* the recommended type of water for shrimp tanks. We'll discuss some of the downsides shortly.

See the Remineralizers section (p93) for more information.

Side note: Osmotic pressure also partially explains how shrimp absorb minerals from the water. By storing calcium in their shells, they remove it from cells in their body so there is a lower concentration of calcium compared to the external water. This lets them absorb calcium from water when it flows through their gills. In addition, there are proteins called **calcium transporters** that can grab calcium from the water and pull it into the cell, even when the cell has a higher concentration. The same is true for other molecules.

Oxygenating RO/DI Water

RO purification and distillation techniques both purify almost everything from the water – including oxygen. As such, if we do a 50% water change with freshly purified water, we're dropping DO levels by 50% until the filter replenishes the DO. In a healthy tank with good filtration, this temporary drop in DO only causes minor stress for a short period of time. If shrimp are already stressed by disease or poor water quality, then a sudden drop in DO can cause additional stress that may lead to death.

We can easily avoid this DO drop and reduce stress from water changes by putting an air stone into purified water an hour before putting it into the tank.

Tap Water

In contrast to the control we get with RO/DI water, **we have no idea what's in tap water**. Water tests can help estimate what's in tap water but there are always unknown components that we don't know the danger of. They can be harmless or extremely toxic. There's also an issue with chlorine and chloramines in many areas, although those are taken care of by dechlorinators like SeaChem Prime.

We can get an idea of what's in tap water by using a TDS test and combining it with GH/KH tests to determine how much of the TDS is unaccounted for contaminants. If tap water is 150 ppm with 120 ppm (~5 dGH) GH and 20 ppm (1 dKH) KH, then 140 of the 150 ppm is accounted for, indicating the tap water doesn't have many contaminants. There are still a few issues though:

- Even if the GH is in the recommended range of a shrimp species, does it include the right ratio of minerals? A calcium test is needed to determine that. Testing procedures are available on p271. Our chapter on Molting Problems (p267) goes more in-depth into mineral ratios.
- What makes up the 10 ppm of unknown TDS? If it's sodium, then it's perfectly healthy. If it's copper, then any shrimp will die.
- Will tap water stay the same? In many regions, tap water parameters vary dramatically depending on the season and current water treatment processes.

Side note: The data we've found suggests that tap water has the right ratio of calcium and magnesium for shrimp *on average*, but it varies widely by region (Pehrsson, Patterson, & Pery, 2008).

Table 1. Mineral content of water samples (mg/100g)

	Mean	Median	Min	Max	Avg mg in 2 liters	DRI* mg/day (male 31-50y)
Ca	3.0	2.7	0.0	10.0	61	1000
Cu	0.0098	0.0017	ND	0.4073	0.20	0.90
Fe	0.002	0.0003	ND	0.065	0.04	8
K	0.5	0.2	ND	20.4	9.8	4700
Mg	0.9	0.8	0.0	4.6	19	420
Mn	0.0002	ND	ND	0.0099	0.005	2.3
Na	3.8	1.9	0.1	39.1	76	1500
P	0.01	0.01	ND	0.2	0.29	700
Zn	0.005	0.001	ND	0.075	0.10	11

http://www.iom.edu/CMS/3788/4574.aspx

Water Recommendations

Neocaridina tend to survive in a wide range of water parameters so they can often be kept in tap water, assuming parameters are in range. Doing so makes problems more difficult to troubleshoot due to unknown elements in the water, but it can work perfectly well. We recommend going to local fish stores or joining local aquarist/shrimp groups to ask if others have success with Neocaridina in local tap water.

In contrast, Caridina do not respond well to additional contaminants. They also need very particular water parameters that can only be achieved with RO/DI water or very clean tap water. If tap water is above 40-50 ppm of TDS, then we don't recommend using it with Cardina. It also will need remineralization to get the right GH just like RO/DI water.

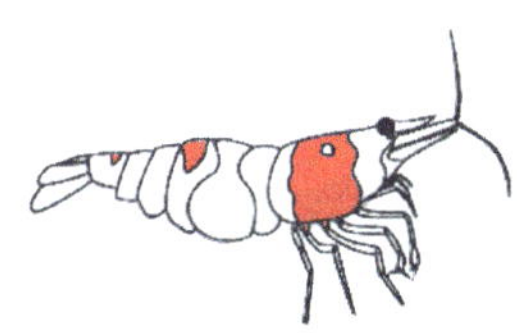

Pro Tip: Water right out of the tap is oxygenated and chlorinated. Let tap water sit for 24 hours before testing to let gases leave the water and get the most accurate measurements.

An additional reason Caridina aren't kept in tap water is because **tap water typically has some KH in it, which exhausts active substrate sooner and leads to unstable pH**. Review the Active Substrates section (p63) for information on extending active substrate lifespan.

Scientific Support For RO/DI Water

One study compared tap water and remineralized RO/DI water to see which one caused the best survival and breeding for Crystal Red Caridina shrimp (Coswatte, Suraweera, Mudalige, & Liyanage, 2022). Despite having the exact same pH, GH, Kh, ammonia, nitrite, nitrate, TDS, and temperature, the remineralized RO/DI water showed significantly better survival and breeding. While all parameters that hobbyists can measure may be the same, these tests do not capture everything that matters. For this reason, we always recommend using RO/DI water for best results.

The Problems With RO/DI Water

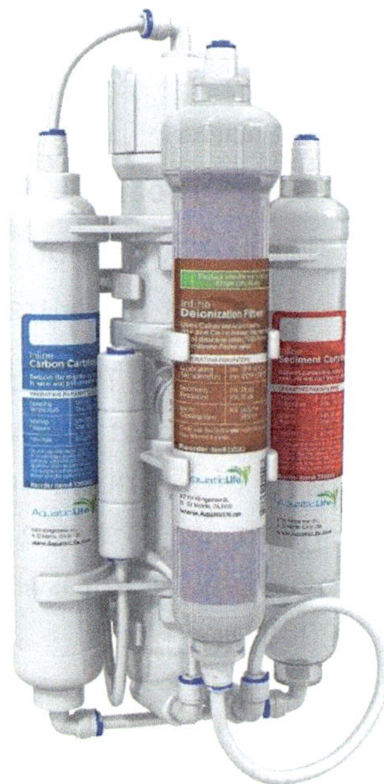

RO Buddie RO System

There are some downsides to RO/DI water, specifically concerning how to get it. Our options are either to:

- Buy an expensive filtration system ($60-300 USD, excluding replacement parts).
- Buy gallons of RO/DI water ($0.75-1.00 per gallon) from the local grocery or fish store and carry the water home.
- As mentioned previously, RO/DI water often has very low DO levels that can cause stress during large water changes.

When keeping cheaper shrimp, it may be worth the risk of using tap water and dealing with potential problems. For more expensive shrimp (like some Caridina), an RO unit is an expensive but necessary investment. A commonly used RO unit is the Aquatic Life RO Buddie[xxxi].

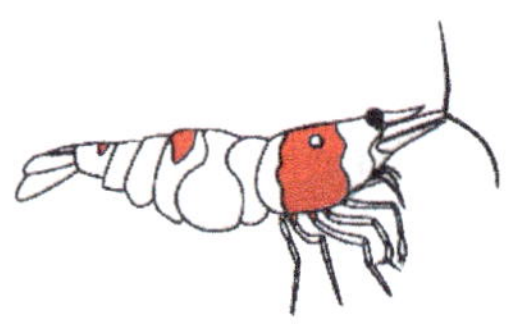

Remineralizers

Shrimp remineralizers are powder or liquid products designed to add the right ratio of calcium, magnesium, and other trace elements that shrimp need in the water to molt properly. As a result, **all remineralizers raise GH**, **while some products also include carbonate to raise KH**, too. Most hobbyists use remineralizers with purified water but *they may be used with any type of water to increase GH/KH*, as long as parameters stay within recommended ranges.

As we've mentioned before, remineralized RO/DI water is the best option for shrimp tanks because it offers complete control over what is in our water. In contrast, tap water comes with numerous potential problems like copper leaching, mineral imbalance, unstable parameters, etc. If the budget allows for it, **we *highly* recommend remineralized RO/DI water**.

Types of Remineralizers

There are three different questions to answer when looking at remineralizers: Does the remineralizer raise KH? Is the remineralizer designed for shrimp? Is the remineralizer liquid or powder?

Does the remineralizer raise KH?

Remineralizers are designated as either GH+ or GH/KH+. **GH+ remineralizers** are recommended for low-pH Caridina tanks with active substrate and RO/DI water OR anyone using tap water that wants to increase only GH and/or balance their mineral content to avoid molting problems. **GH/KH+ remineralizers** are intended for Neocaridina tanks with RO/DI water or anyone using very soft water.

Is the remineralizer designed for shrimp?

Some remineralizers are specifically marketed for shrimp (i.e. Salty Shrimp, Shrimp King Mineral, etc.) while others are intended for general aquariums (i.e. Seachem Equilibrium). In theory, the difference is that shrimp-specific remineralizers have been tested to guarantee the best molting performance and health of our shrimp, while more general remineralizers may not optimize shrimp health.

Is that true?

It's difficult to say because none of the manufacturers provide any sort of data on their product performance and most don't even offer an ingredient list, so we know very little about any of them. Often, the best we can do is take recommendations from other shrimp keepers. The problem with doing so is that hobbyist tests often don't have control groups to reliably say the remineralizer *caused* the difference as opposed to slight differences in temperature, bacterial contamination, etc., between tanks. In the end, the best remineralizer is often whichever one is readily available and cost-effective.

Is the remineralizer easy to use?

Remineralizers are broken down into two types: powdered and liquid. Powdered remineralizers require some measuring to add the right amount to water, whereas liquid remineralizers are pre-measured to add a certain amount of GH per pump. Another issue with powdered remineralizers is that they absorb moisture and harden into unwieldy chunks if they aren't sealed properly after use. This makes liquid remineralizers easier to use overall but powdered remineralizers may be bought in bulk for cheaper.

Remineralizer Efficiency

Here's a breakdown of some common remineralizers based on the manufacturer's instruction for use and the volume of water they treat (more info on next page):

[xxxii]*Product links*	**Price (USD)**	**Volume Treated (gal)**	**GPD (free shipping)**	**GPD (w/o free shipping)**
Salty Shrimp GH+ (230g)	25.55	2300	90	71
Flip Aquatics Shrimple GH (500 mL)	19.99	2500	125	71
Brightwell Caridina GH+ (500 mL)	17.8	1250	70	50
Salty Shrimp GH+ (110g)	17.08	1100	64	46
Seachem Equilibrium (300g)	8.62	375	44	24
Salty Shrimp GH+ (850g)	51.16	8500	166	137

(Gallons treated per dollar) GPD = Number of gallons raised by +1 dGH per dollar spent (example: Salty Shrimp GH+ (230g) is enough to raise GH by 1 dGH for 1100 gallons of water and costs $17.08 1100/17.08=90.02). We have a spreadsheet with additional data[xxxiii].

Shipping costs are determined by Amazon pricing, except for Flip Aquatics Shrimple GH. Flip Aquatics shipping is $14.99, while Amazon (no prime) shipping is $6.99 for all products aside from Salty Shrimp GH+ (850g), which is $11.03. All prices subject to change.

How to Use a Remineralizer

Our goal with remineralizers is to reach a specific GH/KH. To do that, simply add the right amount of remineralizer for the volume of water according to the product instructions. We always recommend doing this outside of the tank to avoid overdosing and eliminate sudden changes. Check the GH/KH/TDS are correct using liquid test kits, then, add the remineralized water to the tank.

What Affects a Shrimp Tank (Summary)

Check out the chart on the following page for a summary of everything we've discussed that may affect the parameters and health of a shrimp tank:

Key

- ↑ Raises
- – No Effect
- ↓ Lowers
- △ Daily Change

What Affects A Shrimp Tank

	pH	GH	KH	NH_3	NO_2	NO_3	DO	Nutrients	Surface Area	Growth	Breeding
Active Substrate	↓ | ↓ △	↓*	↓	↑*	↑*	↑*	–	↑*	↑	–	–
Botanicals	↓	↓*	↓	↑	↑	↑	–	↑	↑	↑	–
Carbonate	↑ | ↓ △	–	↑	–	–	–	–	–	–	–	–
Filtration	↓ △	–	–	↓	↓	–	↑	–	↑	↑	↑
High Bioload	–	–	–	↑	↑	↑	↓	↑	–	↓	↓
Food	–	–	–	↑	↑	↑	↓	↑		↑	↑
Inert Substrate	–	–	–	–	–	–	–	–	↑	–	–
Plants	↑ △	–	–	↓	↓	↓	↑ | ↑ △	↓	↑	–	↑
Remineralizer	↑*	↑	↑*	–	–	–	–	–	–	–	–
RO/DI Water	–*	↓	↓	↓	↓	↓	–	↓	–	–	–
High Temperature	–	–	–	–	–	–	↓	–	–	↑	↑
Water Changes (Tap water)	?	?	?	?	?	?	?	?	?	?	?

*Depends on product

Basic Low Maintenance Tank Setup Recommendations

The following recommendations are for setting up healthy low-maintenance shrimp tanks. These tanks have simple filtration, thick substrate, plenty of plants, and support an average bioload, so they create a well-balanced ecosystem that is great for newer shrimp keepers. We discuss how to set up other tanks (for selling and breeding) later on in the book (p225) because they focus on creating intentional imbalance for the sake of optimizing shrimp growth, reproduction, and coloration. These imbalanced systems are inherently more difficult to maintain so require more knowledge and care.

	Neocaridina	**Caridina**
Tank Size	5+ g (19+ L)	10+ g (38+ L)
Feeding Frequency	1-2 times per week (25-30 mg per 10 adult shrimp per feeding)	
Food Types	Alternate two types of food (Ideal): -Mixed-Source protein (>30% protein) for adults -Powdered food for babies	
Plants	At least three types of plants: -Floating plant -Moss (any type) -Low-care rooted plant (Cryptocoryne or similar)	
Maintenance	Water changes and filter cleanings as needed	
Temperature	65-75°F	
Substrate	Inert	Active (pH-lowering)
Lighting	Based on plant needs	
Filter	Sponge or double-sponge filter (HOB if you hate the noise from air pumps)	
Heater/Cooler	As needed – See Heater/Cooler section (p73) for more info.	

Low Maintenance Tank Supplies

Please see the Tank Types *chapter (p225) for more detailed information on other setups.*

Budget Options

These budget options are based on overall price along with our experience with the brands. For example, NICREW lights are technically a few dollars cheaper, but we've had a few of them fail, whereas our AQUANEAT lights have not failed over the past 4 years of use. The product lists include everything that is critical for a healthy shrimp tank.

- Budget Neo 20-gallon (76L) Tank Supplies (Excluding Tank)[xxxiv]
- Budget Caridina 20-gallon (76L) Tank Supplies (Excluding Tank)[xxxv]

If budget allows, we suggest using test equipment listed in Recommended Options below.

Recommended Options

These lists include upgraded test equipment, lights, filters, and substrates.

- Recommended Neo 20-gallon (76L) Tank Supplies (Excluding Tank)[xxxvi]
- Recommended Caridina 20-gallon (76L) Tank Supplies (Excluding Tank)[xxxvii]

Purified Water Supplies

- Neocaridina RO/DI Water Supplies[xxxviii]
- Caridina RO/DI Water Supplies[xxxix]

See roman numeral references at shrimplyexplained.com/book-links/ for indicated lists.

Plants

Aquarium Co-Op and Flip Aquatics are both great options for aquarium plants. A cheaper option is to join local aquarist groups on Facebook or visit the r/aquaswap subreddit.

Tanks

For a simple rimmed tank, it's hard to beat the Aqueon 20-gallon long available at Petco. Petco has a 50% off tank sale every quarter, which is a great time to buy 1 tank - Oh, maybe two for this deal... I mean, it's basically stealing so why not get 10?

Where To Go From Here

Once the equipment and plants are acquired, we have everything needed to *start* a tank – but we can't add shrimp just yet. Shrimp need a healthy ecosystem to support them, but simply throwing all the equipment and water into a tank isn't enough. The right equipment helps immensely, but we also need time and patience. Let us explain:

This guide focuses on building a shrimp tank, but we're actually creating an ecosystem for more than just shrimp. If we want to help our shrimp thrive, our best chance of success is to create a low-maintenance ecosystem that mimics their natural environment. That includes supporting healthy plants, but also populations of bacteria, fungi, and many small animals called **microorganisms**. The more diverse lifeforms we have, the more stable and healthy our tank is. To support this life, we need to **cycle** a tank, which is explained in the next chapter.

Bladder snail (Physella acuta) *climbing up tiger lotus* (Nymphaea zenkeri). *How many ostracods do you see?*

Setting Up A Shrimp Tank

The Cycling Process

Cycling is the process of developing a healthy microbiome (microbial environment) to support aquatic life in an enclosed system. The "cycle" refers specifically to the **nitrogen cycle**, which is the production and breakdown of nitrogen-based compounds through the ecosystem. Here's how the cycle works:

As mentioned in the Parameters chapter (p12), ammonia is released from decaying organic waste such as dead plants/animals and shrimp poop. It is broken down first into nitrites and then nitrates by different types of **nitrifying bacteria**, which we'll refer to as **beneficial bacteria** (BB for short). Plants use nitrogenous compounds released by BB as building blocks for new growth, which in turn creates food for herbivores. The herbivores release waste (ammonia) after eating the plants, thus continuing the cycle. This is just one example of the many incredibly efficient processes that have evolved throughout billions of years of life on earth – and we get to make it happen in our own tank!

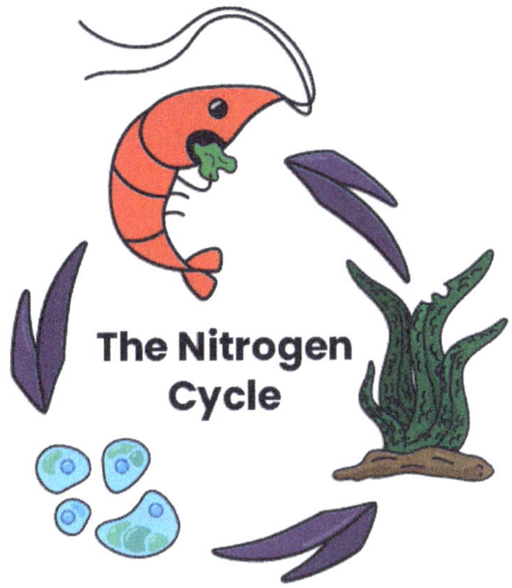

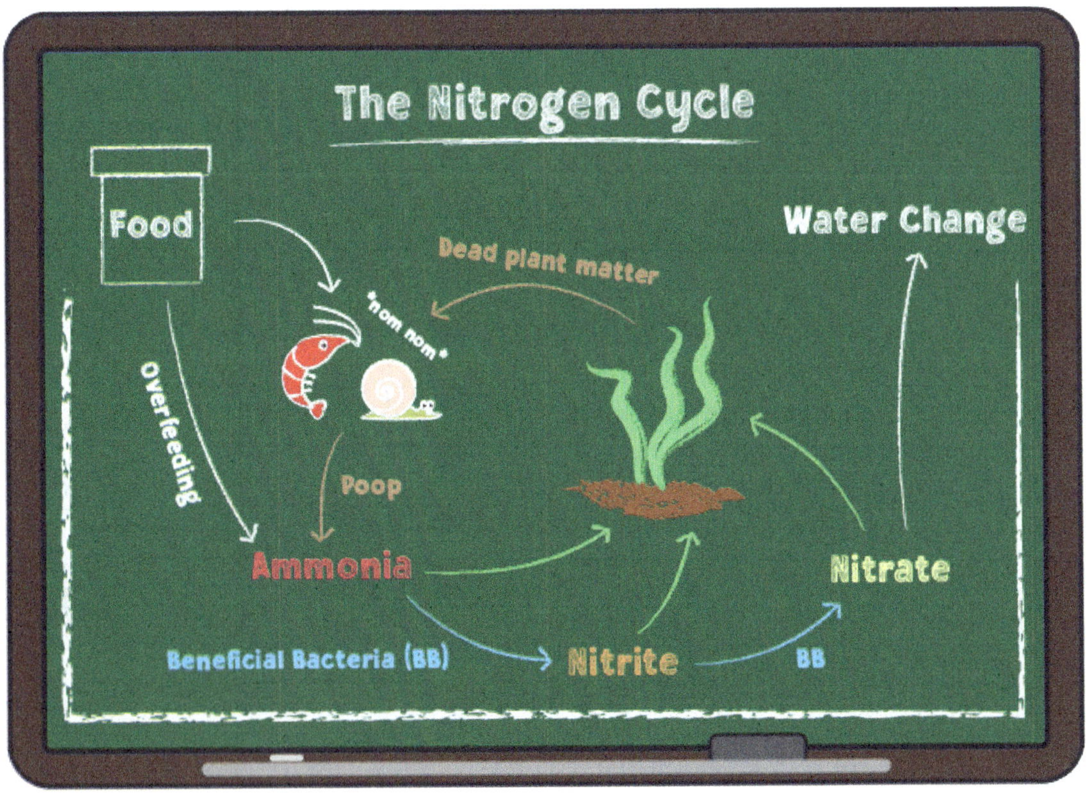

Even though the nitrogen cycle is everywhere, getting it up and running in our tank isn't always easy. There are also many ways to disrupt the cycle in such a small ecosystem. The next few sections cover the easiest steps to start and maintain a foolproof cycle. After that, it's time to add shrimp!

A Word of Warning

Before we get into the nitty-gritty details, we need to go over something.

Tank setup and cycling determine our success as shrimp keepers. The steps aren't challenging, but they take time (6-8 weeks) and patience.

Now, you may be disappointed because that's a long time to wait. You also might be confused because you've probably heard that some people can cycle a tank within a week. This is possible, but it also drastically reduces the chances of keeping a healthy shrimp tank.

You see, a cycled tank just means the nitrogen cycle is present. It does not tell us if there's enough food or if parameters are stable enough for our shrimp. What we want to create is a "**seasoned**" or "**mature**" tank with diverse food sources and nutrient cycles. This requires time. We have to fight the urge to rush through this process because, otherwise, we may end up with a tank of dead shrimp and a lighter wallet. It's an experience we can avoid by following along with the next four steps of cycling:

We have two options during this waiting period – impatience or fascination. We can always sit around tapping our feet and yelling at the bacteria to grow faster. It doesn't do much though.

Instead, take a little bit of time each day and observe the tank. Watch how tiny green dots of algae suddenly grow on the glass; Try to spot the first microscopic white copepods bustling around in the substrate; Appreciate the light green shoots of new growth from plants. There's *a lot* of amazing stuff going on as your tank matures. We'd hate for you to miss it!

Getting familiar with the microbiome also helps us notice when it changes, so we can catch problems before they impact our shrimp. Check out the Microorganisms chapter (p112) to learn more about the tiny critters that may pop up in a tank!

Now, let's get into the four steps to start cycling a tank.

Step 1: Tank Setup

Add the substrate, water, plants, décor, filter, and any other equipment we want to the tank. Adding anything at a later point changes the conditions of the tank and may disrupt the cycling process. That's why it's best to throw everything in the tank at the start, except for the animals. It's a tough environment for plants at first, but they're hardy enough to handle it.

Adding Beneficial Bacteria

We've talked a bit about how important BB are for a tank but haven't discussed how to introduce them. The fact of the matter is that BB are guaranteed to enter our tank because they are found *literally everywhere*. Even chloramines and other disinfectants struggle to eradicate them, as demonstrated by numerous papers discussing the effect of nitrifying bacteria on public water distribution systems[1,2]. Dip a hand in tap water and then into the tank and - BOOM - it's likely been inoculated with at least some BB cells. Given a few weeks and a food source (ammonia), these cells grow into hardy bacterial populations, ready to handle any waste in our tank.

That being said, adding just a few cells at the start means it can take some time for the necessary population to grow, which is why aquarists add them in other ways - sometimes even without knowing.

For example, plants carry millions of BB on their surfaces, even after common treatments to remove pests. Just a small amount of moss is enough to seed the tank and jumpstart the cycle. In addition, fish, snails, shrimp, and other organisms carry BB on their surfaces.

Aside from aquatic organisms, there are two other common ways to seed a tank with BB. The first is to run a filter in a different tank that's already seasoned. After a week or two, the BB and microorganisms from that tank populate the filter, then you just move that filter over to the new tank. This method isn't possible if we don't already have access to an established tank, so we can instead use a commercial product like FritzZyme 7 Nitrifying Bacteria[xl]. This product contains BB that can be poured directly into the tank to help jump-start the cycle. It's very easy but does cost a bit of extra money.

After adding BB, we need to help the bacteria grow so they establish stable populations in the tank. That's where ammonia comes in.

Does Adding BB Speed Up the Cycle?

Hobbyists are always looking for ways to speed up the cycling process and get shrimp in their tank as fast as possible. It is true that adding lots of BB through a pre-cycled filter or commercial products establishes the nitrogen cycle quickly, but remember that nitrogen is just one nutrient in the tank. Other nutrients like phosphorus are not immediately dangerous but their cycles still need to develop so we don't experience a

nutrient imbalance. In addition, a stable shrimp tank benefits immensely from a strong food web that starts with algae, which feeds larger microorganisms like copepods. These microorganisms, in turn, end up in the stomachs of macrofauna like shrimp and fish. This biodiversity helps cycle nutrients, remove waste, and reduces maintenance requirements, so it's good for shrimp keepers and our shrimp - but it does take time to develop.

Ostracod - Magnified - Source: Janek Lass, CC BY 4.0 via Wikimedia Commons

Algae and bacteria reproduce rapidly, so they may only need a week or two to populate a tank, but the same isn't true for larger microorganisms. For example, some ostracod species reach sexual maturity around 30 days after hatching (Brown, 2019). They may only be on their second generation after our tank has matured for 6-8 weeks but that still creates a more established population than what we'd have after only a week of cycling. Allowing these populations to establish themselves before larger animals enter the food web helps to create a thriving ecosystem.

Overall, while adding BB does speed up the cycling time, it doesn't drastically speed up how quickly the ecosystem matures so we still recommend 6-8 weeks for the best chance of success with Neocaridina. They can technically be added soon after the cycle is complete but much more frequent feeding and maintenance is required, which in turn leads to a less stable environment and a higher chance of tank failure.

A mature tank is especially important for Caridina, which are particularly sensitive to an unstable environment. The low pH environment also slows the nitrogen cycle and reduces the rate of bacterial growth, which is why **we recommend giving Caridina tanks 8-12 weeks to establish**.

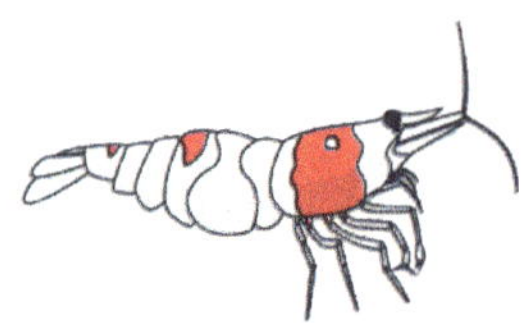

Step 2: Adding Ammonia

BB only grow if they have food – in this case, that's ammonia. Traditionally, the ammonia source came from a live fish that got thrown in the barren landscape of a new tank. The fish was fed, then pooped out some ammonia, then the BB did their thing. This process is called a **fish-in cycle** but is rarely used now because it is extremely stressful for the fish to be in such a high-ammonia environment. The fish may not survive, and a dead animal isn't the best omen for a new tank. This method also requires a lot of work, with frequent water tests and water changes to limit stress on the fish.

Another option is **fishless cycling**, which uses bottled ammonia like Dr. Tim's Aquatics Ammonium Chloride[xli]. This is a very controlled way of increasing ammonia to the right level and keeping it there until the BB establish themselves. Unfortunately, it can be time-intensive, requiring frequent testing of ammonia, nitrites, and nitrates to know when to add more. It's also an additional expense. That being said, **if we intend to add fish to the tank, then fishless cycling is the recommended method**.

The easiest and cheapest way to add ammonia is to do a **snail-in cycle**. Snails are trained survivalists that can handle harsher conditions than fish. Species like ramshorn snails have been shown to survive in up to 20 mg/L of ammonia (total ammonia nitrogen [TAN] technically) (Hung T-C, Stevenson, Sandford, & Ghebremariam, 2018), which is well above the 4-5 mg/L of ammonia seen in most tanks during cycling. Just put a few snails in a tank, feed a little shrimp/fish food, and watch the cycle develop without worrying about constant testing or water changes. **We recommend the snail-in method for shrimp-only tanks (+snails)**. Recommendations for types of snails are provided in the next section.

The last option for an ammonia source is from the substrate itself. Active substrates designed for planted tanks (Like Fluval Stratum or Amazonia ADA) leach ammonia and eliminate the need to add anything ourselves. Even with nutrient-rich substrates, we still recommend adding a few snails to develop a bioload and encourage a diverse ecosystem.

A Word on Bioload

What we are doing with an ammonia source is introducing a **biological load**, or **bioload**, to the tank. The bioload can refer to the amount of ammonia produced by a given species OR the total amount produced by all animals in a tank. Animals like shrimp and small snails produce little ammonia so have a *low* bioload, whereas goldfish are pooping machines with a *high* bioload. **A large population of low bioload animals can still have a high bioload overall.**

The BB and plants in our tank need to acclimate to the bioload. When the bioload is too low (understocked), then the BB and plants don't have enough food (ammonia) so plants may starve while the BB enter a dormant state. If the bioload is too high (overstocked),

then ammonia is being produced too quickly for the plants and BB to keep up, resulting in an ammonia build-up that stresses or kills inhabitants.

When using bottled ammonia, we are conditioning our tank to a high bioload. This is not a problem if we also intend to add fish to the tank, as they tend to have a higher bioload that matches the amount of ammonia being added.

But what happens when we stop dosing ammonia and add shrimp that produce a low bioload?

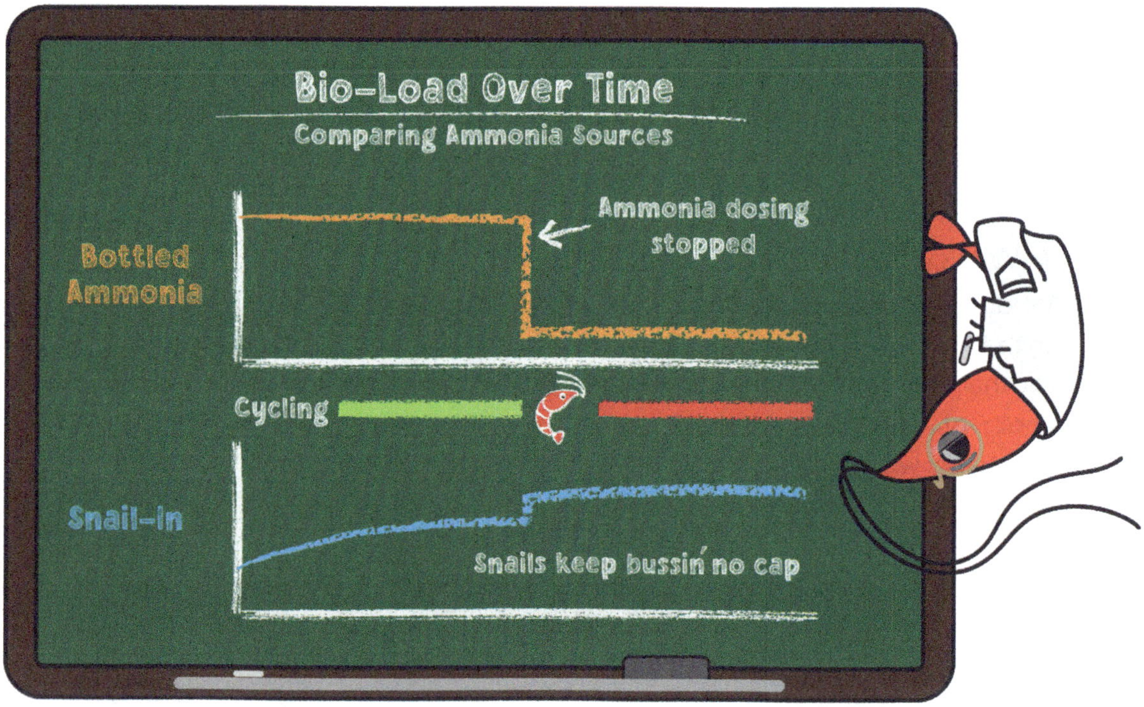

That's a big change to the environment, so the total bioload may be too low and lead to a cycle crash as BB become dormant and plants starve. Even if BB aren't affected, the ecosystem adapted to a high energy input (ammonia) so sudden changes force the system to adapt and be temporarily destabilized. *That isn't a great time to add shrimp.*

The snail-in method prevents this problem because snails create a steady natural bioload. The initial population of shrimp we add to our tank increases the bioload slightly, but it is minimal compared to what the snails already produce so the cycle is much less likely to be disrupted.

As we see from the graphs above, there is a much smaller change in bioload when the shrimp are added to the snail-in tank than the tank which was dosed with ammonia. This reduces the chance of disrupting the natural nutrient cycle that has developed during the cycling time.

Recommended Types of Snails

#1 - "Pest" Snails (i.e. ramshorn and bladder snails)

Pest snails are super cheap, excellent survivalists, and protect our tank from the risk of overfeeding (biggest mistake of new shrimp keepers) by eating any leftover food. Most importantly, they reproduce based on food availability to provide an adaptive bioload that grows as our tank matures. While we can end up with a lot of them, their population won't grow out of control unless they're overfed.

We recommend throwing 5-10 "pest" snails in the tank after setting it up. We technically only need one individual because these species reproduce asexually but having more than one speeds up the process. Many plants come with a couple of these snails (even if they claim to be snail-free) but we can easily buy some or find hobbyists giving them away.

Bladder snail on a rock - Dat doris, CC BY-SA 4.0 via Wikimedia Commons

Rabbit snail doing rabbit snail things

#2 - Mystery, Nerite, or Rabbit Snails

These snails are all slow (or unable) to reproduce in freshwater tanks so many shrimp keepers prefer them over Option #1. They come in a wide variety of appearances and can be a unique addition to your tank, but we will have to go out and buy them.

To provide a high enough bioload for a cycle, we generally recommend adding 1-2 of these larger snails for every 10 gallons of water.

Snails to Avoid:

Apple snails – These snails will eat all but the toughest plants in a tank. Since shrimp love plants, we do not recommend keeping apple snails in the same tank.

Step 3: Grow-Out

Regardless of which method you choose – **Do not do water changes unless ammonia levels are above 4 ppm**! Water changes disrupt the fragile ecosystem being built so they should be avoided until the cycle is complete.

Snail-In Method

(Best for tanks with few to no fish)

If we opt for the snail-in cycling method, then just sit back, throw a pinch of food (0.02g is enough) in the tank every day or two and watch nature take its course. **Skip feeding if most of the old food hasn't been eaten yet or add a little extra if all the food disappears quickly**. Some white fungus-like growth may appear on leftover food, which is a good sign that decomposers are becoming established in the tank.

The other thing to do is pay attention to plant health. Many plants "**melt**" initially, meaning they shed their leaves to grow ones that are better suited for the conditions of your tank. After melting, we should see new growth within a week or two, indicating that they are adapting to the new environment. If you don't see signs of growth, then we recommend researching the plant type to see if the tank meets its light and nutrient needs. At the low starting population, snails may not produce enough ammonia (and other nutrients) for plants initially, so fertilizers may be beneficial for fast-growing plants.

Fishless Cycling with Bottled Ammonia

(Best for tanks that will have fish)

If we decide to use bottled ammonia, then follow these steps:

1. Add BB products to speed up the cycle.
2. Dose enough ammonia to bring the tank to 2 ppm. Record this amount for later use.
    - **Please note that ammonia levels above 5ppm can harm BB and plants**. If levels ever rise above 5 ppm, then perform a water change to bring ammonia down.
3. Measure ammonia and nitrites every 2-3 days. If ammonia levels drop, then add more to keep it at 2 ppm.
    - Ammonia levels drop during this time when ammonia evaporates from the tank.
4. Continue measuring and adding ammonia as needed until nitrites appear. This may take up to 12 days.
5. Reduce ammonia dosing to maintain 1 ppm of ammonia and continue measuring ammonia, nitrites, and nitrates every 2-3 days. Add ammonia as needed.
    - A lower ammonia level is needed because the bacteria that convert nitrite to nitrate do better at lower ammonia levels and this is the point where they need time to establish.
6. A nitrite spike should happen within the next 10 days. At that point, reduce ammonia dosage to stay at 0.5 ppm.
7. Continue measuring ammonia, nitrites, and nitrates until nitrates appear.
8. Increase ammonia dosage to maintain 1 ppm. Continue measuring and dosing until the 1 ppm of ammonia is completely converted overnight. ***There should be no measurable ammonia or nitrites at this point***.

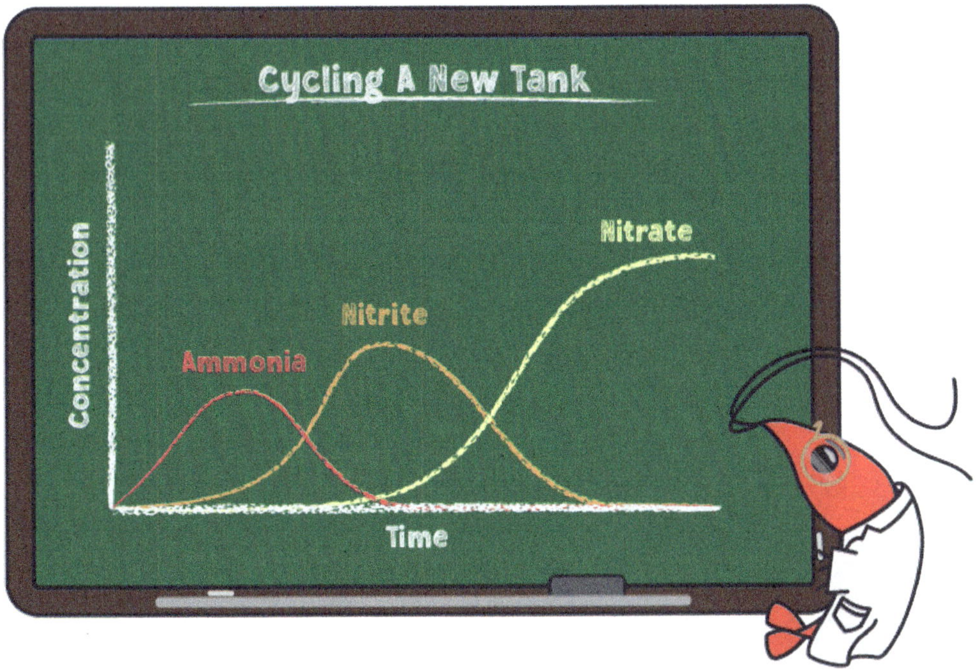

9. Increase ammonia dosage to 2 ppm and wait until the tank completely converts it overnight. There should be no measurable ammonia or nitrites.
10. Keep adding the same amount of ammonia to feed the BB until animals are added.
11. If you add fish to the tank, then now is the time to do it. They are best for replacing the bioload of ammonia that you have been adding. Let the tank adjust to the new inhabitants for **at least 1-2 weeks** before adding shrimp.
12. Shrimp are normally young and small when first bought, so the fish might think they're the perfect food. To reduce the chance of shrimp being eaten on entry, put the shrimp in a net (*after proper acclimation* [p122]) and lower it to the bottom of the tank, into an area where there are plenty of hiding spaces.

What To Expect During the Grow-Out Period

Algae

Regardless of the number of plants, type of substrate, or lighting regimen, a new tank is very likely to have lots of algae growth during the first few weeks and months. It may not look pretty but just know it gets better.

Hair Algae growing on glass (and on the snail) during cycling process

Algae growth happens because a lot of changes occur in a new system. This results in nutrient imbalances that one or more algae species take advantage of. **Algae are actually helping our tank by using nutrients that would otherwise lower water quality**. Another benefit is that algae is an important part of delicious biofilm, which our shrimp love!

It gets easier to control algae as a tank matures and nutrient demands stabilize. This still might require modification of light cycles and fertilizers, but those topics deserve their own book. For more information on handling algae and nutrients in a planted tank, please check out The Aquascape Guide[xlii].

Biofilm/Mold

We may see white fuzz (example[xliii]) growing on leftover food and driftwood, especially in the first week or so after adding it. This is a common mold that shrimp and snails quickly

eat their way through. We can remove it manually, but it is easiest to let it get eaten or let it disappear on its own in a week.

Woods like spiderwood and grapevine are more likely to have visible biofilm growth because they have a lot of stored sugars.

Microorganisms

Quite a few shrimp keepers get surprised when they see little white dots and squiggly lines swimming around or crawling on the glass - Don't worry. **Most of them are harmless little critters that help break down waste in your tank**. Some of the common harmless types are copepods, daphnia, limpets, rhabdo worms, and detritus worms. These are covered more in-depth in the Microorganisms chapter (p112). The main ones we do not want to see are planaria and hydra, which we cover in the Shrimp Diseases chapter (p277). We highly recommend reviewing those chapters to know what to look out for.

Baby shrimp surrounded by copepods

These tiny creatures come in on plants, from the bags that animals are transported in, and occasionally through food. Even when plants and food are treated beforehand, some eggs are extremely hardy and may still make it into a tank. **Again, most microorganisms are perfectly harmless and are very beneficial** because they help stabilize the ecosystem, but some people don't like the look of them. For better or worse, they are part of a healthy ecosystem, and we encourage you to embrace these little critters. Trying to keep a shrimp tank without microorganisms is just asking for an unhealthy tank.

All microorganism populations are controlled by food, which typically includes detritus and any food being added to the tank. If we see a LOT of microorganisms in a tank, or a sudden increase in normally stable populations, then that's a sign of excess food. Once we cut back on food and clean the substrate, we'll see their population naturally decline within a couple of weeks. Another control method is to put fish in a tank to eat the microorganisms. That's easiest to do right after finishing the cycle but, if we leave fish in a tank, just know that some baby shrimp may get eaten along with the microorganisms.

Harmless rhabdocoela worm on glass

Again, check out the Microorganisms chapter (p112) for more information and pictures of different types of common creatures in shrimp tanks.

Step 4: Testing

There are 4 signs to know that a tank is cycled and the ecosystem is healthy after 6-8 weeks (8-12 weeks for Caridina). Ideally, a tank should show all four of these signs before adding shrimp.

Sign #1: Water Testing

At the end of the cycle, we should see ammonia and nitrites **drop to 0 ppm and stay there**. Confirm they stay at zero for at least a few days before adding any larger animals like fish or shrimp.

Nitrates should show up during the cycling process but may not be caught through testing - especially if the tank is heavily planted or has fast-growing floating plants. If we have been adding food (or another ammonia source) consistently during the 6-8 week process and still have zero ammonia, nitrites, and nitrates, then it's safe to say the aquarium is cycled and handles waste properly.

If we do see nitrates building up, then water changes are needed to **bring nitrates down below 10-15 ppm** before adding shrimp.

Sign #2: Plant Growth

Within 6-8 weeks, we should see new growth on plants - typically brighter green shoots of leaves or moss. This indicates they are adapting well so they are ready to help process ammonia and protect our shrimp. If plants aren't growing as well as we'd like, then consider fertilizers or lighting changes.

Sign #3: Microorganism Activity

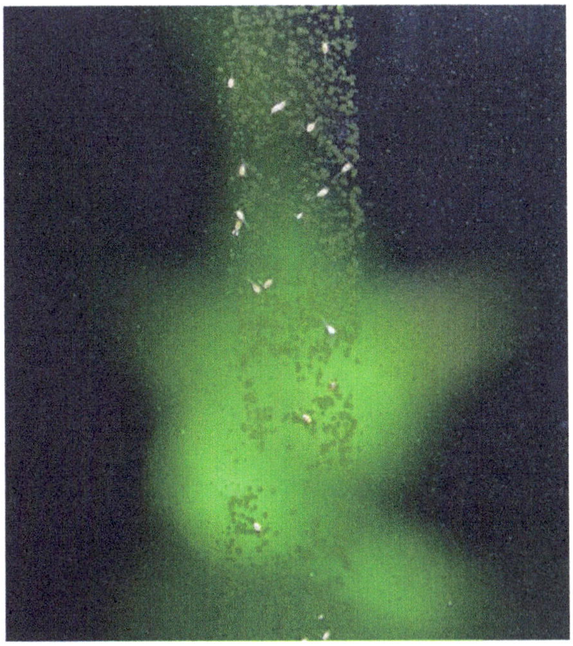
Copepods feasting on green spot algae

There should be tiny white specks (roughly 1mm or less) and possibly small thin worms (a few mm to a cm in length) moving around throughout your tank. This is a great sign that we have a diverse, resilient ecosystem ready for shrimp. Their populations may continue growing as the tank matures but should eventually stabilize. A rule-of-thumb for population size is based on our reaction when looking at a tank. If we look in and think "Wow, there are hundreds of little white dots in there", then that's a good sign. If the reaction is, "Wow, there are THOUSANDS of little white dots in there. It's literally overrun!", then that's a sign to reduce feeding and possibly clean the substrate to remove excess debris.

Sign #4: Algae Growth

Ideally, there should be a bit of algae growing on the glass. Algae create additional surface area and food to support a healthy shrimp colony. If it's on the glass, then we recommend just cleaning off the front wall for viewing purposes, while leaving the remaining walls to develop a beautiful green mat of tasty food. We can encourage this growth by leaving the lights on for longer each day but only do this toward the end of the cycle if there's still no sign of algae after 6-8 weeks.

Algae growth isn't always visible when there are a lot of plants sucking up all the nutrients. If the other signs are present, we should still be fine to add shrimp even if there's no visible algae growth. Some species of algae are *always* present in a tank.

Healthy Microorganisms

The health of our shrimp depends on millions of tiny critters in our tank, many of which we can't see without a microscope. If you can see little dots floating around or some worms crawling around the glass – don't freak out! Most are harmless or beneficial to shrimp tanks since they create a cleaner, stable, and diverse ecosystem for shrimp to thrive in. Many of the organisms help break down waste, grow, and in turn become part of a healthy shrimp's diet. They also tend to be more visible in shrimp tanks because there are no larger predators actively hunting them, unlike in fish tanks. The following list lets us identify harmless organisms to avoid confusion (Is that a baby shrimp?) or concern that may come from seeing something we aren't familiar with. Not all microorganisms are healthy, though, and we'll get to those in Shrimp Diseases chapter (p277).

Understanding the life cycles of microorganisms also helps us detect problems in our tank. For example, some hobbyists may see a bunch of worms in their tank and immediately try to find a medication to get rid of them. The problem with this approach is it attacks a symptom of a problem, not the problem itself. If there are a bunch of worms in a tank, then that means there's plenty of food available due to overfeeding or a lack of maintenance. The worms are there to break down waste and cycle nutrients throughout the tank so killing them with medication simply leaves excess waste around to pollute water and spread disease. If the root cause isn't treated, then there will be more issues down the road - keep this in mind when caring for a tank.

To be clear, there are MANY more hidden microorganisms in our tanks (paramecium, amoeba, stentor, etc.) that are not on this list but play important parts in the ecosystem. The diverse set of life that we get to cultivate an environment for is truly staggering.

Detritus Worms

First up, detritus worms! While not everyone enjoys having wriggling worms in their tank, worms are incredibly good at breaking down waste and churning through the substrate. Think of them like farmers trying to till the soil so it's healthy for the plants and animals.

Detritus worms - Kai Medina (mk170101), CC BY-SA 4.0 via Wikimedia Commons - Arrows added by Shrimply Explained

A "detritus worm" is a loosely defined term including many different types of beneficial or harmless worms like tubifex, bloodworm, or rhabdocoela. They vary in body size, color, and

shape so the main concern when we see a worm is "Does this look like planaria or move like a leech?" (See Pests and Disease section for comparisons [p291]). If not, then it's a detritus worm.

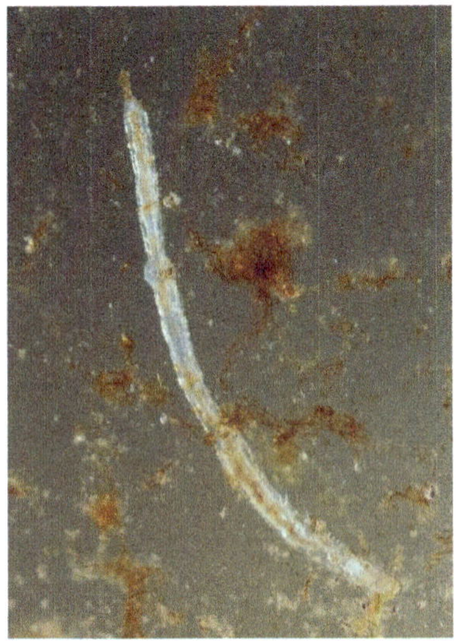

These worms are extremely hardy so they are present in just about every aquatic ecosystem. Their eggs can also survive extreme conditions, including desiccation, methylene blue, formalin, and iodine treatments (Park In-Seok, Baek Se-Won, & Moon Ki Hwan, 2019). As such, **detritus worms can still end up in a tank even when plants are treated and quarantined properly**.

Detritus worms are typically more prevalent in shrimp tanks because there are no predators to eat them. This means their population grows to consume the amount of food available to them. **When a tank is full of detritus worms, it's a good sign of overfeeding**. To control the population, limit food, use a feeding dish, or consider relatively shrimp-safe bottom-feeding fish like kuhli loaches or pygmy Corydoras.

Detritus worm - Macrostomum flatworm - Photo Credit to Nora Quinn (TT/Insta: @missMoshie)

Copepods

Copepods are small crustaceans (about 1-2 mm-long) found in nearly every aquatic environment. They make up a large portion of the tiny white dots we see on the glass or flitting around the tank. With good eyesight, it's possible to identify them via the oar-like arms they use to swim and sense their environment.

Copepod - Andrei Savitsky, CC BY-SA 4.0 via Wikimedia Commons

Copepods spend every second of their day feeding on phytoplankton, eating up to 373,000 per day. This helps keep water clean and helps copepods grow so they can, in turn, be eaten by larger creatures like our shrimp. Shrimp do not actively hunt them, but copepods are so small and numerous that shrimp often pick them up and eat them while scavenging.

Copepods are so numerous because they reproduce quickly, typically reaching sexual maturity after a few days to a few months based on the species. This lifecycle starts with sexual reproduction, then the female carries eggs (sometimes visible to the naked eye) until they hatch as **nauplii** (larva) and quickly grow into adults.

Pregnant copepod - Andrei Savitsky, CC BY-SA 4.0 via Wikimedia Commons

When a copepod experiences environmental stress, it produces **resting eggs**. These eggs are *extremely* resistant to adverse conditions, capable of surviving temperatures as low as 1.5°C and dry periods of 9+ months (Brazil, Caetano, Vargas, Bozelli, & Santangelo, 2022). Even with proper treatment and quarantine of plants, copepods are almost guaranteed in a shrimp tank – and should be appreciated due to their importance in the food chain!

Daphnia

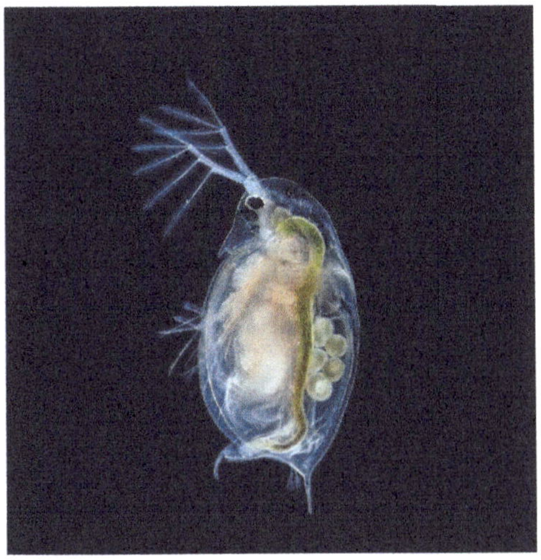

Brandon Antonio Segura Torres & Priscilla Vieto Bonilla, CC BY-SA 4.0 via Wikimedia Commons

Daphnia are planktonic crustaceans commonly called "water fleas". Like copepods, their small size (1-2 mm) makes them little more than white specks in a tank. They are filter feeders that remove algae, detritus, and bacteria from the water to keep water clear.

These little creatures reproduce sexually or asexually and reach sexual maturity 10-15 days after birth, so they quickly populate a tank. That being said, daphnia are not all that common in aquariums because they only feed in the water column. Based on other hobbyists' experience, filters often keep the water clean enough that daphnia do not reproduce much[xliv].

Ostracods

Ostracod - Janek Lass, CC BY 4.0 via Wikimedia Commons

Ostracods (AKA seed shrimp) are bivalve microcrustaceans that make up a portion of the tiny white dots roaming around the tank. Adults reach approximately 1-2 mm in size and eat a variety of microscopic foods ranging from detritus, algae, plant matter, dead animals, bacteria, and others. They are completely harmless toward shrimp, though, and commonly make up a percentage of a shrimp's diet (Weber & Traunspurger, 2016). Again, shrimp do not intentionally hunt them, but ostracods are so ubiquitous that shrimp happen to eat them while foraging.

Ostracods can reproduce sexually or asexually and lay extremely hardy eggs that can last for years in dry or otherwise adverse conditions (Smith, n.d.), which is why they are found in nearly all aquatic ecosystems.

Slime Molds

The slime mold is a name given to a variety of different organisms but only the *Myxogastria* group is visible to the naked eye in our tank. These types of slime molds are single-celled organisms that combine with others of the same species to form (comparatively) extremely large *single-celled* webs with multiple nuclei (see all the yellow growth in the image). The way they grow and move is mesmerizing (BBC, 2013).

Slime molds are uncommon but useful decomposers that may grow on wood and glass in the tank. They are perfectly harmless and not indicative of any problem in a tank. Slime molds appear randomly and disappear within a few days to weeks. They can also be removed manually by scraping them off glass and hardscape then vacuuming up the remaining pieces.

Slime mold on aquarium glass - Credit: u/clammyclemens

Limpets

Freshwater limpets that appear in our tanks are mollusks from the genus *Acroluxus*. Freshwater limpets are slow-moving scavengers that help clean a tank. The one thing they do quickly is reproduce, since they are hermaphroditic and can reproduce asexually. Limpets constantly drop egg capsules as they move, so tanks with limpets often have lots of tiny white spheres on various surfaces. If one is up against the glass[xlv], it looks almost like a bladder snail with its ovoid body, tiny dots for eyes, and a radula scraping against the glass for food. The biggest visible difference is that limpet shells are shaped like tents that extend out past their body on all sides, while bladder snails have a domed shell with heads that extend past their shell.

Overall, they are perfectly harmless to shrimp and are beneficial for a diverse ecosystem.

Limpet - Alexander Mrkvicka, CC BY-SA 3.0 via Wikimedia Commons

Bladder snail (Physella acuta) *with bubble butt*

Exuviotrophic Ciliates

These are an honorable mention because they're not visible to the naked eye but are *very* interesting. These organisms were brought to our attention by Nora Quinn (TikTok @missMoshie). They're weird little potato-like creatures that live on our shrimp and Nora managed to get some incredible footage of them[xlvi].

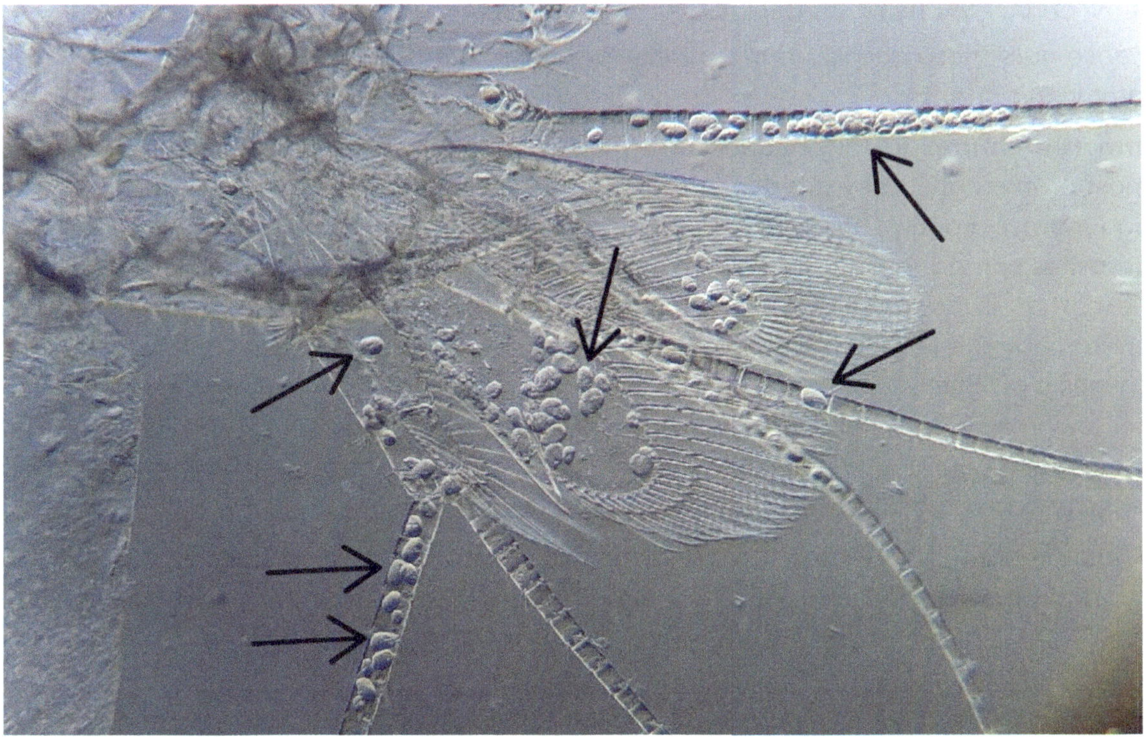

Exuviotrophic ciliates inside a shrimp molt - Every one of those potato-shaped things is a ciliate - Photo credit to Nora Quinn (TT/Insta: @missMoshie)

Basically, these little organisms are ciliates (they have cilia - little hair-like appendages used for feeding and moving) that live symbiotically with our shrimp and other crustaceans. They hibernate on the shell until a shrimp molts, after which the ciliates come to life and eat the fluid that shrimp produce to help them get out of their old shell (the exuvial fluid). They reproduce rapidly inside the leftover molt before dispersing to find other hosts to hibernate on until they molt and restart the cycle.

Nature is wild. Thanks again to Nora for sharing her knowledge with us!

Where To Buy Shrimp

As we've discussed, shrimp do best with stable parameters. If the breeder's conditions are extremely different from our tank, then shrimp are likely to have a harder time acclimating, in which case some deaths are normal in the first few weeks.

It Matters How Shrimp Are Bred

Please note these definitions are mainly for Neocaridina shrimp. Caridina are discussed later in this chapter.

Imported shrimp are the cheapest and most widely available option but they are also the most problematic. They are often bred in parameters that are very different from home aquariums, are kept in overcrowded and unhealthy conditions where diseases and parasites spread rapidly, and the shrimp are stressed from repeated travel from breeder › exporter › importer/seller › buyer. This frequently leads to deaths when imported shrimp are bought and introduced to new tank. Lastly, imported shrimp are sorted by size, which causes a problem. Females are larger so are more likely to fit the size requirement, so it is not uncommon to get entirely females when purchasing a bag of 10 shrimp, which means we need to go buy more to hope for a male. For all these reasons, we generally recommend avoiding imported shrimp when possible.

If you do want imported shrimp, we recommend Flip Aquatics because of their 30-day quarantine policy that also reduces stress on shrimp thanks to the break in travel before sale.

Note: The type of water in a tank also affects the best choice in shrimp. Tap water often has unique properties (pollutants, mineral ratios, etc.) so imported shrimp tend to do worse in it. They are more likely to adapt to remineralized RO/DI water.

The other choice is **homebred shrimp**, meaning any shrimp bred locally or within the same country. Extremely local breeders are best because the shrimp are already adapted to the local tap water (if the breeder uses tap water), so the transition to their new home is stress-free. They also don't have to travel as far and are typically bred in better conditions with fewer diseases. To find local breeders, join aquarist groups in-person or online and find someone you trust.

If local breeders aren't available, a good domestic home breeder in the United States is Mark Shelly Aquatics. **Get 5% off your order using code "explained5" at checkout**. A problem with home breeding is that breeders often have trouble keeping up with demand, eventually turning to importing. Some of them are honest about it, but not all of them, so do some research and look at reviews to avoid bad experiences. If the seller doesn't specifically say "homebred", then the shrimp are almost certainly imported.

Please note we know Mark Shelly personally and do receive a commission if the code is used. We recommend him because we know how well he cares for his shrimp.

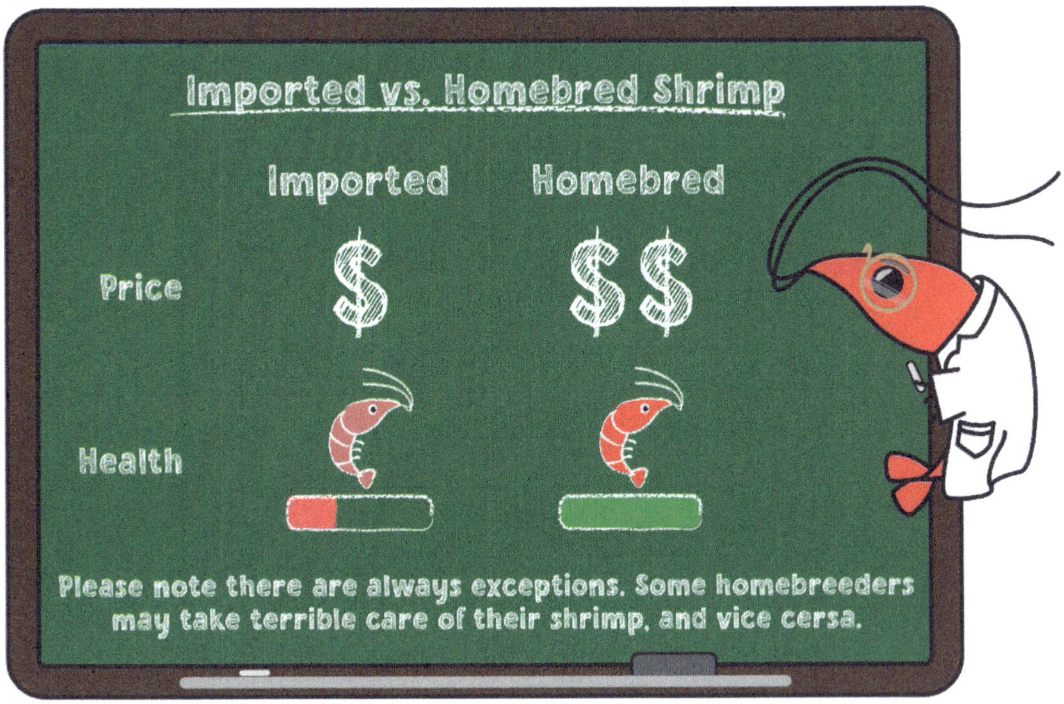

Where To Get Caridina

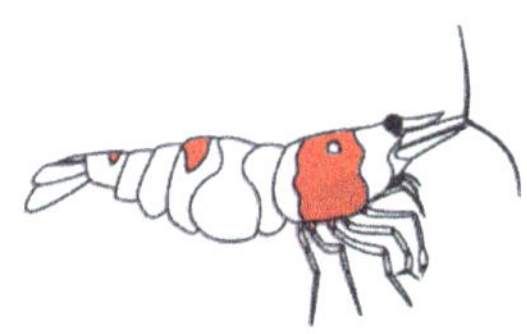

Caridina are a slightly different story than Neocaridina because they are less hardy. This means that they have to be bred in cleaner and more consistent conditions (typically remineralized RO/DI water) no matter where they are bred in the world. This makes the transition to a new tank far easier, while also cutting down on issues with disease and parasites. Imported Caridina still deal with stress from repeated transit from importers to sellers to their new home, but this is generally not as problematic. Homebred Caridina are still preferred if you can get them directly from the breeder but imported Caridina are often perfectly fine.

We've had good experiences with Flip Aquatics and Shrimply Fintastic.

Understanding Shrimp Grades

A "grade" of shrimp refers to their appearance – how bright and opaque their coloration is or how closely they match the "perfect" pattern of what's seen in shrimp shows (p248). For example, Cherry shrimp grades are "Red Cherry" (low grade), "Sakura", "Fire Red", and "Painted Fire Red" (highest grade), with the higher grades having a higher percentage of their bodies covered in red, along with lower shell transparency.

The price of a shrimp breed generally increases with higher grades. There are some exceptions to this though. Grades can be applied to an individual but also need to be

thought of from a colony perspective. A single high-grade shrimp may appear in a colony of otherwise dull cherry shrimp. Is that shrimp worth as much as a high-grade shrimp from a colony of entirely high-grade shrimp? No, because that single high-grade shrimp from a dull colony is very likely to have dull offspring because it is not a pure-bred. Be suspicious if high-grade shrimp are being sold for cheap, as this often suggests they are outliers rather than consistent pure-breds.

Each breed of shrimp has their own grading system based on somewhat arbitrary preferences, so do some research on each specific breed to understand how to evaluate their quality. We cover grading and breeding more in the Shrimp Breeding chapter (p191) and Caridina Shrimp Breeding section (p251).

Adding Animals

Once the tank is cycled, it's time for the most exciting part – Adding our shrimp and other animals!

Adding new organisms to the tank, especially larger ones, is a huge change that this new ecosystem needs to adapt to. The tank can be delicate at this stage, so it's best to take things slowly. If you choose to add animals other than shrimp, then we recommend introducing them one group at a time, prioritizing the hardiest ones with the highest bioloads first.

For example, let's say we want a tank with 6 ember tetras, 6 otocinclus (otos) catfish, 6 Amano shrimp, and 10 Red Cherry shrimp. The fish have higher bio-loads than shrimp, so we want to start with them. Out of the two fish, otos tend to be more sensitive to environmental changes. Out of the shrimp, Amanos tend to be hardier. This means we should introduce the ember tetras first, then the otos, then the Amanos, and the Cherry shrimp last. When adding shrimp, try placing them in a sheltered corner of the tank where they can quickly hide from fish.

After each new addition, we recommend carefully monitoring the parameters to look for changes over the next week, especially in ammonia and nitrites. If the tank seems stable after a week, then add the next group of animals. Rinse and repeat until all the animals are added. To be clear, this order of animals is a recommendation. There are differing theories on this topic.

Introducing new animals is always a bit stressful – for both us and the animals. The next section explains how to limit that stress and give our animals the best start to a healthy life.

How Many Shrimp Should We Buy?

We recommend getting 1-2 shrimp per gallon and starting with a minimum of 10 shrimp. Many online sellers add a couple extra to the bag, so if we order 10 shrimp we typically end up with 11-12 shrimp. If we order fewer than 10 shrimp, then there's a risk of three things happening:

- We may not get males and females, so they won't breed.
- A few shrimp may not adapt well to the new tank, especially if we get adult imported shrimp. Having 10-12 shrimp increases the likelihood that both males and females survive to create a stable colony.
- A smaller starting population increases the risk of inbreeding issues. At least 10 shrimp offer a better chance at genetic diversity to reduce inbreeding problems. More on that in the Inbreeding Problems section (p213).

How Many Shrimp *Can* We Have?

Each female produces anywhere from 20-50 offspring every month or two (Nur & Christianus, 2013), depending on age and size. Not all babies survive but their short breeding cycle creates an exponentially growing population when they don't have any predators. Within 3-6 months, you might be worried about your tank getting overrun with shrimp. Talk about a good problem to have!

In reality, overpopulation isn't a huge issue for shrimp tanks because their population is limited by food availability. If we let the population grow to the amount of food naturally produced by the tank, then shrimp eventually reach the maximum amount the tank can support (AKA carrying capacity), at which point breeding slows or stops. **The natural carrying capacity for a tank is roughly 5-10 adult shrimp per gallon**, depending on the amount of surface area for biofilm and algae to grow on, along with light and nutrient availability.

If our goal is to breed as many shrimp as possible in a tank, then we can push the population limit higher by adding more food. This does increase the risk of poor water quality but consistent testing, gravel vacuuming, and water changes let us push the population limit far above what's possible naturally. This works well for Neocaridina tanks since Neos aren't as susceptible to poor water conditions or the fluctuations that occur during maintenance. This strategy is a bit more dangerous for Caridina though. Good water quality is absolutely critical for Caridina so lots of additional food often causes more harm than good.

With more food and maintenance, Neocaridina tank populations can reach up to 20 adult shrimp per gallon. According to master shrimp breeder Mark Shelly (Mark Shelly Aquatics), breeding typically drops off significantly past this population density, regardless of how much food is added. In contrast, Caridina breeders like Kelvin Wong rarely keep more than 5-7 adult shrimp per gallon (Shrimp Mania, 2024) since they don't want to risk water pollution from the amount of food required to maintain a larger population.

In the end, it all comes back to what our goals are as a shrimp keeper. If we just want to enjoy a low maintenance tank, then let the population naturally find its maximum limit the ecosystem can maintain. If we're trying to sell a bunch of shrimp, then it may be worth the extra effort to feed more and perform more maintenance.

What To Do With All Those Shrimp

If shrimp are breeding like crazy and you're worried about having too many shrimp, then congratulations! That's a great opportunity to give a few shrimp away to friends, offer them as a gift to a love interest (after helping them with the romantic process of tank setup), or sell them to a local store.

We're getting ahead of ourselves though. First, let's get our shrimp into their new home.

Acclimation

The tanks where our animals come from are completely different environments from any tank we have set up. Dropping them directly from their bag into our tank is like teleporting someone from a balmy spa into a snowstorm while still in their bathing suit—shocking, to say the least. To avoid stressing out our animals, we always want to **acclimate** them, otherwise we may lose some or all of our new pets.

Acclimation is the process of getting the parameters within an animal's current environment (the bag) as close to the parameters of the new environment (our tank). We recommend **drip acclimation**, which involves introducing tank water into the bag at a slow, consistent rate. Having a smooth transition into a tank is an important first step for successful shrimp keeping, especially if the shrimp are stressed from shipping or a previous poor environment.

Here are the steps for successful drip acclimation:

1. **Make Sure The Tank Is Cycled**
2. **Plan The Day**
3. **Drip Acclimate**
4. **Add Shrimp To Tank**

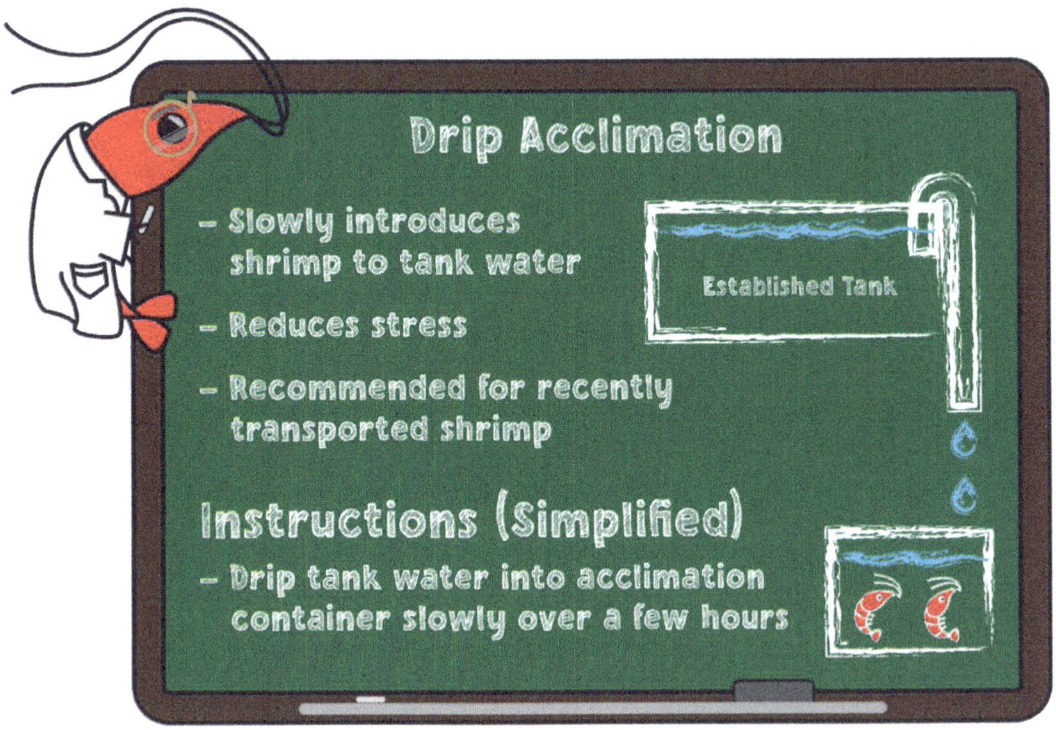

Side note: If you're familiar with fish acclimation, the drip acclimation procedure may seem like a slow and unnecessary process. Shrimp are more affected by water quality so acclimation is more stressful. That being said, healthy shrimp can handle large changes as long as they are *temporary*. If the shrimp were cared for and shipped properly by the seller, then following the fish acclimation procedure (p128) may work perfectly fine. The same is true when transferring healthy shrimp between tanks. The problem with shipped shrimp is we often don't know how healthy they are, so it is generally safest to do a slower drip acclimation.

Step 1: Make Sure The Tank Is Cycled

Double-check that the tank shows Successful Cycling Signs discussed in the Cycling section (p99). If everything looks good, then continue to the next step.

Step 2: Plan The Day

The acclimation process takes a few hours, so we want to plan for it to avoid stressing ourselves or our shrimp. That means having all the equipment necessary for proper drip acclimation, along with 3-4 hours to pick up and acclimate the shrimp without being rushed. We can certainly do other stuff during this time, but we recommend sticking around the tank to check acclimation setup every 30-60 mins. It's surprisingly easy to overflow a container.

Depending on how we buy our shrimp, there are a few different things to be aware of:

When Picking Up Shrimp In-Stores...

Our goal is to protect our shrimp from significant temperature changes or excessive movement during transport.

Avoid Temperature Changes

If buying shrimp on a very cold or warm day, then the small amount of water in their bag may change temperature rapidly during the trip home. To avoid this, put the bag in a cooler (that's not cold), Styrofoam box, or another insulated package. Adding ice or a heating pad may seem like a good idea, but they could cause excessive temperature changes, so we recommend not using them.

Avoid Excessive Movement

Just as we dislike earthquakes, our shrimp dislike being sloshed around in the bag on the way home from the store. Don't be a Darla[xlvii]!

To prevent excessive movement, bring a clean piece of cloth (something that doesn't leach chemicals into the water) or moss to place in with the shrimp. This gives them something to hold onto and reduces their stress levels. Also, it's a good idea to remove all but a small amount of air from the bag. More air means more room for the water to move around. If the trip home is less than an hour, then an inch or so of space at the top is all we need to ensure they have enough oxygen. Lastly, keep the bag somewhere that it won't roll or shift during the journey home, whether that is in our lap (if you're not driving) or in a container that won't slide around.

When Ordering Online...

This is the easiest option that requires little planning, as most online sellers already take precautions to avoid stressing your shrimp. All we need to do is be around for the delivery so the package doesn't sit outside and get too hot or cold. This may not be a big problem if the seller uses insulated packaging for shipping, which most do.

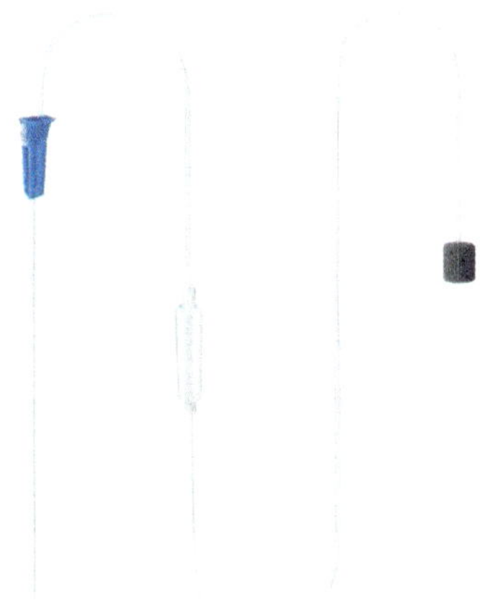

Drip acclimation tubing

Once our shrimp arrive safely, it is time to drip acclimate!

Step 3: Drip Acclimation

Equipment

1. **Drip acclimation kit** – For slowly introducing tank water into the water our shrimp are in. Allows for a safe and stress-free transition into a tank. Can be DIY or store-bought for around $10[xlviii].
2. **Fish net** – After acclimation, we always recommend using a net to move shrimp into the tank instead of pouring them in with water from the store. This reduces the chance of contaminating a tank with unwanted organisms that may cause disease.
3. **Large container** – Must be able to hold at least 4 times the volume of water in the bag of shrimp. The fish (shrimp) net should also be able to fit inside to net our shrimp out at the end of the acclimation process. We also want something that can float in the water. A large Tupperware container normally works well.
4. **Smaller container or glass** – Must hold at least 1/4 cup (or 50 ml) of water.
5. **Cloth covering** – Must be large enough to cover the top and sides of the large container and thick enough to block light.

Instructions

1. Inspect the shrimp for any signs of health problems and check the bag for pests.
    a. Familiarize yourself with common diseases by reading the Shrimp Diseases chapter (p277).
2. If the shrimp were in transit for a long time, then check ammonia levels in their bag.
    a. **If ammonia is high (>5 ppm) and the tank has pH above 7.5**, then switch to the fish acclimation method (p128).
3. Put a small amount of water (1/4 cup or 50 ml) from their bag into the small container.
4. Pour the rest of the water—with the shrimp—into the large container. We recommend swirling the bag slightly, then dumping quickly to keep the shrimp from sticking to the inside of the bag.
    a. If any stragglers are stuck to the sides of the bag, pour the water from the small container back into the bag to free the shrimp. After that, pour the poor straggler shrimp into the large container.

5. Position both containers below the water level of the tank. Make sure they are out of range of pets or children.
6. Place the hooked end of the drip acclimation kit into the tank and the other end in the *small container*.
7. Squeeze the bulb on the acclimation tubing to get tank water flowing. This may require multiple squeezes.
8. Adjust the flow rate to roughly **1 drop per second**. This is generally done with a wheel on the tube for store-bought kits.
9. Place the dripping end of the acclimation tubing into the large container with our shrimp.
    a. The smaller container is no longer needed.
10. Cover the large container with a cloth or move it to a dark area.
    a. Less light leads to lower stress on shrimp.
11. After the volume of water has doubled in the large container (roughly 1 hr.), then adjust the **flow rate to 2-3 drips per second**.
12. Remove the drip acclimation tubing once the volume is 4 times what it was initially – roughly 2-3 hours.
13. Measure the temperature of the tank and the large container to verify they match (within 1-2°F or 1°C ideally)
    a. If the temperature does not match, then float the container in the tank until the temperature is close.
14. Turn off the tank lights to reduce stress.
    a. This is not *necessary*, but it can help.
15. Net shrimp out and place them in the tank.
    a. Try to avoid adding any water from the bag they were shipped in. The bag water may have excess ammonia or unwanted hitchhikers.
16. Leave the lights off for an hour or two while our shrimp adjust to their new home.

Drip acclimation setup to Step 9.

Fish Acclimation Method

The fish acclimation method is for fish, but also for shrimp when the bag they are in has high ammonia levels. In this case, the slower drip acclimation method is not suitable because it exposes them to high ammonia levels for a long time. In addition, if the tank has high pH, then the tank water will raise the pH of the bag water and make ammonia more toxic. On the other hand, if the tank pH is below 7.5, then the ammonia stays relatively non-toxic (for a couple of hours exposure) so we recommend doing drip acclimation.

Equipment

1. Measuring Cup (1/2 cup or 100ml)
2. Fish Net

Instructions

1. Turn off tank lights.
2. Open the bag your animals are in.
    a. Don't cut it – keep the bag whole.
3. Lower the bag into tank water until it is fully supported by the water but <u>do not let go of the bag</u>.
4. Roll the edges of the bag to create a flotation ring at the top that holds the bag steady in the water. Keep the rolls loose to trap more air.
    a. At this point, you may let go of the bag.
    b. If you are doing this method because of high ammonia, then skip to step 9.
5. Add approximately ½ cup of <u>tank water</u> to the bag.
6. Wait 15 minutes, then add another ½ cup of <u>tank water</u> to the bag.
7. Check if the pH of the water inside the bag is ±0.5 pH. If not, repeat Steps 5-6 until it is.
8. Check the temperature of the bag to see if it is ±2°F (±1°C) of the tank. If not, continue floating the bag until the temperature equilibrates.
9. Use the fish net to remove the animals and place them in the tank.
    a. This reduces the ammonia/contaminants in the bag water that enter the tank.
10. Leave the lights off in the tank for an hour or two while animals adjust to their new home.

Post-Acclimation

Congratulations on safely adding animals! The stressful part is done, but we recommend paying careful attention over the next few days to check that animals settle in well. It can be tough to tell sometimes because most aquatic animals aren't very expressive (except for Shrimply - although we're not always sure what he's trying to express...). Also, **shrimp tend to hide for the first few days so it's normal to not see them for a while if the tank is heavily planted**.

We don't know what Shrimply is trying to communicate here...

Here are some signs to indicate that acclimation is NOT going well:

1. **Aggression between tank mates**

   If this happens, we may need to put aggressive inhabitants into another tank or return them. Dwarf shrimp rarely ever show aggression.

2. **Frantic swimming**

   If most shrimp are frantically swimming around instead of scavenging as normal, then check water levels for parameter changes. If nothing is out of the ordinary, then look for any dead bodies or molts. When females molt, they release chemicals (**pheromones**) that cause the males to swim frantically in search of a female. If there are no dead bodies or molts and parameters are fine, then the animals may just be feeling healthy and active.

3. **Elevated ammonia or nitrites**

   Ammonia and nitrites should always be 0 ppm in a cycled tank. Elevated levels are a sign of a cycle disruption or overload. We explain what to do and how to avoid cycle disruptions in the next chapter.

4. **Staying at the waterline or trying to climb out**

   If *most* shrimp (75%+) are at the waterline or trying to climb out of the tank, then there may be water quality issues. Check parameters and make sure your filter is working well. Consider if any changes have been made in/around the tank, like new cleaning spray, flea treatment of dogs/cats, etc. If nothing is out of the ordinary, then they may just be exploring or eating dead algae near the waterline.

Cycle Disruptions

Identifying A Cycle Disruption

If we detect ammonia or nitrites in the water at any point after cycling, then the nitrogen cycle may have been disrupted. This is also sometimes referred to as a cycle crash.

Preventing Cycle Disruptions

A cycle disruption happens anytime the current combination of plants and BB are unable to handle the bioload in a tank, resulting in ammonia or nitrite spikes. High ammonia levels prevent BB from breaking it down efficiently, which makes the problem worse. Common causes include:

- **Overstocking**
    - Having too many animals in the tank can overload the nitrogen cycle, leading to ammonia spikes and subsequent crashes.
- **Overfeeding**
    - Feeding too much at once, or letting leftover food build up, can result in a large release of ammonia that crashes the cycle.
- **Insufficient Maintenance**
    - Neglecting routine aquarium maintenance tasks like water changes, filter cleanings, and substrate vacuuming (Covered in the Maintenance chapter [p151]) can lead to a buildup of waste and toxins. Over time, this can disrupt the nitrogen cycle and lead to a crash.
- **Contamination**
    - Chemicals from cleaning products or pet medication can stay on our hands long after use and may disrupt the cycle. Always wash hands thoroughly before putting them into the tank to avoid contamination.
- **Insufficient filtration**
    - Inadequate or improperly maintained filtration systems may not effectively remove waste and toxins from the aquarium. This can disrupt the nitrogen cycle and lead to a crash.
- **Rescaping**
    - During a rescape, you may remove plants and décor with BB growing on them. Removing too many plants or BB may leave too few in the tank to handle the current bioload.
- **Substrate Disturbance**
    - Cleaning too deeply into substrate can release trapped ammonia and bacteria, resulting in an ammonia spike or bacterial infection. To avoid this, only vacuum the top layer of substrate when possible.

- **Chlorine Toxicity**
    - If the water used for water changes or during the initial setup of the aquarium is not properly treated to remove chlorine or chloramines, it can harm the ecosystem and cause a crash.
- **Medications/Chemicals**
    - Certain medications, such as antibiotics or copper-based treatments, can harm or kill BB. Similarly, the use of excessive chemicals or additives can disrupt the balance of the nitrogen cycle and lead to a crash.

Re-Establishing the Cycle

To re-establish a cycle, we need to keep ammonia levels low to support the growth of the BB that break down nitrites. This is best done through frequent testing and water changes. If ammonia levels are above 1 ppm for an extended period of time and the tank is above 7.5 pH, then we may consider lowering the pH of our tank to reduce the amount of toxic ammonia present.

If ammonia levels are kept under control and there's enough surface area for BB, then our BB will re-establish themselves. **We can help them establish faster with additional filtration and by adding bottled BB to the tank**. It also may help to remove animals or excess food to reduce the bioload.

After doing what we can, just keep doing water changes and testing until the cycle re-establishes.

Orange Rili (Neocaridina davidi *female*) *carrying eggs – AKA "berried".*

What is Normal Shrimp Behavior?

See the Anatomy section (p166) for a refresher about shrimp body parts referenced in this chapter.

Eating Behavior

Shrimp need to eat *a lot* of biofilm so they spend most of their time grazing. This is why their claws on the front two pairs of walking legs (chelae) constantly scrape the ground in search of food, in a sort of tap-dancing motion. During this time, they also smell the water column with their antennae for signs of nutrient-dense food sources. This is how they sense food when we put it in the tank.

When we add food, shrimp quickly sense it with their antennae and start to move toward it. How quickly they approach is a good indicator of how hungry they are, as well-fed shrimp often meander toward food instead of bolting straight for it. Regardless of their hunger levels, shrimp go into a sort of feeding frenzy when they finally find the food, tearing into it and fighting over scraps by waving their tiny limbs at each other. This is the only time we ever see shrimp be aggressive and most hobbyists think it's *probably* not a problem because their claws are too small to harm each other. It's like stuffed teddy bears trying to fight each other.

That being said, we disagree that feeding frenzies cannot cause problems. See the Food Size section (p144) to learn more about how food aggression impacts shrimp health.

Another normal eating behavior that may cause concern is when shrimp gather at the waterline of the tank after a water change or top-off. You'll often see this because shrimp now have access to dried algae that they didn't beforehand.

Cleaning Behavior

Aside from eating, we may also see shrimp preening by rubbing their legs across their face or along their abdomen. Their antennae may spread out while shrimp try to clean thoroughly, as well. It looks a little weird but is perfectly normal. Interestingly, there's some evidence that shrimp use their claws to stimulate the production of antifouling compounds at the base of each antennae, which they then rub over their antennae to keep them clean (Schmidt, Chien, & Tadesse, 2006).

Shrimp sometimes curl their abdomen tightly underneath themselves, forming a sideways U-shape (see image to right). They do this to help clean their shell and stretch it out. During this time, a small crack may be visible in the middle of the shell. Don't worry – it's normal.

Pre-Molting Behavior

When a shrimp prepares to molt, it often forms a U-shape repeatedly to stretch out its shell. During this time, it may stay in one place and fidget or twitch its limbs more than normal.

When a shrimp is very close to molting, a thin clear line often develops between the head and abdomen. This is where the shell breaks so the shrimp can squeeze out of it. Many shrimp keepers panic when they see this break in the shell because they've heard about the **White Ring of Death** (WRoD). This is the name of a molting problem where the shell breaks and the shrimp can't get out, eventually dying of stress after a few days to a week.

The pre-molt break is fairly distinguishable from the WRoD in appearance because the WRoD normally is a larger crack (2+ mm vs 1 mm) and the edges are scrunched up around the break, indicating the shrimp is struggling to escape. Even these symptoms are not signs of a serious issue unless they stick around for more than 24 hours. Check out this

Short[xlix] for a clear comparison of the Pre-Molt Ring vs WRoD. If one or more shrimp develop the WRoD, then please see the chapter on Molting Problems (p267).

An additional cause of concern during the pre-molt period is when a lot of shrimp suddenly start swimming around. **When a female prepares to molt, male shrimp swim around frantically to try finding her.** This is because sexually active females release chemicals called pheromones into the water that attract males to her. One or more males may try clinging onto the female at this time, which can be cute, but it can also be dangerous and stressful if too many males gather at once. **To avoid stressing pre-molt females, we recommend trying to have at least a 50:50 male:female ratio** and have plenty of plants or rocks for females to hide in. Having more females than males is even better, especially for shrimp breeders.

To be clear, having a few more males than females isn't a death sentence for a molting female. There's no need to go out and buy another batch of shrimp if the ratio turns out to be 60:40 M:F. When the ratio starts skewing 70:30 or more is when more females may be necessary.

Molting Behavior

To escape their shell, a shrimp usually pulls their head out first, then quickly flicks their tail to pop out of the abdominal molt (Marine Things, 2023). This whole process takes just a few seconds, so be on the lookout for pre-molt behavior to get a chance to see the process. It's very easy to miss!

If a shrimp is struggling to molt, we may see them twitching or jumping around by flicking their tails – similar behavior to when they get scared but repeated consistently without stimuli. This is extremely stressful for them, so they also may stop and lie down on their side. After multiple attempts to break out of the shell, it is very unlikely they will succeed but it is best to leave them alone. Any attempts to intervene tend to stress them out more. **The only intervention that may work is adding an air stone to the tank so they have optimal oxygen levels**.

The stress of molting is one of the main causes of death for older shrimp. If a shrimp is 1-2 years old (again, lifespan depends on tank temperature) and has a molting issue, that's not necessarily a reason to worry about all the other shrimp. If molting problems happen repeatedly though, then please see the chapter on Molting Problems (p267) to try protecting remaining shrimp from the same fate.

Post-Molting Behavior

After a successful molt, shrimp have a very soft shell that is vulnerable to predators. Even if there isn't a predator in the tank, they typically go hiding for up to a day because that's their survival instinct. During this time, they'll often be found hiding in crevices or hanging underneath a branch, staying mostly stationary while their shell hardens.

Right after molting is also when mating occurs. If a sexually active female molts while she has a well-developed saddle, then we may get lucky enough to see a male maneuver itself underneath her and press himself against her for a few seconds to fertilize her eggs. After this, the female hides for around 24 hours, but we may see her moving eggs into her swimmerets. This is extremely cool but very rare to see. The mating process is covered in more detail in the Lifecycle chapter (p184).

Once a shrimp's shell is mostly hardened, you'll see them venture out again. **Their first stop is often visiting their old molt, which they eat to recover minerals** (Truong, Hines, & Emerenciano, 2023). This is why it is so important to leave molts in the tanks instead of removing them.

Abnormal Behavior

There are a few stress indicators to look out for, especially after making changes to the tank. These include the following:

1. Staying still for more than 30 minutes without pre- or post-molting behavior
2. Laying on their side or upside down, or even upright without being supported by their legs
3. Repeated flicking of their tail and jumping around without an obvious reason to be startled (molting problem or toxin)
4. Floating down in the water without swimming
5. Multiple shrimp gathering at the waterline and trying to agitate the water (sign of low oxygen)
6. Multiple shrimp trying to climb out of the water (bad parameters or toxin)

Red Cherry shrimp (Neocaridina davidi female) eating a bloodworm snack.

Food

What Do Shrimp Eat?

Our shrimp are classified as **opportunistic detritivores**, meaning they eat just about anything they can get their claws on that's not alive and healthy. Their natural diet includes **biofilm**, dead plant/animals, and microorganisms like copepods that shrimp happen to pick up while scavenging (Weber & Traunspurger, 2016). In our tanks, there aren't many dead plants or animals (hopefully), so shrimp diets are often supplemented with various commercial foods.

Commercial shrimp food

As a hobbyist, learning about shrimp food can be incredibly frustrating because there are tons of different sources claiming that "X food is the *best* for growth!" or "You should be feeding your shrimp seven different foods on a rotating schedule each week!". Unfortunately, very few foods have been evaluated for their performance, so these claims are rarely backed up by any data.

So, what are we supposed to do? Buy every food and test them out?

That's outside of most people's budget and time constraints, so the rest of this chapter is dedicated to a scientific look at a shrimp's dietary needs to create some basic guidelines for what's *sufficient* versus what's *optimal*, depending on our goals as shrimp keepers.

A Closer Look at Biofilm

Without a doubt, the most common food anyone lists for shrimp is **biofilm**. Biofilm is a complex community of algae, bacteria, and fungi found on just about every wet surface, whether in a kitchen sink or an aquarium. These communities are held together by a sticky substance called **extracellular polymeric substance**, or EPS, which is secreted by many different species. EPS is made of sugars, proteins, and other organic compounds that, when combined with all the cells held by it, turns biofilm into a relatively nutritious and readily available food source for animals like shrimp.

Biofilm takes time to develop in our tanks, which is why we recommend letting a tank sit for 6-8 weeks before shrimp are added. This time allows diverse biofilm communities to develop, as shown in one study where researchers compared changes in species diversity for biofilm grown on different materials over 63 days (Viau, Marciano, Iriel, & Lopez-Greco, 2016). After letting biofilm develop, researchers moved the biofilm-covered materials to containers with *Neocaridina davidi* and observed the tanks for an additional 63 days. During this time, biofilm was the *only* food source available for the shrimp. This experiment

tracked various health indicators and found that **biofilm was sufficient to support Neocaridina growth, survival, and reproduction**. Water quality also remained consistently healthy (High DO, consistent pH, safe ammonia/nitrite/nitrate levels), indicating the organisms in biofilm successfully sequestered nutrients. The material with the highest surface area (agrovelo) was found to harbor the largest biofilm biomass.

Based on this research, we conclude that **biofilm, by itself, is *sufficient* for shrimp to grow and complete their life cycle**. The surface area in a tank determines how much biofilm grows and, therefore, how large a shrimp population can be supported. By including plants, substrate, and sponges, we drastically increase surface area and provide more food for shrimp. If we simply leave a tank alone with clean water, sufficient lighting and surface area, then we get an ecosystem that supports a small population of shrimp.

Many hobbyists don't just want *some* shrimp though – We want ALL THE SHRIMP! That's where complications come in.

Finding Balance in A Shrimp Tank

The previous experiment proved that biofilm provided the *necessary* nutrients for shrimp. It did not demonstrate that biofilm offered *optimal* growth, breeding, and coloration – factors that tend to be incredibly important for shrimp breeders. If our goal is to breed as many shrimp as possible OR to breed shrimp as fast as possible, then we want to consider how best to supplement their diets while balancing shrimp and tank health.

Balancing Food Availability and Cost

One consideration is food accessibility. We've shown that biofilm has enough nutrients to support shrimp and it's abundant in ecosystems. The problem is that shrimp need to spend their entire day scraping it up to get enough food, which wastes energy that could otherwise go toward growth. This is natural – but not efficient. **Commercial foods are designed to be accessible and digestible, so less energy is wasted on foraging and more energy goes to growth. More food also supports a higher shrimp population than a biofilm-only diet.** Downsides to commercial foods include the cost and effort to buy.

Balancing Protein Content In Food

Another consideration is protein content. Biofilm has been shown to contain around 35% protein (Shaikh, Rashid, Onwusogh, & et al., 2023) but, as we mentioned before, it takes energy to obtain enough biofilm to survive on, which is why we turn to commercial foods. Numerous studies have shown that food with higher protein content increases shrimp growth rate, going up to a maximum of 58%. One

downside is that higher protein foods also increase ammonia production since protein contains lots of nitrogen (Gauquelin & Cuzon, 2007). This points to a necessary balance of food quantity and maintenance. If we feed a lot of high protein food, then we need to be concerned about elevated levels of nutrients like ammonia and nitrates which need to be removed through regular maintenance.

There's also the consideration of where protein comes from. Some breeders emphasize plant-based foods. Others feed mainly animal protein. Let's take a look at the research.

The Science of Protein

One study tested diets containing various amounts of animal and plant protein on the growth and survival of Homarus americanus (American lobster) (Tlusty, Myers, & Metzler, 2008). They found the lobsters fed 100% animal protein had initially higher molting/growth rates, but all the subjects contracted shell disease and died... Not ideal. Lobsters fed mixed diets (including *Spirulina* algae or soy-based proteins) showed lower growth but higher survival. Shrimp hobbyists have seen similar results, as shrimp fed solely on leftover artemia and bloodworms (from fish feedings) occasionally develop molting problems.

> **Side note**: Spirulina powder contains 50-70% protein[i].

Another study on Neocaridina compared animal-based foods to one mixed-protein food that included *Spirulina* (Türkmen & Karadal, 2016). The results told a slightly different story than the previous study, since Neocaridina fed the mixed-protein food demonstrated a higher growth rate than the groups fed solely animal-based food and no difference in survival.

These studies show that feeding some plant protein seems to improve overall survival and health for shrimp and other crustaceans. What if the diet is *entirely* plant-based?

Our final study tested the effect of entirely plant-based diets compared to diets with varying levels of animal protein on *Litopenaeus vannamei* (Soares, Peixoto, Davis, & Davis, 2021). The results showed that 10-40 g/kg of additional animal protein led to weight gains of 9-16%, respectively, with no change in overall survival. The increased animal protein also led to a higher consumption of food.

Based on these studies, we conclude that **a balanced diet containing both animal and plant protein is likely to optimize shrimp growth and survival**. If our goal is to maximize growth and breeding, then one food containing both is sufficient, or we can alternate feeding a plant-based food with an animal-based food. If the goal is simply to have a healthy, low-maintenance shrimp tank, then biofilm is sufficient to support a population of shrimp without needing additional food.

Food Categories

Mixed Protein (Plant and Animal) Foods:

If only feeding a single type of food, these are good options with diverse ingredients.

Ingredient lists on products are based on the quantity of the ingredient included in the product, meaning there's more of the first ingredient than the second, more of the second than the third, etc. To qualify for this category, we look for animal AND plant protein in the first three ingredients. This is an arbitrary cutoff to more easily categorize foods so may not fully represent the nutritional content.

- Fluval Shrimp Granules
- Sera Shrimp Food
- TetraColor XL Tropical Granules
    - High carotenoid content to enhance coloration
- Dennerle Shrimp King Complete
- GlasGarten Shrimp Dinner
- Hikari Mini Algae Wafers

Animal-Based Foods

1. Fluval Bug Bites
2. GlasGarten Artemia Shrimp Snacks
    a. Limit feeding to once a week to avoid health issues
3. Hikari Frozen Bloodworms (Best to buy in-person at a store)
    a. Limit feeding to once a week to avoid health issues

Plant-Based Foods

1. GlasGarten Snowflake
2. Shrimp Envy - Growth Envy
    a. Full disclosure – we are not paid to promote Shrimp Envy's products but we do know and like the owners
3. Vegetables (after blanching)

These lists are NOT comprehensive. There are hundreds of foods on the market that are healthy. Visit ShrimplyExplained.com/book-links/ for an updated list with links[li].

Other Considerations for Food

- **Coloration**

  Pigments like **carotenoids** are responsible for shrimp shell and body coloration. Shrimp cannot produce pigments themselves so must obtain them from foods like spirulina. Diets with higher carotenoid content lead to greater concentrations in the body and stronger coloration (Tomas, Sganga, & López-Greco, 2020).

- **Food Size**

  Smaller shrimp have been shown to avoid clusters of larger shrimp. This causes them to miss out on food and have reduced growth potential (Obaldo & Masuda, 2006). To support the growth of baby shrimp, breeders use powdered foods that spread around the tank to incorporate additional nutrients into biofilm, which is the main food source for baby shrimp. You can buy commercial powdered foods or grind up larger pieces of food to accomplish the same thing. Check out our Food Size section (p145) to learn more about how food size impacts feeding recommendations.

  Powdered Food Options:

  - Glasgarten BacterAE
  - GlasGarten Shrimp Baby
  - SL-Aqua Shrimplet Food
  - Shrimp King Baby Food
  - ***Visit ShrimplyExplained.com/book-links/ for an updated list with links[lii].***

- **High Mineral Foods**

  Various foods are promoted as "calcium-rich" to prevent molting problems and improve shell health. In general, research suggest that the majority of calcium and other minerals be obtained from the water (Truong, Hines, & Emerenciano, 2023). Admittedly, this research is on saltwater shrimp but we believe this is also the case with freshwater shrimp. In our experience of helping hobbyists with molting problems, calcium-rich food *can* help, but only if the water lacks the right mineral ratios. If the water has the right mineral ratios, then adding calcium-rich food can offset the mineral balance in shrimp and *cause* molting problems. As such, **our recommendation is to only feed calcium-rich food when experiencing molting problems and to stop feeding it if molting problems do not improve within a week**. Please see the chapter on Molting Problems (p267) to learn more.

- **High Cholesterol Foods**

  Cholesterol is used to create hormones that control molting and reproduction in shrimp. Shrimp cannot produce their own cholesterol so they must get it from food. Foods like artemia, bloodworms, and egg yolk have high cholesterol content and may be fed occasionally (once a week) to promote breeding and molting. See the chapter on Molting Problems (p267) or our video about bloodworms[liii] to learn more.

Feeding Rules of Thumb

When adding any food to the tank, consider what type it is to determine how cautious to be and when to remove leftovers.

1. **Feed cautiously**
    - **a.** We can always feed more but can't always take food out
2. **Food quantity depends on shrimp population**
    - a. Larger population can handle more food
    - b. Larger populations also increase the bioload and need for maintenance
3. **Feed powdered foods extremely cautiously**
    - a. 1/4 or less than recommended by manufacturer initially
        - i. May increase over time if there are no water quality issues
    - **b.** Excess cannot be removed easily so mistakes can be fatal
4. **Leftovers of small/soluble foods should be removed after 1-2 hours**
    - a. Includes small granules (1-2 mm), and sugary foods like banana
    - b. Overfeeding is extremely dangerous
5. **Leftovers of larger foods should be removed after 2-3 hours**
    - a. Includes larger granules (3+ mm)
    - b. Overfeeding is less dangerous
6. **Insoluble foods can be left in for 4+ hours**
    - a. Includes soy bean hulls (Snowflake) and fibrous foods like carrots
    - b. We recommend removing leftovers within 12 hours regardless of manufacturer recommendations
    - c. Overfeeding these foods is relatively harmless but may eventually cause issues if left in for multiple days

Guidelines To Optimize Food for Growth and Breeding

- Feed a diet of diverse protein sources. This can be a single type of food with plant and animal protein, but multiple food sources are likely to offer a more varied and well-rounded diet.
- Natural food sources like Indian almond leaves improve diet diversity and biofilm availability. Indian almond leaves and other botanicals are covered in-depth in the Botanicals section (p82).
- Feed as much as shrimp can eat within 2-3 hours (depends on food size – See Food Size section [p144] for more information).
- Feeding smaller amounts multiple times per day can reduce the impact food has on water quality (See Food Size section [p144] for more information).
- Add powdered food 1-2 days a week to help with baby shrimp survival and growth.
- Use a feeding dish to remove leftover food.
- Cut back on feeding if shrimp do not attack new food quickly or if nitrates start to accumulate.
- Increase maintenance when feeding more often.

The most important aspect of feeding is understanding how to balance the input and output of nutrients in an ecosystem. We can feed a larger quantity of food to increase growth and breeding, but we also need to understand what our ecosystem can handle. Increased feeding is balanced by an increase in plants, filtration, and maintenance. Without those adjustments, overfeeding is extremely dangerous.

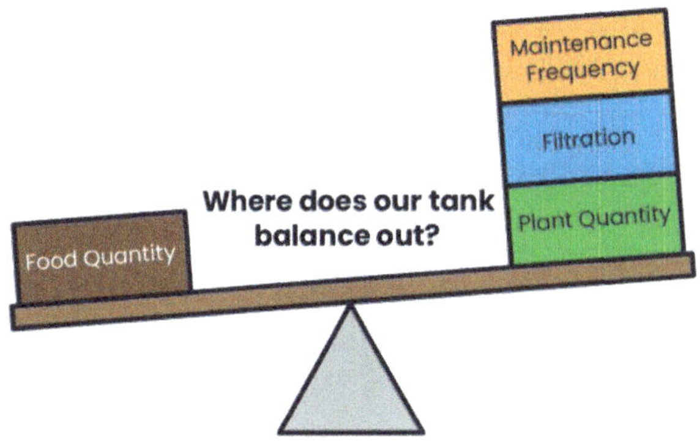

With that information in mind, let's go over how much to feed.

How Much Should We Feed?

Most of the papers say that shrimp are fed *ad libitum*, meaning the food is available at all times, but at least one paper specified 4% of a shrimp's body weight per day (Vazquez, Delevati-Colpo, Sganga, & Lopez-Greco, 2017).

It's hard to weigh shrimp though. Luckily, we can reference research studies that weighed Neocaridina at different ages, so we don't have to do that work.

One study weighed male and female Neocaridina at 30, 60, and 90 days after hatching, finding them to be approximately 20mg, 45mg, and 60mg respectively (Vazquez, Delevati-Colpo, Sganga, & Lopez-Greco, 2017). Another experiment tracked their weight for longer periods of time and found that females weighed approximately 60mg at the time of sexual maturity (roughly 75 days) days and up to 165 mg at 7 months of age (Tropea & Lopez-Greco, 2015). Each female shrimp gained an average of 10mg per molt and molted 2 times a month. This means an average increase of about **10-15% per molt**. The males were not tracked in the second experiment.

While it's not much data, it's enough for our purposes because most shrimp are sold before sexual maturity at around 60 days of age. This means we can estimate what 10+ shrimp weigh and calculate food requirements from there.

Calculations

Let's say we buy 10 Neocaridina for a tank. Those 10 should weigh 45-60mg each. We'll use the lower number for a conservative estimate to avoid overfeeding, bringing the total body weight to 450mg. If *ad libitum* feeding is 4% of total body weight, then we should feed **18mg of food per day for every 10 shrimp in the tank**. *That's not very much!* Assuming they all survive and molt properly, then increase the amount of food per day by 15% every 15 days while paying attention to how quickly they find and eat the food.

To be clear, **this math provides an estimate of the amount of food to add if there are no other food sources**. Every tank and shrimp population is different, so try observing how

quickly the food is consumed and adjust accordingly. There's also a lot of flexibility with feeding schedules so we may choose to only feed every other day or every few days and shrimp will be perfectly fine. In fact, less feeding is highly recommended if the goal is to maintain a low-maintenance ecosystem. **If goals include fast growth and breeding, then more frequent feeding is recommended, especially if shrimp gobble up food within an hour or so.**

25-30 mg of food in a weigh boat. Liquid test kit bottle for scale.

Remember, **it's best to be on the safe side and avoid overfeeding**. If the tank has matured for at least 6-8 weeks, then the biofilm, algae, and microorganisms should be able to sustain shrimp for at least a few weeks without any food being added, unless it's a breeding tank with a huge population of shrimp. Take a look at the Signs of Overfeeding section (p148) to know when there's excess food in a tank.

Please see The Danger of Active Substrate (p64) for more information about how overfeeding creates a ticking time bomb in a low-pH tank.

Food Size Matters

A lot of shrimp food is sold as relatively large pellets. When we toss one in, all the shrimp swarm around these pellets in a feeding frenzy, which gives us some pretty incredible visuals. That being said, there are a few concerns with feeding these pellets that are worth discussing.

How Food Size Affects Water Quality

One team of researchers performed research on feeding different sizes of food to farmed Pacific whiteleg shrimp (*Litopenaeus vannamei*) (Obaldo & Masuda, 2006). A key concern of this group was testing how different food types affected water quality. To test this, they placed 2g of food of different sizes in an aquarium for 2 hours, then removed the food, dried it, and weighed the leftovers. They found that **smaller food particles leach more of their mass – more nutrients – into the water, which decreased water stability.** This makes sense because smaller food exposes more surface area (by volume) to the water, so more of the food gets soaked and starts to dissolve.

$$Volume = \frac{4}{3}\pi r^3$$

$$Surface\ Area = 4\pi r^2$$

Explanation: Here, we have the equations for Volume and Surface Area (SA) (r = radius). In the SA formula, r is *squared*, whereas r is *cubed* in the Volume formula. This leads to an exponentially larger increase in volume as the size of food particles increases. For example, a food particle with a 2 mm radius has 50 mm^2 SA but only 34 mm^3 volume, while food with a 4 mm radius has 4x the SA and 8x the volume. A much lower percentage of the larger food is touching the water as compared to the smaller food, as demonstrated by the Surface Area to Volume Ratio.

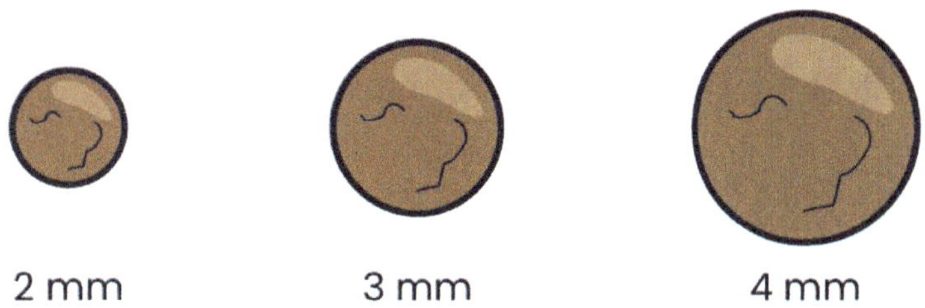

Food Radius	Surface Area	Volume	Surface Area-To-Volume Ratio
2mm Food	50 mm^2	34 mm^3	1.47
3mm Food	113 mm^2	113 mm^3	1
4mm Food	201 mm^2	268 mm^3	0.75

What we can take from this research is the understanding that **feeding smaller food particles increases the risk of water pollution** – *but only if it isn't eaten*. The way the researchers tested the impact of food size on water pollution doesn't take into account how shrimp actually eat. For example, is more waste produced by shrimp when they eat small pellets in a controlled way? Or when shrimp crowd around larger pellets and frantically tear into them? There are pieces flying off larger food during feeding frenzies (see video) and that almost certainly increases the amount of small debris entering the water and substrate.

How Food Size Affects Shrimp Interactions

Another consideration of food size is how it affects the way shrimp interact. Smaller foods let each shrimp take away a portion for themselves. On the other hand, larger foods that stay together force shrimp to gather and stay closer to other shrimp than they may otherwise want to.

Amano shrimp and cherry shrimp enjoying soybean hulls.

Think about it this way – would you rather eat in a popular sit-down restaurant or a popular bar that's packed shoulder-to-shoulder? The restaurant may have every table filled but still leaves enough space for everyone to enjoy their meal, whereas you might find yourself in the middle of a fistfight at the bar.

It's kind of the same with shrimp.

According to Zheng et al., a higher stocking density leads to a greater number of injuries among juvenile tiger prawns, as determined by antennae breakages (Zheng, Mao, Su, & Wong, 2020). Another study on Pacific whiteleg shrimp demonstrated that larger food sizes led to more incidents of aggression (Obaldo & Masuda, 2006). Injuries (and subsequent infections) are very common problems when too many animals of any species are kept together in a small space – and this can happen to our dwarf shrimp too.

While our dwarf shrimp species are relatively peaceful and lack the claws of tiger prawns, they can certainly tear into dead shrimp, proving their claws are strong enough to cause injuries during feeding time. Smaller individuals know this and avoid getting into the thick of the fray, so feeding larger pellets that shrimp are forced to gather around limits the amount of food going to smaller shrimp.

By feeding a large pellet that doesn't break apart easily, we're creating an artificially high density of animals and increasing the risk of injuries occurring. While the risk isn't all that high for dwarf shrimp, it is present and easily avoidable by breaking up larger food

into more manageable portions that reduce crowding. It may be worth feeding larger food to inexpensive shrimp if you really want to see a feeding frenzy but, if you have expensive shrimp, then why take the risk?

How Feeding Frenzies Affect Energy Usage

A seemingly irrelevant study on light intensity and growth rate in *Penaeus merguiensis* (banana prawn) brought up an interesting theory stating the following:

> *"Less movement means more energy for somatic growth and may contribute to the growth improvement of prawns..."* (Hoang, Barchiesis, Lee, Kennan, & Marsden, 2003)

Most shrimp keepers want the largest and most colorful shrimp, which means we want their energy focused on growth and color production. **The research suggests that fighting for food around a larger pellet uses more energy than peaceful eating of personal-sized pellets**. If we're trying to optimize growth and coloration, then there's a good reason to consider feeding smaller pellets.

Summary

To summarize, larger foods may lead to more waste from shrimp frantically tearing into them. It also forces shrimp together, potentially leading to wasted energy or injuries. On the other hand, smaller foods can break down faster and are harder to remove from the tank if they're overfed. **This information suggests that smaller food size may be better for the health, growth, and coloration of shrimp. The increased risk of water pollution with smaller foods can be mitigated by removing excess food sooner (1-2 hours) and feeding smaller quantities more frequently**.

Various Caridina grouped up around food - Source: Shrimp Mania (YouTube)

Signs of Overfeeding

Overfeeding is arguably the biggest problem new and experienced shrimp keepers face. We completely understand the desire to make sure the animals in our care don't starve but many shrimp keepers drastically overestimate how much food shrimp need.

The truth is that a tank which has been left to mature for 6-8 weeks (8-12 for Caridina) already has plenty of food in it. Shrimp are not going to starve unless we decide to add a large population (5-10 shrimp per gallon) immediately. It does depend on our goals though, since commercial foods can increase shrimp growth and breeding. If we choose to feed frequently, then look out for the following signs of overfeeding:

- Visible leftover food after more than a few hours
- Increase in nitrates not associated with added fertilizer
- Large increase in microorganism populations like copepods or detritus worms
- Shrimp lacking interest in food

If any of these signs appear, then cut back on feeding for a week or more until the signs go away. We can even stop feeding entirely during this time, because there are still large amounts of biofilm, algae, or decomposing organic matter in most tanks.

Feeding Non-Commercial Foods

The main concern with non-commercial food is whether the food is safe for shrimp. A good way to check is to ask shrimp keepers or search for their experiences online, but this isn't always easy. We've simplified this process for you by breaking it down into three safety questions:

1. Could the food contain copper?
2. Could the food have any pesticides on it?
3. How messy is the food?

Could the Food Contain Copper?

Copper is extremely toxic to shrimp. It's rare that any commercial fish food has large amounts of copper, but it is possible. To be safe we recommend only feeding fish food that says it's shrimp-safe, buying shrimp-specific food, or carefully reading reviews.

One important note: After hearing this, you may check the ingredients of your shrimp food and find it contains *copper sulfate*. Don't panic though – copper sulfate is much less toxic than copper and is safe for shrimp at the levels found in these foods. In fact, shrimp actually require a very small amount of copper for a healthy circulatory system. Copper acts like iron does for our blood cells, binding to oxygen at the gills to deliver it to other parts of the body (Zhou, Davis, & Rhodes, 2014).

Could The Food Have Pesticides on It?

Shrimp enjoy fresh fruits and vegetables - if they aren't poisoned. Unfortunately, many insecticides used on crops also hurt shrimp because, well, they're aquatic insects. The best way to avoid pesticides is to buy food with the USDA Organic label on them. This label indicates the food has stricter standards for pesticide use and, if any pesticides are still

used, they are typically less harmful to shrimp. One study found that organic foods are 5x less likely to have pesticides on them but they may still be present (Montiel-Leon, Duy, Munoz, Verner, & et al., 2019). To further reduce the risk of pesticide exposure, always be sure to wash vegetables thoroughly and blanch them, which also kills any bacteria in the food (Ceylan, 2019).

Some shrimp keepers choose to feed canned vegetables since many of them are already pre-washed and sterilized. That's totally fine, too! Just check that those vegetables are unsalted/unseasoned, so we don't add a bunch of extra stuff to the tank.

How Messy Is the Food?

If we add food that dissolves and clouds the water, then we may be doing more harm than good. Any dissolved nutrients will hopefully be filtered out, but they may change water parameters and affect the ecosystem. It's best to add food that holds together well like Snowflake (soybean husks), carrots, peas, cucumbers, etc. **Soft foods like bananas or peaches have lots of sugar that quickly dissolves into the water column and causes water quality problems**.

Shrimp are messy eaters, so they often drop small bits of food into the substrate, no matter what consistency the food is. These leftovers can accumulate and affect water quality just like dissolved food. To avoid this, some shrimp keepers use **feeding dishes** that trap any leftover food and remove it from the tank. These dishes are also an easy way to prevent overfeeding by removing uneaten food within 2-3 hours. Many shrimp keepers have perfectly healthy colonies without feeding dishes though, so they're by no means a necessary purchase.

Feeding Vegetables

Including peas, carrots, green beans, stinging nettle, spinach, kale, zucchini, cucumber, and broccoli. We may also feed leaves from Indian poplars, alders, oaks, and Indian almond trees (Schoolman & Arndt, 2017).

Shrimp eating a blanched carrot.

Just like humans, shrimp benefit from a variety of different foods to get important micronutrients. As such, many shrimp keepers feed different vegetables 1-2 times a week. We don't want to just throw raw veggies into the tank though, as they can have chemicals and bacteria on them which may be harmful to our shrimp. Follow the recommendations in the previous section to avoid pesticides and follow the blanching instructions below to remove any bacteria:

Blanching process

1. Boil enough water to cover the vegetables.
2. Add vegetables and boil according to the times on this chart[liv]. If the vegetable isn't on the chart, follow the time of the vegetable that is most similar to your vegetable.
3. Place vegetables in ice water.
4. Eat some of the vegetables if you're hungry, then give the rest to your shrimp.

To store blanched vegetables for future feeding:

1. Pat dry the vegetables.
2. Spread them out on a baking sheet or wax paper and place in the freezer. Spreading them out prevents them from turning into an unwieldy frozen clump.
3. Once frozen, transfer to a freezer bag and store. They last a year or more.

Shrimp Tank Maintenance

Tank maintenance is a controversial subject. Some people subscribe to the idea that 10-15% water changes every week are necessary to maintain good water quality. Others never do water changes and just top off their tanks when evaporation occurs and are also successful. What we need to understand is that both approaches work to accomplish the same thing, which is to create the most stable environment for our shrimp. The approach that is best for you depends on our tank setup and goals.

Types of Maintenance

There are a few different types of maintenance that we may need to do. In case you haven't kept aquariums before, let's cover the terminology for them:

Water Changes

A water change refers to the replacement of old tank water with new tap or remineralized RO/DI water. This is done to replace minerals and remove waste products that may build up in the water. How quickly waste builds up depends on how balanced we choose to make the tank. Again, the entire goal of this book is to help explain how our actions affect tank balance. Here's a list of major factors and how they affect water change frequency/volume:

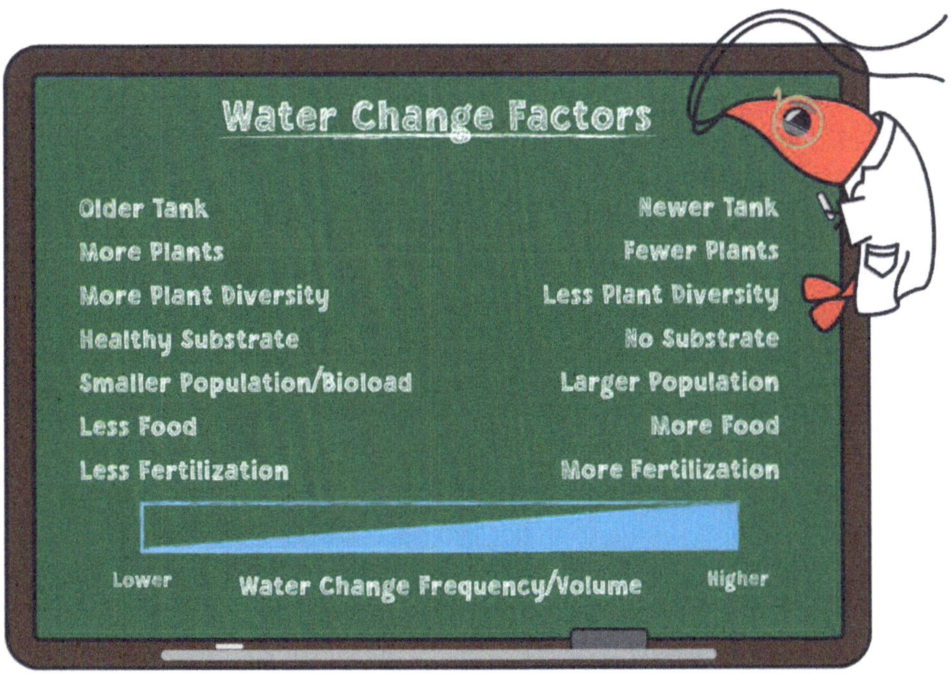

Shrimp keepers generally do smaller water changes (10-15% of tank volume) than fish tanks to avoid sudden parameter changes. That being said, small and slow water changes may be necessary for a tank - or they might not be. It's an extremely controversial subject among shrimp keepers, partly because every piece of content recommends something

different. We understand that's incredibly frustrating to hear but it depends on the goals for the tank, how healthy the water going into the tank is, and how healthy the shrimp are.

Can Shrimp Handle Large Water Changes?

Healthy shrimp can handle large (30-40% or more) water changes every week without an issue. We know this because it is standard routine for many Neocaridina *and* Cardina breeders focused on breeding the most shrimp[lv]. These large water changes work because breeders always use healthy water (typically remineralized RO/DI water that may be pH-matched) so they know the water going into the tank is safe and has very consistent mineral and nutrient levels compared to the old water. **Any small stress from temporary parameter changes due to water changes is typically less than the stress caused by waste buildup in crowded tanks**.

In contrast, doing such large water changes with tap water can be dangerous because tap water can vary so much throughout the year (Culligan Water of Kansas City, n.d.). Even if water has the same pH, GH, KH, and any other measurable parameters, there may still be differences that hobbyists can't detect. That's why **we recommend smaller (10-15%), slower, and less frequent water changes when using tap water**.

In addition, **how the water gets added into the tank matters**. With tap water of questionable quality, slowly dripping the water in over 30-120 minutes is a safe option. In contrast, remineralized RO/DI water can often be dumped in over a matter of minutes unless the shrimp are a very fragile variety. The exception being if dumping the water in disturbs the substrate, which increases disease risk. **It's always a good idea to pour water in slowly enough that it does not disturb the substrate or kick up debris**. Placing a bowl or plate on the substrate and pouring water onto it is a good way to reduce substrate disturbances, since the dishware redistributes force from the falling water.

If shrimp have just been introduced to the tank and are stressed from acclimation, then doing a large water change is an unnecessary stressor that we recommend avoiding until they are established in the tank for at least a week. On the other hand, if there's a contaminant in the tank (insecticides, a massive powdered food spill), then a large and stressful water change may be the lesser evil. Every situation is different so it's important to understand how a water change affects our tank.

How Water Changes Affect Tanks

Did we just disturb the substrate? Well, then it might be a good idea to do one large water change to remove any debris that got kicked up. Doing so reduces the risk of shrimp getting a disease from substrate bacteria, which is often worse than the stress from a water change. A hydrogen peroxide treatment (p295) after substrate disturbance is also a good idea to reduce disease risk.

Is there a thick substrate layer with lots of plants to create a healthy and diverse ecosystem? If so, then a weekly water change routine is less likely to be necessary and may even be harmful if the water change disturbs the substrate and kicks up debris.

Does the tank have a thin substrate with few plants? Then the tank may not cycle nutrients well, so larger and more frequent water changes are a good idea.

Survival of The Fittest

For hobbyists who just want healthy shrimp tanks, it's entirely possible that water changes aren't frequently required IF the tank is set up like we recommend in the Low-Maintenance Setup (p238). If, on the other hand, the goal is to be a seller with a reputation for hardy shrimp, then water changes may be highly beneficial.

Think about it this way: Would you rather buy shrimp that have been proven to survive large water changes or ones that haven't been tested?

Doing water changes weeds out weaker shrimp so only the fittest survive. Larger water changes are more stressful so have the strongest selection pressure for hardy shrimp. If a breeder does regular large water changes and their shrimp survive, then that's a great indication their shrimp are fit and healthy. On the other hand, a breeder who never does water changes may have beautiful shrimp (and more of them) but there's less of a selection pressure for hardy shrimp, so their shrimp may struggle to transition to a new tank with different water parameters.

Side note: Breeders typically have very consistent routines to maintain stable water, which reduces the risks associated with water changes.

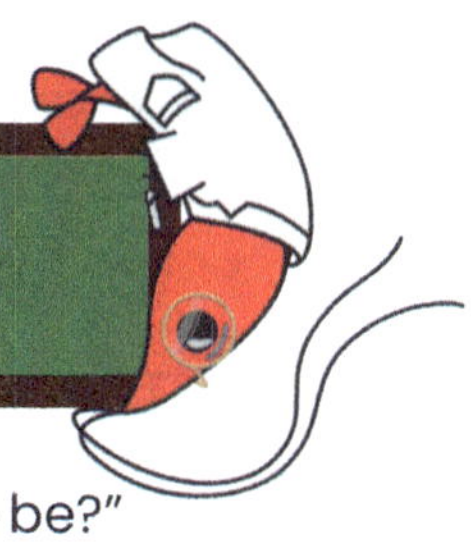

To summarize, there is no one answer to "What should my water change routine be?"

Water changes can be healthy or unhealthy depending on how they are done. What matters is understanding our goals, the overall health of our tanks, and how the ecosystem is impacted by our actions.

Top-Offs

A top-off is the process of replacing evaporated water with new water to keep water volume steady. Always use RO/DI, distilled, or spring water for top-offs to avoid GH creep.

The best way to reduce evaporation and top-off frequency is to put a lid on your tank. One of the cheapest options for a lid is to use two or more panes of glass (with edges sanded down - wear a mask) from a hardware store.

Filter Cleaning

Debris builds up in a filter over time, reducing water flow and performance. The exact procedure varies depending on the type of filter but, in general, simply wash and squeeze out the filter sponges and media in water to remove the debris.

It's best to do filter cleaning in old tank water but it can be done in tap water *even if the water isn't dechlorinated*. Chlorine takes time to kill bacteria - anywhere from a few minutes to never, depending on pH and temperature (Center for Disease Control, 2024), *and that's for free-floating bacteria*. Nitrifying bacteria like our BB stuck in biofilms in our

filter are well-protected (Zhu, Shan, Xinyun, Zhang, & et al., 2021) on top of being hardy so can withstand chlorine levels in tap water for 30 minutes or more. That being said, it's still a good idea to be on the safe side, rinsing with room-temperature water and limiting chlorine exposure to a few minutes since we don't know how sensitive other microorganisms and bacteria are to adverse conditions.

Substrate Cleaning

Substrate cleaning, AKA gravel vacuuming, is the process of sucking up leftover food and animal waste from the floor of a tank. For planted tanks with rooted plants and thick substrate, we only want to vacuum the very top of our substrate, as going deeper disturbs bacteria and may disrupt the cycle or cause a bacterial infection. Yes, this means there will be build-up of debris under the substrate, but leaving the debris helps establish a strong community of decomposers to keep your tank healthy. If the substrate is disturbed, we recommend doing a tank-wide hydrogen peroxide treatment (p295) to kill any unwanted bacteria in the water column.

In contrast, a thin-substrate tank with few rooted plants may not break down debris as well, in which case more frequent substrate cleaning is recommended. For example, Caridina breeders who use undergravel filters and a thin substrate layer (or no substrate) often clean their substrate every week or two before each water change (Shrimp Mania, 2023).

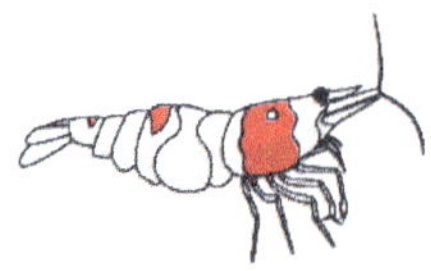

Again, **substrate cleaning depends on how the tank is set up**. With lots of plants, a strong ecosystem, and no overfeeding, substrate cleaning is rarely required and may actually be detrimental, whereas a tank with few plants and no substrate needs frequent cleaning regardless of how much food is added.

Pro Tip: To safely clean the substrate without sucking up shrimp, we recommend putting fine netting around the intake of the vacuum. After that, let the removed water settle in a bucket for a day or so, then shine a light in it to check for any baby shrimp that might have been sucked up. Don't worry – They just went on an exciting roller coaster ride and will be fine once they get back in the tank!

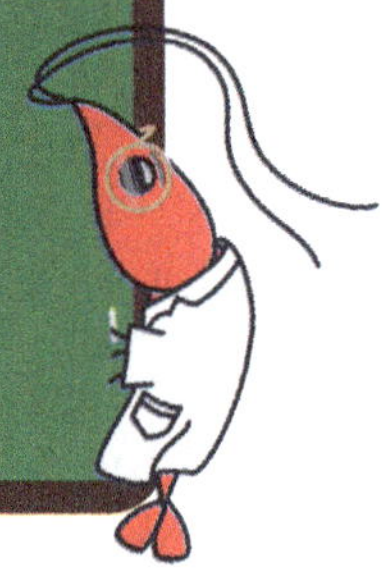

If we use a feeding dish and/or don't overfeed, then substrate cleaning is rarely required except for aesthetic purposes.

Glass Cleaning

Glass cleaning involves scraping off algae using a scrubber or razor blade. It is never required and is only done for aesthetic purposes. Since algae benefits shrimp, we recommend only cleaning the front pane of glass and letting nice green algae build up on the other panes.

Signs Maintenance Is Required

Now, the question is: when do we *need* to perform maintenance?

Even with a low-maintenance tank, some maintenance may still be required. Every ecosystem is unique so there's no "best" maintenance frequency, but here are some signs to look out for:

Out-of-range parameters

What we suggest—at least initially—is to test pH, GH, KH, ammonia, nitrites, and nitrates once a week. If anything goes outside of the recommended ranges, then maintenance is required. As we gather more data on the tank, we'll get a feel for when to do maintenance without having to test as often.

Please see our Parameter Control (p261) chapter to deal with a specific parameter being out of range.

Low Water Volume

Evaporation reduces water volume and increases GH and TDS levels in a tank. We recommend doing top-offs every week or so (**with un-mineralized purified water**) to maintain stable parameters. Getting a tank lid reduces the amount of evaporation and the need for top-offs.

Floating Debris/Colored Water

If there's a lot of debris in the water column, then the filter is not keeping up with the tank's bioload. This is a good sign that cleaning or a stronger filter may be required.

Low Water Flow

The flow rate decreases when a filter gets clogged. This is a good sign to squeeze out any sponges or clear debris from the filter.

Flow rate is only visible for HOBs and canister filters, so this is not a useful sign for most sponge filters. For sponge filters, pay attention to the bubble flow and do water changes when fewer bubbles come out. Take a video when the sponge is new to the tank to have a reference point for expected bubble flow.

Abnormal Behavior

Any abnormal behavior (covered at the end of What Is Normal Shrimp Behavior? [p132]) is an indication that maintenance may be required. Test all parameters to see if any have changed. If so, follow the instructions in the Parameter Control section (p261) to fix them. We should also consider if any changes have been made in or around the tank area (i.e. pesticide spray). In addition, check if the microorganism population is increasing (excess food) or decreasing (sign of poisonous water quality).

If nothing seems to be wrong, then keep an eye on the tank to see if shrimp behavior returns to normal in the next hour or two. Shrimp are weird and some may just want to scare us. They also can't read calendars so often get the date wrong for Halloween.

Accumulated Debris

If there's a lot of uneaten food on top of the substrate, then it may break down and cause an ammonia spike. Remove it with a gravel vacuum, or a kitchen syringe for more control. Remember that disturbing the substrate (especially in deeper substrates) may kick up bacteria that cause illness, so try to avoid disturbance as much as possible unless the substrate is new or cleaned frequently. If you do disturb the substrate significantly, then we recommend a hydrogen peroxide treatment (p295) to kill bacteria in the water column.

Avoiding Contamination

A serious but often undiscussed issue is tank contamination. Humans are constantly introduced to new microbes in the outside world and the last thing we want to do is bring those into our tanks. A good rule of thumb is that **if our hands aren't clean enough to put in our mouth, then they aren't clean enough for our shrimp tank**. We never know what may be enough to disrupt our carefully created ecosystems.

Our own health may also be in danger if we don't follow safe hygiene practices. It is possible to contract diseases like *Mycobacterium* infections from aquariums (Committee, 2021).

Prevent Cross Contamination

Our goal when keeping multiple shrimp tanks is to maintain separate ecosystems. Using the same tools in two tanks without sterilization between them may lead to cross contamination and health issues. This is due to the fact that **each breed of shrimp is typically raised in separate ponds or tanks and therefore exposed to different microbial ecosystems** (Foysal, 2023). One breed may develop immunity to certain bacteria over time, but that doesn't mean other breeds have the same immunity.

It's like how humans develop immunity to bacteria where we live but often can't drink the water in foreign places because our gut microbiomes aren't used to the unique combination of foreign microbes. Shrimp raised in different farms - or even different ponds in the same farm - may not be exposed to the same microbes due to differences in water parameters and local ecosystems (Yanting, Cuilan, Xu-xiang, & et al., 2018). **Cross contamination is a serious concern for all breeds**, but especially so for Neocaridina and hardy breeds of Caridina (mainly Crystal) that are farmed and imported on a large scale. Sensitive breeds of Caridina, like Taiwan Bees, are more often bred at smaller scales in controlled settings, so they may have fewer differences in the microbial colonies they're exposed to. Regardless, they are often more inbred with worse immune function, so it's still a good idea to avoid cross contamination whenever possible. Here are some hygiene recommendations to protect our shrimp and ourselves:

- Good hand-washing technique is extremely important. Wash thoroughly with vinegar or soap and water and make sure to get under those nails! Wash all areas of the body that may enter the tank.
- When using soap, thoroughly rinse hands (more than you normally do) to remove any residue. Soap is toxic to aquatic life.
- Always wash before *and after* tank exposure to prevent the spread of disease.
- Avoid exposing orifices or open wounds to tank water. If wound exposure does occur, thoroughly wash and cover with antiseptic ointment.
- Avoid touching phones and other infected surfaces before putting hands into tanks.
- Always sterilize equipment after placing it in a tank. This includes any aquascaping tools, pH/TDS meters, water testing vials, nets, gravel vacs, or hands. See below for sterilization options.
    - Run sterilizing solution (discussed below) through gravel vac and drip acclimation tubing to kill off any bacteria.
    - Don't use your mouth to start siphons. There are better ways[lvi].
- When possible, have duplicate sets of equipment for each tank, or at least two sets of equipment so one set may be getting sterilized while the other is used. A cheap way to do water testing without worrying about cross contamination is to use dedicated 20ml plastic syringes[lvii] for each tank, then squirt water into a single set of test vials that never touch the tanks themselves.

Sterilization Solutions

- **3% Hydrogen Peroxide**
    - Soak tools in hydrogen peroxide for at least 10 minutes to sterilize.
    - Hydrogen peroxide is a strong oxidizing agent that rips apart cell membranes and destroys DNA. It breaks down quickly in water so it is not anywhere near as much of a threat to aquatic life as bleach. A small amount of hydrogen peroxide left on tools is not a problem for shrimp, making it the preferred method for sterilization during regular maintenance.
    - Leaving hydrogen peroxide exposed to air causes it to break down and become less effective over time. Keep containers closed to preserve potency.
- **1:10 Bleach Dilution**
    - For tools that have only been briefly exposed to tank water, soak in bleach solution for at least 10 minutes. For tools/tanks that have been exposed long enough for biofilm to grow on them (24+ hours), soak in bleach solution for 1 hour (California Fish and Wildlife Services, 2024).
    - Bleach is an extremely effective sterilizing agent. It kills just about anything it comes into contact with, including shrimp. **After dipping tools in bleach, let them dry fully before next use so any residue evaporates**.

- Only use bleach with sodium hypochlorite and water as the ingredients. Scented thickened bleach may have additives harmful for aquatic life.
- Leaving bleach exposed to air causes it to break down and become less effective over time. Keep containers closed to preserve potency.

- **Other Cleaning Solutions**
  - Vinegar is commonly used to disinfect but it does require up to an hour to kill bacteria and is not effective against all types.
  - Isopropyl alcohol does work but residue requires at least 24 hours to break down into compounds that do not harm aquatic life.

Red Cherry (Neocaridina davidi *juvenile) mid-swim.*

Troubleshooting Shrimp Deaths

Even the most experienced shrimp keepers have problems with shrimp deaths periodically. It may be with new shrimp that aren't acclimating well to tank parameters or to a well-established colony that eventually experiences inbreeding issues. Knowing how to identify and fix causes of death is invaluable to shrimp keepers. The rest of this section acts as a detailed reference guide to organize information in *The Science of Shrimp Keeping* and simplify the troubleshooting process.

To keep things light and fun, let's pretend that we're investigating a serial killer with a gruesome modus operandi (MO). What's the first thing to do? Determine a pattern of killing. Were several victims attacked suddenly or were the killings spread out? Were there any signs to identify the murder weapon? Actually, maybe this isn't the best tone to take in the situation. Let's switch gears.

In the graph below, we've organized causes of death based on **Death Speed** (Sudden vs. Gradual) and **Rarity** based on **Water Type** (Tap vs. Remineralized [Remin.] RO/DI Water), to help identify what's most likely affecting a tank. Below the graph, there are alphabetically organized tables containing detailed causes, solutions, and additional references throughout the book to make information readily available. Additionally, there's a free tool[lviii] to help evaluate parameters and detect likely causes of death.

Please note the following information assumes acceptable parameters based on our Parameter Recommendations *(p46), along with more than one shrimp death. While any death is sad, a single death does not necessarily indicate an issue. Once there's a pattern of multiple sudden or slow deaths, then we can start sorting through potential culprits. Consider any recent changes to the tank, including the following:*

- *New Plants/Animals/Food/Equipment*
- *Recent maintenance (water change, substrate vacuuming, etc.)*
- *Changes around the tank (pesticide spray, flea medication applied to pets, children, etc.)*

Use the graph on the next page to identify likely causes of death based on death conditions.

Causes of Death

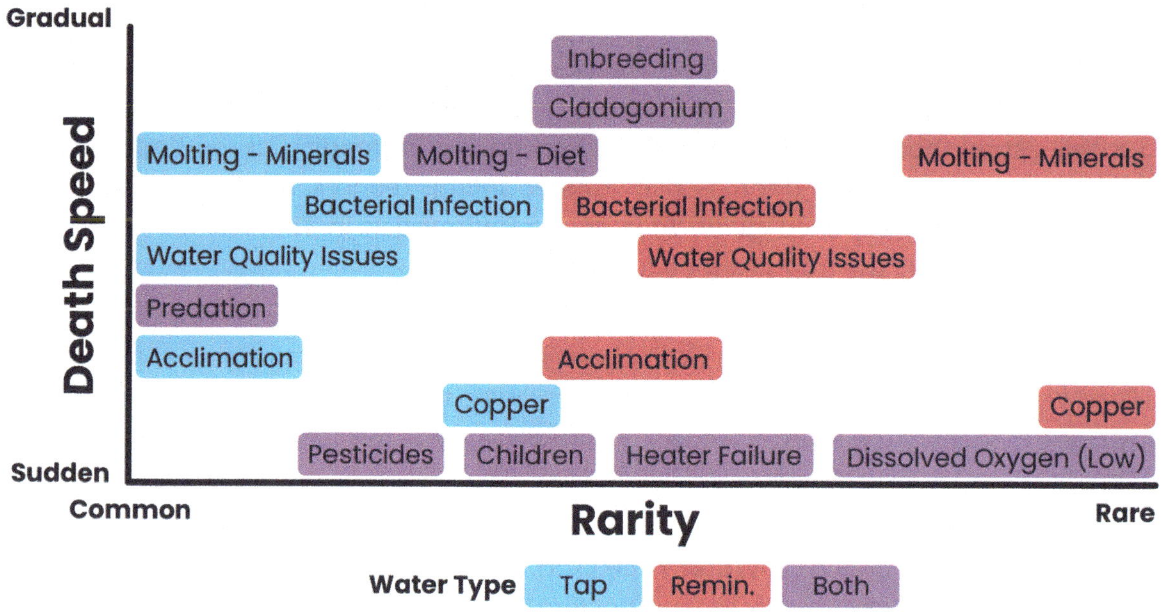

Acclimation

Water Type	Tap	Remin.
Death Rate	Gradual throughout first 1-3 months of adding shrimp	
Symptoms	• White Ring of Death (may be inconsistent) (see p267) • Lack of breeding after 3 months • Random deaths despite healthy parameters	
Rarity	Common	Rare

Shared Causes › Solutions	• Fast/inadequate acclimation (p122) › Keep tank stable so they adjust
	  ○ › Match parameters (+temp) with future shrimp
	• Shrimp were bred in different parameters (especially common for imported shrimp) › Keep tank stable and hope they adjust
	  ○ › Buy healthy homebred shrimp from reliable breeders (not always easy depending on location)
	• Unhealthy shrimp › Keep tank stable and hope they adjust	

Other Causes	• Poor water quality • Different Ca:Mg ratio (p267)	• Different remineralizer from breeder (only a potential issue for very sensitive Caridina breeds like BOAs)
Solutions	• Add CrystalPro Shrimp Minerals for trace minerals (cheap) • Add remineralizer	• Check if remineralizer matches breeders
References	Acclimation (p122), Remineralizer (p93), Parameter Stability (p95)	

Bacterial Infection (Muscular Necrosis + Rust Disease)

Water Type	**Tap**	**Remin.**
Death Rate	Gradual	
Symptoms	• Gradual deaths over weeks or months • Lethargy • Some live shrimp may have cloudy/milky abdomens • Redness around shell edges and pitting in shells (Rust Disease - p282)	
Rarity	Common	Uncommon
Shared Causes › Solutions	• Overfeeding › reduce feeding • Insufficient Filtration/Aeration › add filtration + hydrogen peroxide treatment (p295) • Debris build-up › substrate cleaning + hydrogen peroxide treatment • Dead or dying plants/animals › remove culprit	
Causes	• High nutrient/bacterial load in water • Stress from pollutants that increases chance of infection	• RO Filter system failure
Solutions	• Reduce feeding • Increase filtration/oxygenation • Filter cleaning • Substrate cleaning • Hydrogen peroxide treatment (p295) to kill bacteria in water column • Consistent water changes	• Replace RO filters
References	Maintenance (p148), Feeding Rules of Thumb (p139), Bacterial Infections (p282)	

Children

Water Type	**Irrelevant**
Death Rate	Sudden or gradual – Issue-dependent
Symptoms	• Loud splashing noises • WAY too much food in tank • Random things in tank • Grubby hands in tank water
Rarity	Common to Rare (Depends on life choices)
Causes	• Sexual relations
Solutions	• Adoption
References	My parents

Cladogonium (Green "Fungus")

Water Type	**Irrelevant**
Death Rate	Gradual (Cladogonium takes months to kill a shrimp)
Symptoms	Filamentous green growth underneath shrimp
Rarity	Somewhat Common (for imported Neocaridina)
Causes	Shrimp was bred in overcrowded and unhealthy conditions
Solutions	Cladogonium Treatment Schedule (p288)

Copper

Water Type	**Tap**	**Remin.**
Death Rate	Sudden – 1-2 days after exposure	
Symptoms	Sudden death of **majority** of population after a water change or introduction to new tank	
Rarity	Common	Extremely Rare
Causes	• Water pollution • Leaching from copper pipes (affected by pH)	• Filter system failure
Solutions	• 100% water change with safe water (Ideally Remin. RO/DI water) • Add SeaChem Cuprisorb (most effective) or activated carbon	
References	Remineralized Water (p93), What Copper Is Safe? (p148)	

Dissolved Oxygen (Low)

Water Type	**Irrelevant**
Death Rate	Sudden - often happens at night when plants don't produce oxygen
Symptoms	• *Majority* of shrimp near water line or attempting to leave water ○ This symptom may appear after water changes or top offs when shrimp gain access to algae that was previously above the water line, in which case it can be ignored
Rarity	Extremely Rare (in tanks with filters)
Causes	Inadequate filtration
Solutions	• Add filtration or air stone • Hydrogen peroxide treatment (p295) (It breaks down into oxygen)
References	Understanding Dissolved Oxygen (p31)

Heater Failure

Water Type	**Irrelevant**
Death Rate	Sudden
Symptoms	• Tingling feeling when contacting tank water • Random mass deaths of tank inhabitants
Rarity	Uncommon
Causes	• Incorrect heater size for tank • Manufacturing defect
Solutions	Unplug and remove heater

Inbreeding

Water Type	**Irrelevant**
Death Rate	Very gradual – typically 2-4 years after starting a colony
Symptoms	• Gradual decline in population • Low reproduction rate • Low baby survival rate • Short skirt • Scrunched head
Rarity	Uncommon – becomes more common with older and smaller colonies
Causes	Lack of genetic diversity
Solutions	• Add new shrimp from reliable breeder ○ Add shrimp of different breed to maximize genetic diversity ○ Get same breed if you wish to maintain that line
References	Inbreeding Issues (p213)

Molting – Diet

Water Type	**Irrelevant**
Death Rate	Gradual – weekly/monthly deaths
Symptoms	• White Ring of Death (p267) • Molts stuck on appendages • Bodies next to molts • No visible successful molts within 2-4 weeks of adding shrimp
Rarity	Somewhat Common
Causes	Insufficient cholesterol, iodine, or other nutrients
Solutions	• Add CrystalPro Shrimp Minerals for trace minerals (cheap) • Feed small amount boiled egg yolk for cholesterol (p274)
References	Understanding Molting Problems (p267)

Molting – Minerals

Water Type	**Tap**	**Remin.**
Death Rate	Gradual – weekly/monthly deaths	
Symptoms	• White Ring of Death (p267) • Molts stuck on appendages • Bodies next to molts • No visible successful molts within 2-4 weeks of adding shrimp	
Rarity	Common	Rare
Shared Causes › Solutions	• Minerals leaching from rocks/substrate (p87) › Remove culprit + single 50% water change • Overfeeding of mineral supplements (Mineral Junkie, cuttlebone, crushed coral) › Remove/reduce usage + single 50% water change • Imbalance caused by plant preference for Ca or Mg (only seen months or years after tank is established) › More consistent water changes	
Other Causes	• Natural geology differences • Water treatment processes	• Bad remineralizer (poor QA, not shrimp-focused)
Solutions	1. Add CrystalPro Shrimp Minerals for trace minerals (cheap option) 2. Adjust Ca:Mg ratio 3. Add remineralizer	• Check if remineralizer meets stated GH/KH ratio and whether it's designed for shrimp
References	Remineralizers (p93), Understanding Molting Problems (p267)	

Pesticides

Water Type	**Irrelevant**
Death Rate	Sudden – Deaths hours or days after exposure
Symptoms	• Sudden death of majority of shrimp population – any fish are unaffected • Lack of copepod or ostracods in tank (p112)
Rarity	Uncommon
Causes	• Introduction of new plants grown with pesticides • Pesticides (roach killer, ant spray) applied nearby outdoors or indoors • Flea/tick medication applied to cats/dogs • Feeding unwashed or non-organic vegetables (p148)
Solutions	• 100% water change with safe water (Ideally Remin. RO/DI water) • Add activated carbon

Predation

Water Type	**Irrelevant**
Death Rate	Typically gradual unless predator is prolific
Symptoms	• Presence of fish, aggressive shrimp, planaria, hydra, leeches, (p289) or dragonfly/damselfly nymphs[lix] • Dwindling population • Shy shrimp (hiding from predators)
Rarity	Common
Causes	• Introduction of new plants/animals • Exposure to outdoor insects
Solutions	• Hydrogen peroxide treatment (p295) for hydra • No Planaria or dewormers for hydra & planaria (p297) • Removal of fish, aggressive shrimp, leeches, nymphs (there are no effective chemical treatments that don't also affect shrimp)
References	Shrimp Diseases (p277)

Water Quality Issues

Water Type	**Tap**	**Remin.**
Death Rate	Gradual – weekly/monthly deaths	
Symptoms	• Growth of Vorticella, Scutariella, Monodiscus, Holtodrilus (p277) • Rust Disease (p282) • Water Mold on eggs (p284) • Lethargy • Multiple shrimp crawling out of water (trying to escape) • Seemingly random deaths • Deaths after water changes • High TDS (400+ ppm) with in-range GH and KH	
Rarity	Common	Rare
Shared Causes	Excess nutrients/bacteria in water column from overfeeding, insufficient filtration, debris build-up, or dead/dying plants and animals	
Solutions	• Reduce feeding • Increase filtration/oxygenation • Filter cleaning • Substrate cleaning • Hydrogen peroxide treatment (p295) to kill bacteria in water column • Consistent water changes • Salt dip individuals with advanced rust disease	
References	Understanding TDS (p32), Feeding Rules of Thumb (p142), What Water Parameters Don't Tell Us (p47)	

Shrimp Biology

Anatomy

Without having studied arthropods (insects) before, shrimp like Neocaridina may look extremely foreign - especially close up. They've clearly got a shell, eyes, a tail, and some legs, but there are a lot of appendages without clear purposes. Admittedly, we don't need to know all 30+ appendages to be good shrimp keepers but these animals have some fascinating adaptations. To organize everything, we'll start with the basic segments of their body then discuss specific body parts, moving from the very tip of a shrimp's antennae to the end of their telson (part of the tail). Let's dig in.

External Anatomy

The Basic Parts

Shrimp have three body segments called the **head**, **thorax**, and **abdomen**. Their head and thorax contain vital organs so they are covered in a hard mineralized plate called the **carapace**. Because this region appears to be one body part, scientists often refer to it as the cephalothorax, a combination of the Latin term for head (cephalo-) and, obviously, thorax. The thorax is defined as the area of the body between the head and abdomen. We'll start with anatomy at the tip of the cephalothorax.

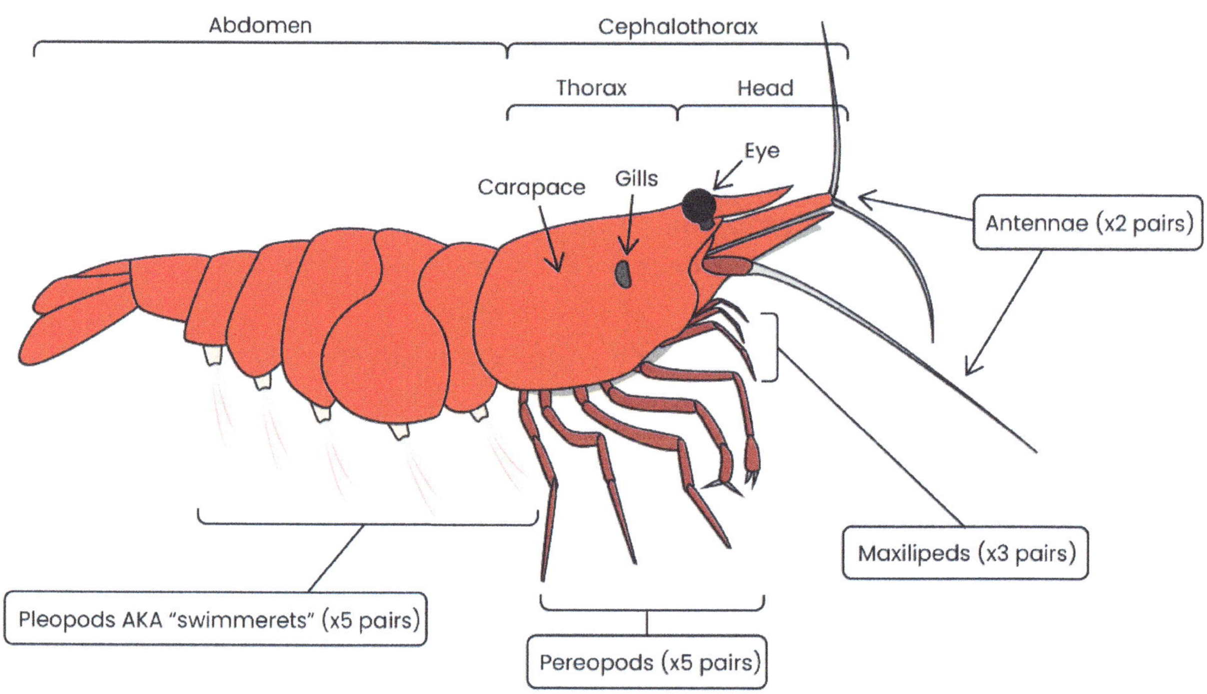

Antennae

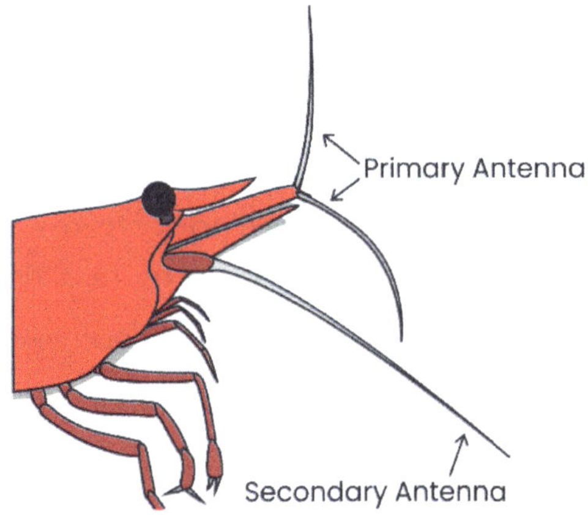

At the tip of the head are two pairs of antennae, referred to as **primary** and **secondary** antenna – That's right, only *two* pairs even though there are clearly *three* pairs of long, thin appendages coming out of their face. Technically, the primary antennae split into flagella, with one pair of flagella oriented up and the other down (Derby, 2021).

The proper function of antennae is critical to a shrimp's survival. As such, they have three methods to keep them working optimally:

1. **Grooming**
    a. Shrimp use their third pair of maxillipeds to keep their antennae clean. Small, stiff hairs on the maxillipeds act as combs to help remove debris and attached organisms.
2. **Molting**
    a. A shrimp replaces its entire exoskeleton during every molt, including the surface of the antennae. This removes old sensory cells on the antennae so new ones may take their place (Derby, 2021).
3. **Antimicrobial secretions**
    a. Glands at the base of the antennae release chemicals that kill bacteria when rubbed by a shrimp's claws (the third maxilliped, specifically) (Derby, 2021) (Schmidt, Chien, & Tadesse, 2006).

Scaphocerites

Scaphocerites are little-known, barely visible "fins" below the primary antennae. Scientists believe their purpose is to help with stabilization and hydrodynamics, specifically during backwards movement (Nakatsuji & Sonobe, 2004). Basically, like tail fins on their face. They have little white hairs on them that are sometimes mistaken for epibionts like *Scutariella japonica*, *Monodiscus kumaki*, or *Vorticella* spp. (p267).

Rostrum

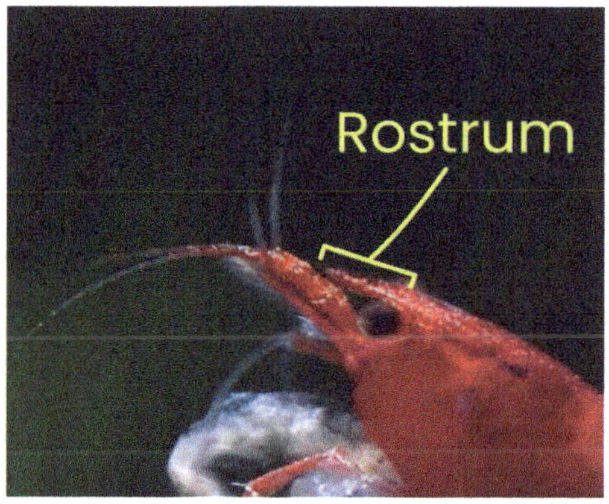

Next up is the rostrum, which is the horn-like structure directly in front of their eyes. In close-ups, the rostrum looks quite sharp and jagged. It has this shape because the rostrum is used for attack and defense by closely related species (Ocasio-Torres, Giray, Crowl, & Sabat, 2015). Our dwarf shrimp have small and unthreatening rostrums which do little except make them slightly more difficult to swallow.

No, we don't have personal experience with swallowing shrimp - stop asking.

Carapace

This extremely tough part of the shell acts as armor to protect most of the shrimp's internal organs and muscles. It tends to be translucent in all but the highest-grade shrimp, so we often see a shrimp's stomach and digestive tract inside (especially when they eat carrots).

The carapace – as with the entire shell – is made of calcium carbonate with varying levels of magnesium to control how flexible the shell is (Raz, Hamilton, Wilt, & et al., 2003). In addition, a material called chitin is also incorporated. Chitin is a similar but harder version of the material fingernails are made of (Anissimov, 2024). It is tough enough to provide great defensive capabilities but lacks much flexibility. This forces shrimp to periodically shed their carapace (and the rest of their shell) so they can continue to grow in a process known as **molting**.

A shrimp eating their molt.

Molting is extremely energy-intensive, resource-intensive, and stressful for shrimp to go through. They give up their defenses and the minerals that went into it, becoming very vulnerable to predation and disease for a period of time before their shell re-hardens. This molting cycle occurs many times throughout a shrimp's life as newly born shrimp molt 7 times in the first 20 days, then slowly reduce that frequency as they get older (Pantaleão J. , Barros-Alves, Tropea, & et al., 2015).

A whole molt. Even the fragile antennae are pulled out of the old shell!

Unnecessary but interesting science!

Some shrimp keepers worry about GH/KH levels because they believe a lack of minerals and carbonate in the water causes the carapace and other parts of the shell to dissolve. This is unlikely because the shell is made up of multiple layers known as the **epicuticle**, **exocuticle**, and **endocuticle**, along with numerous sub-layers.

The epicuticle is the outermost portion of the shell that consists of lipids and proteins – it does not have any calcium/magnesium carbonate that might dissolve in soft water (Li, Han, & Chi, 2022). Instead, the likely problem with soft water is low pH or pH fluctuations that affect the internal pH of shrimp bodies, as mentioned in the pH section of the Water Parameters section. Caridina have various adaptations that let them handle low pH/GH/KH conditions better than Neocaridina.

Eyes

Shrimp are part of an order of animals known as decapods, which have stalked compound eyes (Cronin & Porter, 2008). Compound eyes are made of many individual optical units - called ommatidium (singular) - which are bunched together and look like a disco ball. Each optical unit captures a point of light, then those points are sewn together to form a useful image of the surroundings. This is a similar concept to how pixels on a screen form an image.

Interestingly, shrimp molting is partially controlled by their eyestalks. They release molt-inhibiting hormones so, if a shrimp loses their eyes, they also lose the ability to regulate their molts (Nakatsuji & Sonobe, 2004).

This concludes the appendages attached to the head of the shrimp. Onto the 20 appendages attached to the thorax!

Maxillipeds

Shrimp have three pairs of **maxillipeds** ("jaw legs" in Latin) located around their mouth. These legs are responsible for maneuvering food into their mouths and helping to break it down. In addition, the third maxilliped pair also grooms the antennae with small but stiff hairs (Alexander, Hindley, & Jones, 1980).

Pereopods

> *"These pereopods are made for walkin',*
> *and that's just what they'll do."*
> – Jessica Shrimpson

All shrimp have five pairs of pereopods ("walking legs" in Latin - also spelled as periopods). Jessica is correct that a majority of them are made for walking but two pairs are designed for feeding. The question of whether she struggled to capture the subtleties of shrimp anatomy in song or if she was just ignorant has been lost in the annals of history.

The first two pairs are the ones involved in feeding and tap-dancing around constantly in search of food. These pairs are unique because they have claws that are used for scraping and grabbing anything edible. Due to the presence of claws, these limbs are sometimes referred to as **chelae**.

Thelycum

The thelycum is an opening between a female's fifth pair of pereopods and is where the male delivers their sperm during mating. It is very small and difficult to see on shrimp so it is not a well-known body part.

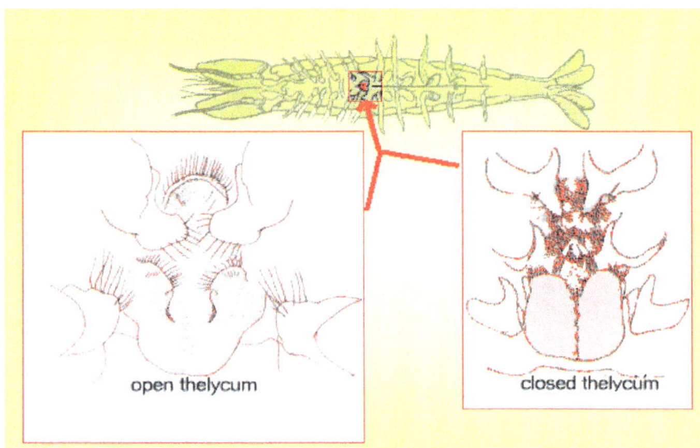

Some related shrimp species can store sperm for long periods of time in their thelycum (Rasch & Bauer, 2016) but this does not seem to be the case with Neocaridina or Caridina.

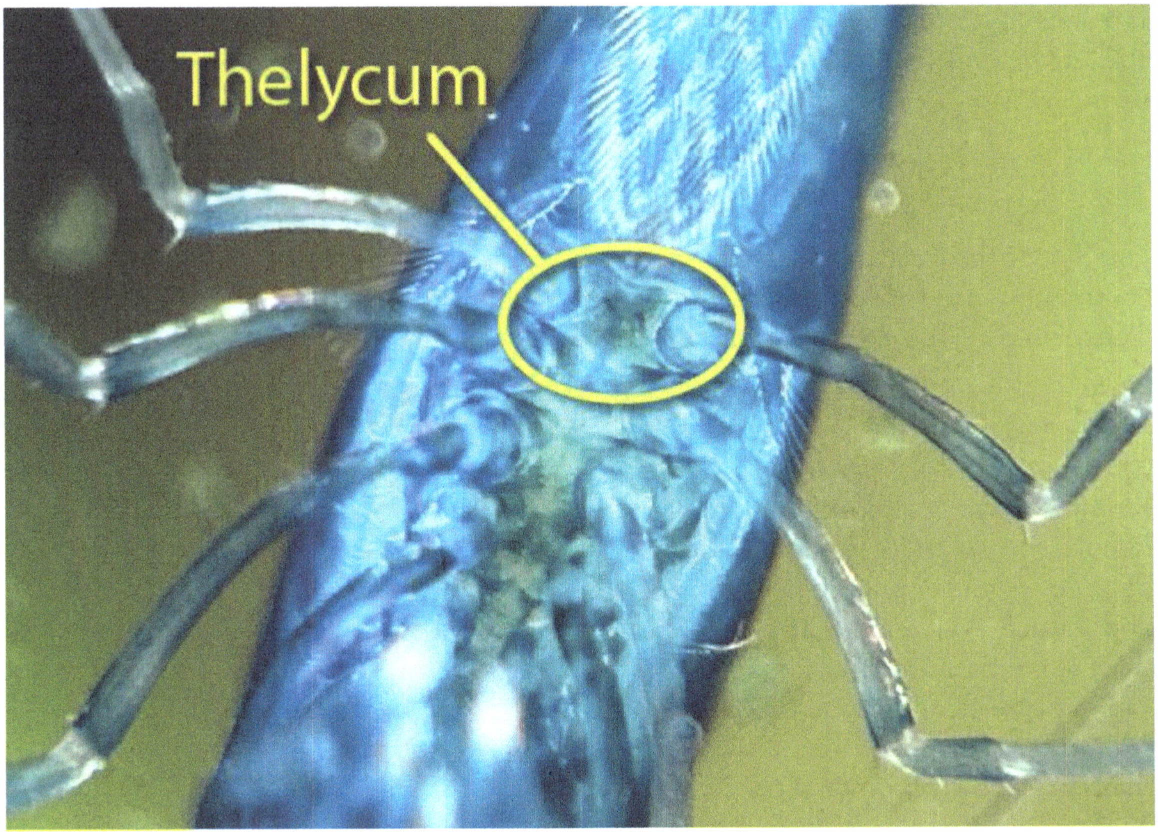

Closeup of thelycum - Credit: c.tiensivu@yahoo.com

The Abdomen

The abdomen refers to the entire region behind the cephalothorax. This is an extremely muscular region which the shrimp uses to swim forward or dart back in an action called the **caridoid escape reaction** (Wikipedia, N.D.). This is a critical escape mechanism from predators that's sometimes observed when shrimp see us approach the tank or get disturbed by objects like aquascaping tools.

The muscles responsible for swimming and the caridoid escape reflex are protected by six distinct segments of the abdomen that overlap with each other. The second segment overlaps with the first and third segments, which happens to be one of the defining physical characteristics of shrimp versus prawns.

The second abdominal segment in female shrimp is larger than the segment for males. This difference creates more room for eggs and is visible once shrimp reach sexual maturity.

Pleopods

Attached to the first five abdominal segments are an equal number of **pleopod** pairs, more commonly called **swimmerets**. These swimmerets are flat appendages that act like flippers to push the shrimp forward through water.

While swimmerets are important for movement, they also serve a critical role in the shrimp life cycle. After mating, female shrimp move fertilized eggs down to their swimmerets and attach the eggs with a glue-like substance. The eggs look like little elongated berries, so females are said to be "**berried**" when they have eggs on their swimmerets. To keep the eggs healthy until they hatch, the mother moves and pulses her swimmerets up and down. This behavior is called **fanning** and it helps circulate fresh water around the eggs, which prevents mold development.

Petasma & Appendix Masculina

Key anatomical differences between male and female Neocaridina include the presence of the petasma and appendix masculina on males. The petasma is located in front of the first pair of pleopods, while the appendix masculina is circular and located at the end of the second pleopod pair. The petasma is a tube holding a sperm packet that the appendix masculina helps deliver through a pumping action during mating.

Tail

What we might think of as the tail is technically the sixth abdominal segment. It has two pairs of modified pleopods that fan out backwards to form the tail fins. These fins are called uropods, with the inner fins called endopods and the outer fins called exopods.

In addition, a part called a telson provides support in the middle of the tail fins (Wikipedia, 2023). An easy way to remember is by thinking of the telson as the "Tail sun" since the sun is the center of the universe (allegedly) and the telson is the center of the tail.

Internal Anatomy

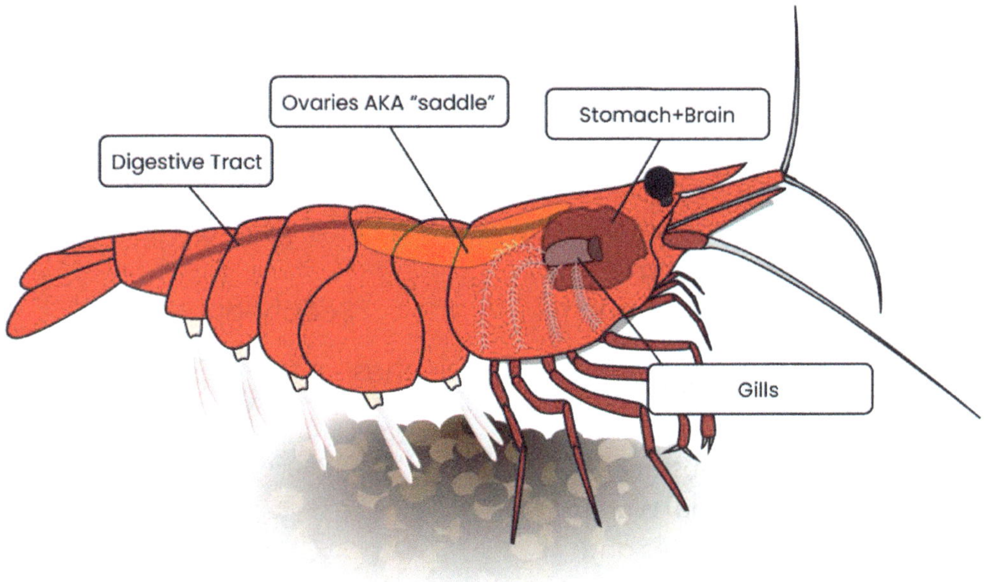

Stomach & Brain

Many shrimp have transparent shells that let us see an internal mass of organs through their carapace, including their brains and stomach. These two organs blend together so they are difficult to tell apart. Some shrimp keepers worry when they see the stomach movement and color change that happens when shrimp eat intensely colored food (like carrots) but that is normal. The only time to be worried is if there's a long white tube crisscrossing around inside of them, as that is a sign of a horsehair worm[lx]. These parasites are more common in ghost shrimp than Neocaridina or Caridina, but it never hurts to be aware of them.

Digestive Tract

The digestive tract is one long tube running along what would be the shrimp's spine - if they had one. This tube ends right below their tail and is often confused for a horsehair worm, as well.

Gills

Gills are respiratory organs that promote gas exchange between the water and blood of aquatic animals. They take in oxygen and minerals while releasing CO_2 and other waste left in the bloodstream. Shrimp gills are extremely complex branching structures housed underneath their carapace. A small ovular opening in the carapace lets water into the **scaphognathite**. The scaphognathite expands and contracts to pump water through eight gill branches (SoniaBajaj10, 2021), all of which have veins running through them to facilitate the transfer of substances between the water and blood. Water flows through these gill branches and exits below the carapace between the pereopods (Preethi N., 2021).

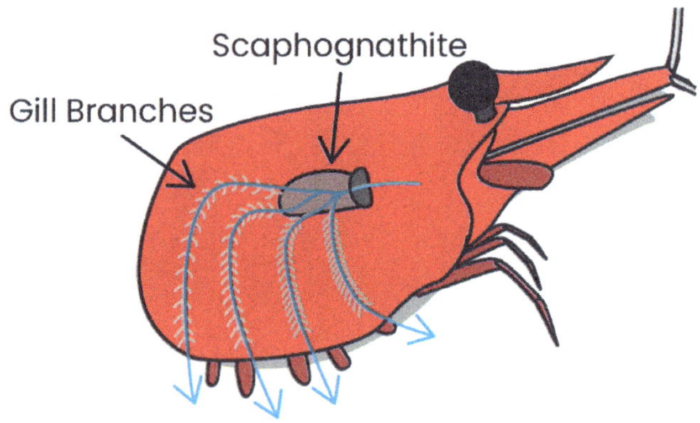

Ovaries

Lastly, an extremely important organ – the ovaries. These are invisible when empty but appear as green, yellow, or white areas when they store unfertilized eggs. The ovaries span part of the thorax and abdomen and look like a horse saddle so the area is commonly called a **saddle** and females with them are said to be **saddled** (see image on next page).

Once a female molts and mates with a male, she transfers the eggs from her ovaries and through the sperm packet, fertilizing them and placing the eggs on her swimmerets until they hatch. Seeing a berried female for the first time is one of the most exciting moments in shrimp keeping. We can't wait until you get to experience it!

That concludes the visible anatomy of shrimp. Now, let's go over the unique differences between males and females to learn how to sex them!

Orange Rili (Neocaridina davidi *female) with developing ovaries (AKA "saddle").*

Sexing

Shrimp have distinct physical differences between sexes - a phenomenon called **sexual dimorphism**. Knowing what these differences are helps us pick out a good mix of shrimp from the seller - at least if they are old enough to sex. The rest of this chapter covers the visible differences between male and female dwarf shrimp, along with how reliable each difference is and why these traits evolved. We've also created a handy Shrimp Sex Determination infographic on page 183.

Please note that sexual dimorphisms become most obvious once shrimp reach sexual maturity (roughly 75 days after birth). Stores often sell shrimp before they reach this point, so it may be tough to tell the sex of your shrimp until they get older.

Caridina have less exaggerated/visible sexual dimorphisms and more opaque shells so they tend to be more difficult to sex. Many breeders rely on antenna length, as discussed at the end of this chapter.

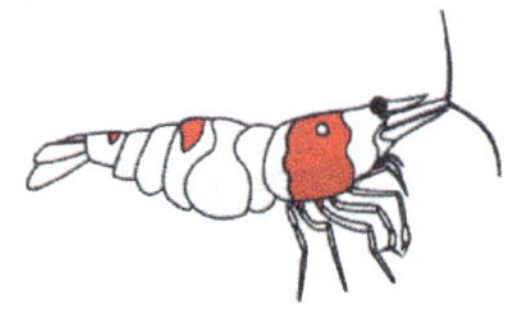

Fertilized Eggs

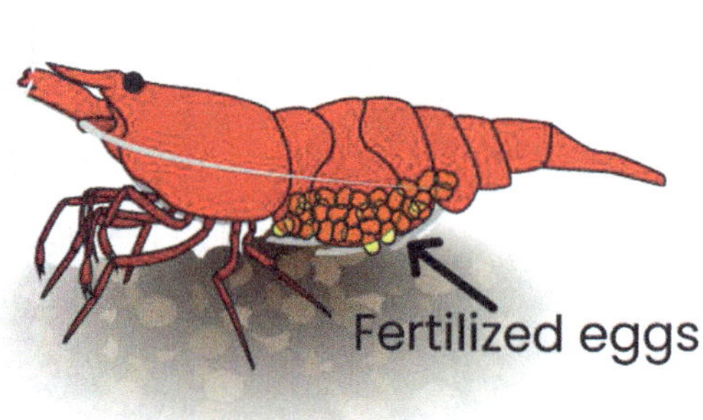

Reliability: 100%

The presence of fertilized eggs is the most reliable method of sex determination because females are the only ones to carry eggs. Fertilized eggs are held under a female's abdomen for 15-30 days and look like distinct jellybeans or berries (where the term "berried" comes from) kept in the abdomen of the female shrimp. *If they aren't distinct shapes, then it may be a sign of a* fungal infection (p285).

Blue Velvet Neocaridina carrying eggs of different colors - Credit: c.tiensivu@yahoo.com

Egg color varies greatly, ranging from white to a light yellow, green, or even blue! The color does not indicate anything about egg health or how long until the eggs hatch. The only way to tell if the eggs are close to hatching is to look for tiny black dots on them. This indicates late-stage eye development.

The presence of eggs is the most obvious way of telling which shrimp are female because eggs are almost always visible either through or underneath their abdomen. In addition, females frequently "fan" their swimmerets back and forth to oxygenate the eggs, which makes the eggs easier to spot. Obviously, this sex determination method fails when none of the shrimp at the store or in a tank are berried, which is why the chapter doesn't stop here.

Evolutionary Explanation: Necessary for reproduction.

Ovaries

Reliability: 100%

The ovaries are where sexually mature female shrimp keep their undeveloped eggs before mating. After mating, the fertilized eggs are moved onto their swimmerets. When the ovaries are visible, they are in the middle of the shrimp's back and often look like a saddle. This is why females ready to mate are said to be "saddled". The saddle color is typically yellow but can be white, green, or a variety of other colors.

"Saddled" Female

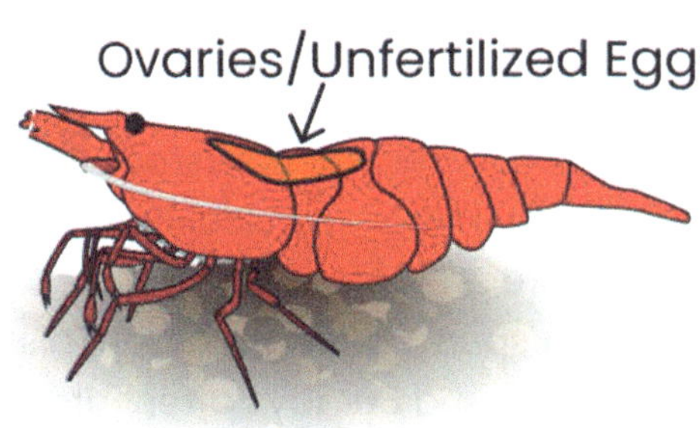

The saddle is the second most reliable method of telling whether a shrimp is male or female because it is often visible through the shell, but that is not always possible with more opaque shells. In addition, the saddle may not be present if the shrimp is young or if it recently had babies.

Evolutionary Explanation: Necessary for reproduction. (Don't worry – The explanations get more interesting.)

Swimmerets

Reliability: High

When a mature shrimp is up against the glass, it's easy to determine whether they are male or female by looking at the first and second pair of pleopods (swimmerets). The pleopod shape and pattern remains the same along the entire length of a female's underside, whereas the first two pairs of pleopods are modified on male shrimp, creating a visual break in the pattern. As mentioned in the Anatomy section (p166), these two pairs of pleopods are the petasma and appendix masculina, which are responsible for delivering sperm during mating. This difference is only visible as shrimp approach sexual maturity around 60 days after hatching, depending on temperature.

Evolutionary Purpose: The visual differences between male and female are due to the male sex organs, which are critical for the species to survive. Shrimp species wouldn't last long without them...

Abdominal Segments

Reliability: High (Less obvious on Caridina)

For female shrimp, the second abdominal segment has a noticeably wider base than males. The difference develops during their juvenile stage and becomes more distinct once they reach sexual maturity. Seeing the difference requires a keen eye--especially when the shell is transparent--but is a reliable differentiator even when eggs aren't visible.

> **Fun fact alert!** This overlapping of the second segment on top of the first and third segment is one of the key traits that distinguish shrimp from prawns.

Evolutionary Purpose: The expansion of the second segment creates room for the female to carry more eggs, resulting in more offspring with the same trait.

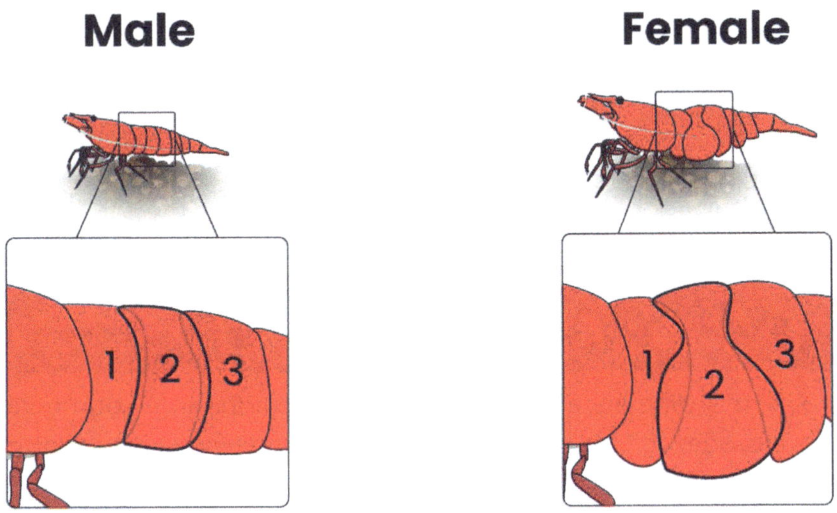

Abdomen Shape

Reliability: Medium

One of the most noted differences between male and female shrimp is the curve of their abdomen. As they mature, females develop a more rounded undercarriage due to the larger second abdominal segment, whereas male abdomens remain relatively straight. **This difference becomes even more pronounced after a female has been pregnant at least once** because her abdomen expands outward to carry as many eggs as possible. This is part of why female shrimp mate after molting, as their shell is most flexible post-molt to expand and hold more eggs.

We could combine this difference with the section on abdominal segments, but it is a slightly different characteristic to look for. The second abdominal segment also is more reliable at a younger age than body shape is, at least until females have their first offspring. After that, abdomen shape is often the most obvious indicator, although it is less pronounced for Caridina.

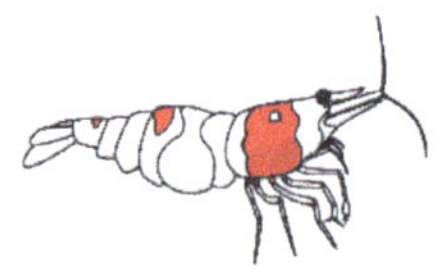

Evolutionary Purpose: A larger abdomen lets females hold more eggs and therefore have more offspring to pass on the trait. Males, on the other hand, have no need for a larger abdomen.

Size

Reliability: Medium-Low

Healthy female shrimp are typically larger than males of the same age. For example, adult female Neocaridina reach up to 1.5 inches (3.8 cm) while males tend to be less than 1.0 inches (2.54 cm). Caridina are slightly smaller, with adult females reaching 1.25 inches (3.2 cm) while adult males reach 0.75 inches (1.9 cm).

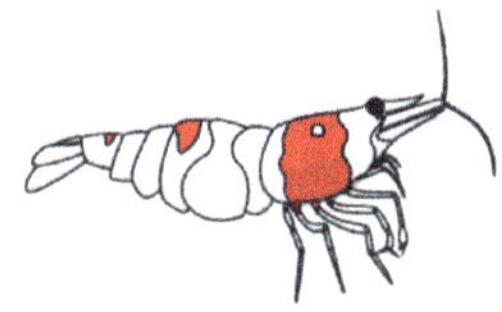

The problem with this method is that we don't always know the age of the shrimp we're looking at, so this is an unreliable sex determination method unless we know their age and can compare a shrimp to another one of the same age. Often, evidence of eggs or a saddle is applicable earlier and more reliably.

Evolutionary Explanation: The size difference between males and females is often seen in species where the females produce many offspring at once (Northern Arizona University, n.d.). It is theorized that this phenomenon is seen because larger females tend to be able to produce more eggs, which in turn leads to more offspring that mate and pass on the larger female genes. Females being larger than males is a common characteristic among insects (which shrimp technically are).

In addition, smaller males may swim faster, which lets them reach recently molted females sooner. This creates evolutionary pressure towards smaller males because they are more likely to pass on their genes.

Color

Reliability: Medium-Low

Male

Female

Female shrimp tend to have more **color coverage** and **brighter/more solid colors**. We say "tend to" because some males are just as colorful as females, especially with high-grade shrimp. As a result, this color difference is only semi-reliable. An additional problem is that shrimp may lose color under stress, making it more difficult to sex them. **We recommend using color in combination with other differences to identify sex reliably**.

As a side note, this color difference is a common reason why shrimp keepers may accidentally pick out all females from their local fish store. They look the biggest and most attractive, so shrimp keepers that aren't aware of this difference pick all females out of the tank and wind up disappointed when their shrimp don't breed.

Evolutionary Explanation: We were unable to find any research that provided a hypothesis for the color difference between male and female shrimp. In other animals, males are typically more brightly colored than females because females choose the most colorful males. They do this because color intensity is typically a sign of good health, so.

There are only a few examples of species where females are brighter than males. This typically happens when the roles are reversed, with males spending resources caring for their offspring (Harvey, 2002) and therefore being the most selective of mates. Since female shrimp are the ones caring for the eggs, there is not a clear explanation for the color difference in shrimp.

Antenna Length

Reliability: High with good eyesight and lighting

One reliable way to determine the sex is by looking at the length of the upward-pointing set of antennae, at least after the shrimp reaches sexual maturity. For males, this pair of antennae is longer than ¾ of their cephalothorax length, whereas this pair is generally shorter than ¾ of a female's cephalothorax length. It takes patience to view the antennae at the right angle to judge how long they are in comparison to the cephalothorax but is a very useful way to distinguish sex for shrimp with opaque shells that don't show the saddle or even eggs. It's especially useful for Caridina since other sexual dimorphisms aren't as obvious as they are with Neocaridina.

Technically, every pair of male antennae are comparatively longer than female antennae. The reason we recommend using the upward-pointing antennae is because they don't move and are in the easiest position to compare to the carapace/cephalothorax length.

Evolutionary Purpose: Researchers found significantly more aesthetascs (tiny hairs that sense the directions of chemicals) on male antennae (Bauer & Caskey, 2006). They proposed that male antennae evolved to have more sensing hairs because it allows them to pick up the pheromones of females quickly and mate more, thereby passing on their genes to more offspring. Longer antennae have more surface area for aesthetascs, so males with longer antennae are more likely to find mates.

Sexing (Summary)

The infographic on the next page summarizes all the ways to distinguish males and females.

Sexing Dwarf Shrimp

For identifying male and female *Neocaridina*, *Caridina*, Ghost, and Amano Shrimp, unless otherwise specified

Differences	Reliability*
All species **Presence of Eggs** Only females produce and carry eggs	100%
All species **Abdomen Shape** Females have more curved abdomens than males	Med-High
All species **Second Abdominal Segment** Females have larger 2nd segment that overlaps 1st and 3rd segment more	High Only easy to see on shrimp with opaque shells
Neocaridina and Caridina **Antenna Length** Females have shorter upper antennae *compared to their head length* vs males	High Takes some practice and good eyesight
Only Neocaridina & Caridina **Color Coverage & Intensity** Females typically have more color coverage and intensity	Med-Low
All species **Body Size** Females are typically larger than males of the same *age*	Med-Low

*After reaching sexual maturity

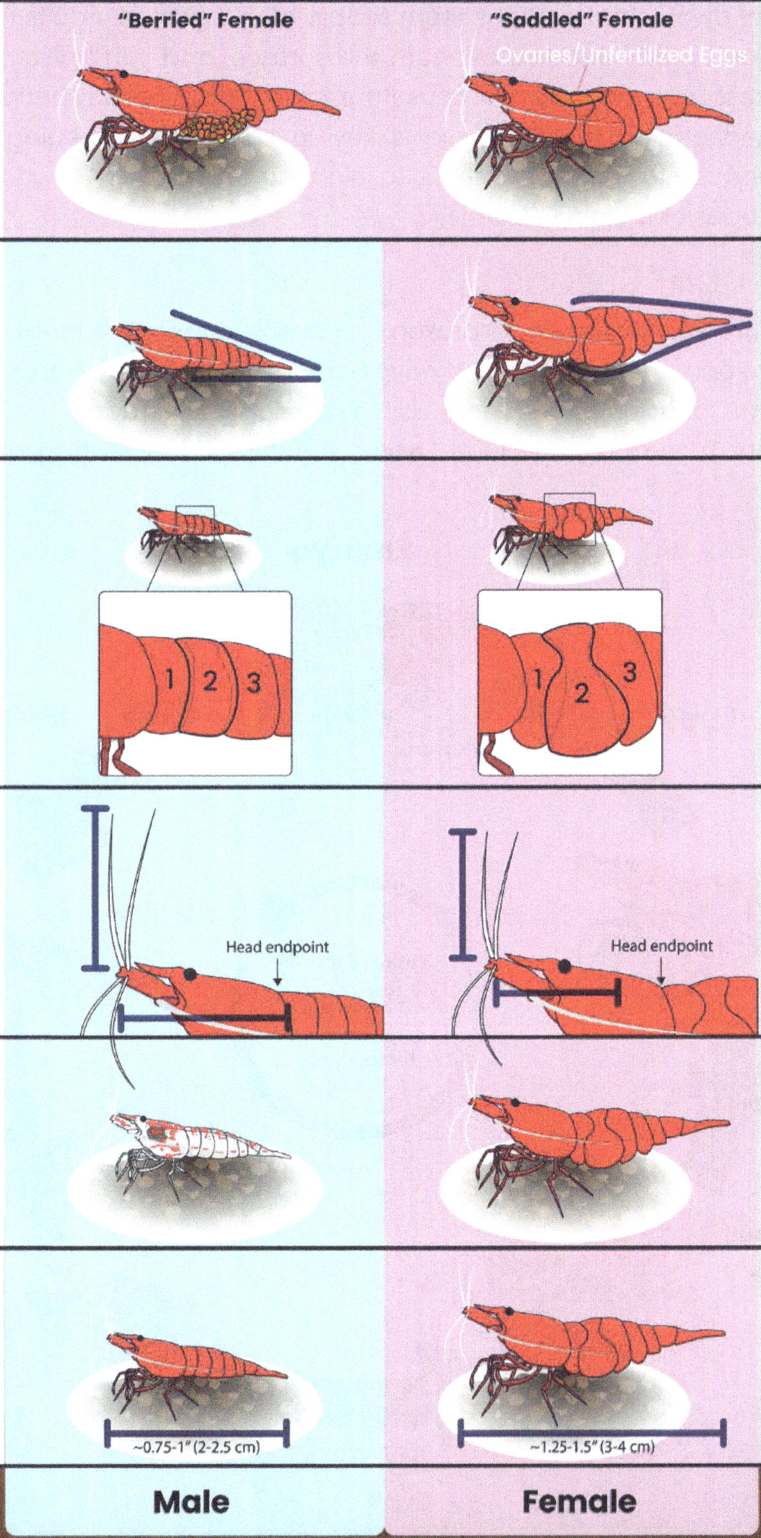

Shrimply Explained

www.shrimplyexplained.com

Lifecycle

The lifecycle of dwarf shrimp might seem simple at first glance. Adults mate, eggs hatch, babies grow up... right? But dive beneath the surface, and you'll discover a surprisingly complex process that uncovers a lot of useful information to help us become better shrimp keepers. Understanding their lifecycle is key to successful breeding, troubleshooting issues, and learning to appreciate these fascinating creatures on a whole new level. Let's explore the stages of their development.

Development Stages

Technically, shrimp growth can be broken into *twelve* different stages based on very fine details of development that we'd only have hope of seeing with a very good microscope. Instead, we'll be focusing on the more visible development stages that are important for shrimp keepers. These stages include: **Larval**, **Juvenile**, **Adult**, **Mating**, and **Embryo**.

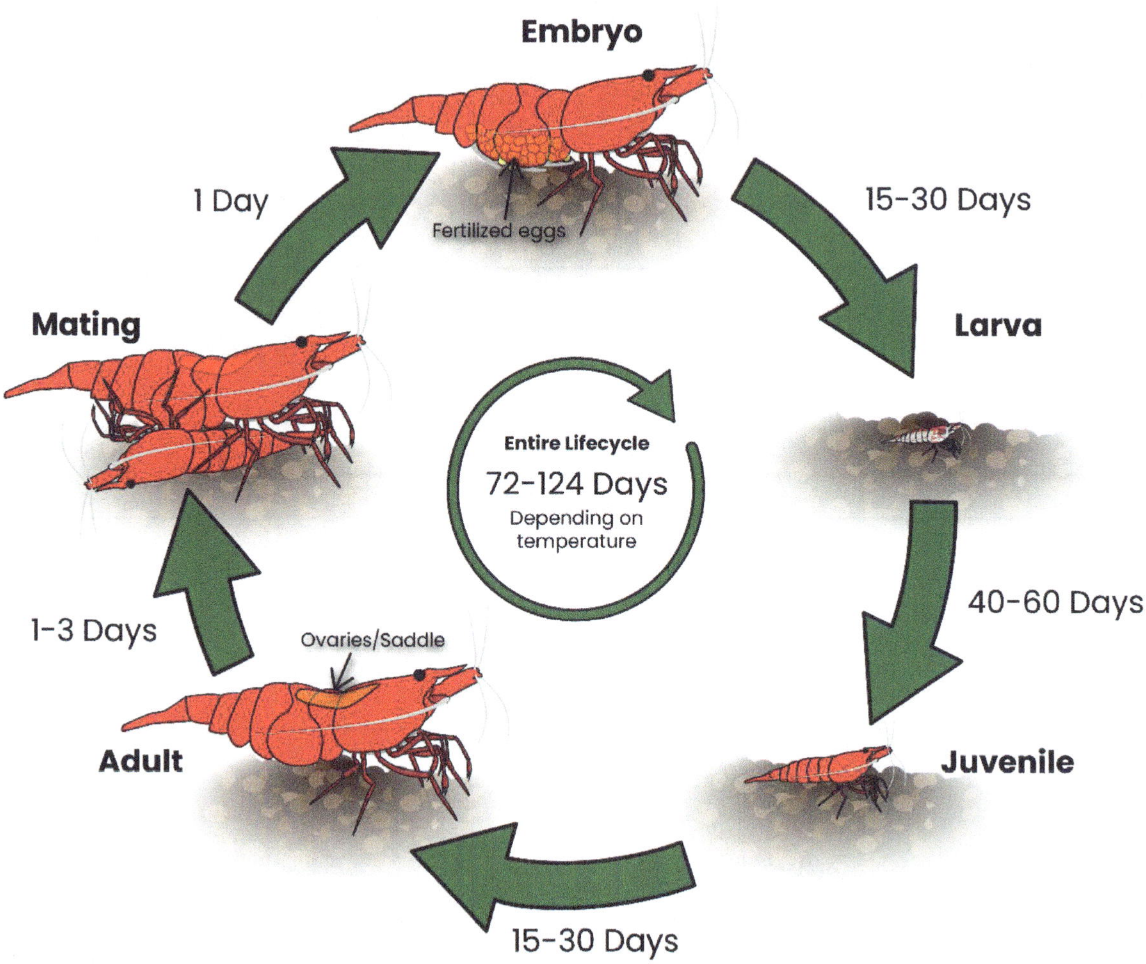

Stage	**Length (Days) [Total Age]**	**Development Signs**
Larval	40-60 [same]	Small (1-12 mm), rapid growth, color development, males can be sexed at ~16-20 days
Juvenile	15-30 [55-90]	12-15mm, ovary development
Adult	1-3 [56-93]	Ovary maturation
Mating	1 [57-94]	Female molting, ~10 second copulation, eggs move to female pleopods
Embryo	15-30 [72-124]	Female fans eggs with pleopods, eggs turn from opaque color to transparent

Please note the time ranges are dependent on temperature, with the longest times estimated for development at 65°F (18°C) and the shortest times for development at 82°F (28°C). Most information in this chapter is from this study (Pantaleão, Gregati, da Costa, Lopez-Greco, & Negreiros-Fransozo, 2017).

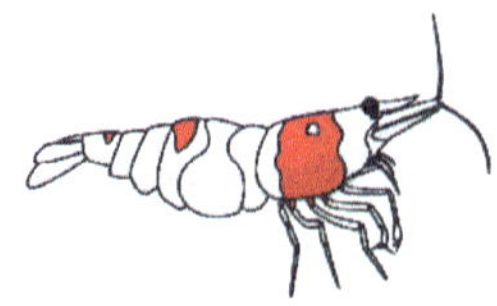

The stated times are for Neocaridina. Caridina develop at a slower rate so may take 4-6 months to reach sexual maturity (Yam & Dudgeon, 2005).

Larval Stage

Timeframe

Length (Days): 40-60
Days Post-Hatching: 40-60

Appearance

The **larval stage** covers the period from hatching to the initial development of female ovaries. At hatching, larval shrimp (shrimplets) look like miniature versions of adults. They are approximately 1-2mm but grow and molt frequently, eventually reaching about 12mm in length after 40-60 days. Molting happens every day, at first, then becomes less frequent as larvae age. Shrimp start out with a transparent body and slowly develop their color over time, although they do not reach the intensity of their final adult coloration at this stage. We can cull shrimp based on their developing color at the later parts of this stage, but some larvae don't show their true coloration until later.

A week-old baby shrimp. Photo Credit: c.tiensivu@yahoo.com

Technically, males start to develop differences in their 1st and 2nd pair of pleopods at 16-20 days after hatching, which is visible when looking at shrimp from below with a good set of eyes or a magnifying glass/camera. Other visible differences noted in the Shrimp Sexing section (p176) develop later and become more prominent as shrimp age.

Behavior

Larvae instinctually understand how vulnerable they are, so they spend the first couple of weeks hiding near the bottom of the tank. They'll also be found buried in plants like Java moss. With their transparent colors and small size, larvae are incredibly difficult to see for the first few weeks.

It's common for hobbyists to worry at the larval stage because we see the eggs disappear from a female but don't see any larvae in the tank. Don't worry – They're just great at hiding. Give it 1-3 weeks and they'll start to appear. If they don't appear after that time, then start thinking about potential problems, which may include a lack of biofilm, unstable parameters, or inbreeding issues causing premature death.

One fascinating fact we found in the research is that shrimp do not need to eat until 6-8 days post-hatching (Pantaleão J. , Barros-Alves, Tropea, & et al., 2015). Instead, they hatch with an egg yolk that lets them hide and grow without taking risks for food. If larvae in a

tank die within six days post-hatching, then it is a sign of a problem *other than biofilm availability.* Many sources recommend powdered biofilm supplements to improve shrimplet survival, but we can take more informed actions by understanding when deaths occur.

Juvenile Stage

Timeframe

Length (Days): 15-30

Days Post-Hatching: 55-90

Appearance

The juvenile stage is bookended by the start of ovary maturation in females, and it ends when ovaries are fully developed. In shrimp with transparent shells, visible ovary maturation starts out like light wisps or spots growing on the back (spanning cephalothorax and abdomen) above the intestines. These spots are the first eggs that develop in the ovaries. By the end of the juvenile stage, ovaries are full of eggs and shrimp are ready to reproduce, at which point females are said to be "saddled".

Juvenile shrimp range from about 12-15mm. Females put their energy into egg production so they stop molting at this stage, while males continue to molt.

Behavior

Juvenile shrimp are larger and less cautious than larvae, so juveniles tend to be more active and visible.

Adult Stage

Timeframe

Length (Days): 1-3 (Technically for the rest of their life)

Days Post-Hatching: 56-93

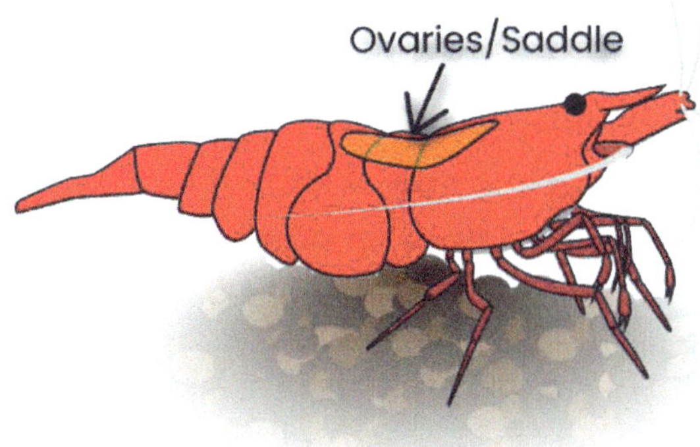

Appearance

Adult shrimp reach sexual maturity with complete development of the male petasma + appendix masculina and female ovaries. At this stage, sexes are more easily distinguished by all characteristics discussed in the Shrimp Sexing chapter (p176). Adults range from 0.5 inches (1.3 cm) (young) to 1.5 inches (3.8 cm)

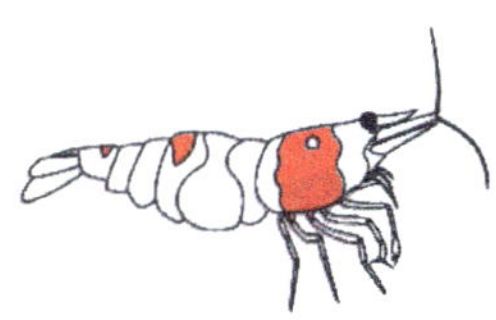

(fully grown female), depending on age and sex. Adult Caridina tend to be slightly smaller than Neocaridina, ranging from 0.5-1.25 inches (1.3-3.2 cm).

Behavior

Behavior remains the same as larger juveniles. Adults live for between 1-2 years, depending on temperature.

Mating

Timeframe

Length (Days): 1
Days Post-Hatching: 57-94

Appearance

No change from adult stage.

Behavior

Females enter the mating stage by releasing pheromones into the water that attract males. This is a common cause of concern because the tank suddenly comes alive with restless males searching for the waiting female. Males eventually find the female and may group around her or clasp onto her back while waiting for her to molt. Excess male attention can be stressful for females and may kill them, which is why having a good balance of male-to-female shrimp is important (ideally >50% females) in any tank. During this process, females try to reduce stress by hiding, which is why plant cover or rock stacks can be beneficial.

Once the female molts, a male quickly positions himself under her, thrusts rapidly, then leaves. The whole process takes anywhere from 10-60 seconds. Some shrimp pairs recover quickly while others spend 30-40 seconds appearing stunned on their sides. It's easier to understand the process with a video (Shrimply, A Scientific Dive Into The Cherry Shrimp Lifecycle, 2024).

After mating, the female shrimp goes back into hiding for up to 24 hours, during which time she moves eggs from her ovaries, through the sperm packet, and onto her pleopods (swimmerets). The eggs are held together with a thin sticky filament.

Embryo Stage

Timeframe

Length (Days): 15-30
Days Post-Hatching: 57-94

Appearance

Female shrimp carry fertilized embryos on their pleopods until they hatch into larval shrimp. Embryos look like oblong berries so a female shrimp carrying eggs is said to be "**berried**". The embryos are initially filled with yolk, just like a chicken egg. Over time, embryos absorb the yolk and eggs become transparent, at which point we can often see small black eyespots through the shell. **Embryos typically hatch 2-4 days after eyespots appear**.

> **Side note**: Eggs can be blue, green, white, peach, and a variety of colors in-between due to different yolk colors (see image below). Some shrimp even have multiple different egg colors.

Blue Velvet Neocaridina carrying eggs of different colors - Credit: c.tiensivu@yahoo.com

Behavior

During this time, the parent constantly moves her pleopods up and down to create water flow that cleans the eggs – a process called "**fanning**". If females don't fan their eggs and if water quality is bad, then *Saprolegnia* water molds can grow on eggs and kill them. At that point, female shrimp are also in danger so may need to be salt dipped. Check out the Shrimp Diseases chapter (p277) to learn more.

Occasionally, shrimp drop their eggs instead of carrying them until they hatch. There are a few reasons eggs may drop:

- Stress – Shrimp that are moved to tanks with different water conditions are more likely to drop eggs – Caridina especially so. Neocaridina can often be moved

between tanks with similar water parameters without dropping eggs, whereas Caridina can (on rare occasions) be overly stressed just from being netted. Eggs dropped from stress are often left in molts.

- Age – Young shrimp are less likely to carry eggs the entire time, especially if it's the female's first clutch. This may be due to the female's inexperience or the result of improper sperm deposition by males, leading to unfertilized eggs that must be dropped. If eggs aren't fertilized, then they are more likely to be dropped soon after the female becomes berried.
- Health – Dropped eggs are a common problem with inbred shrimp, which is why dropped eggs are more common with newer lines of shrimp.

If dropped eggs are fertilized, then they can be saved using an **egg tumbler**. An egg tumbler is a container with air pumped slowly into the bottom of it, which lifts eggs up to keep water flowing around them, just like females do while fanning. Dropped eggs found in molts or on dead shrimp are more likely to be viable. Check out this video for more information on how to save dropped eggs (Mark's Shrimp Tanks, 2017).

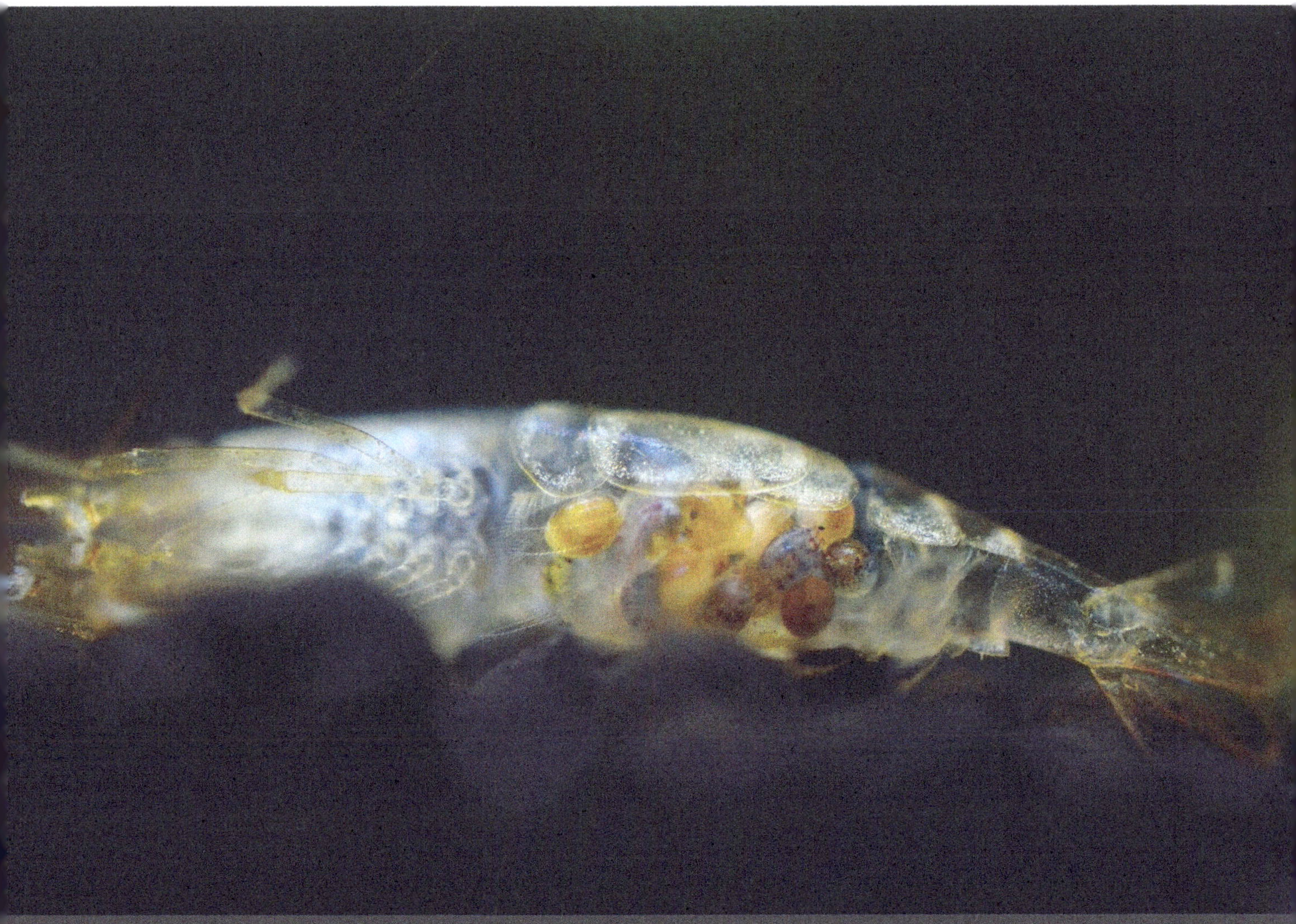

Early molt with remaining eggs. Black dots on eggs indicate those eggs are developing eyes and close to hatching.

Breeding

One of the reasons shrimp are so popular is because they are quite easy to breed once we set up a tank properly; We don't have to collect the eggs and move them to a special tank; We don't have to worry about parents eating the offspring. Just keep the tank healthy with some routine maintenance and let the shrimp do their thing!

It may take shrimp a few months to settle in and have their first offspring but, once they do, we see another reason shrimp are so popular – There's *a lot* of variety in their appearance. From different shell opacities, shell colors, patterns, and even flesh colors, no shrimp is quite alike. They can be similar - which is the goal for shrimp breeding competitions - but even professionals struggle to get perfectly matching shrimp.

Some shrimp keepers want to create a new breed to sell or to breed **high-grade** shrimp for competitions. Understanding the basics of genetics is critical for successful breeding. On the other hand, others just want to enjoy a peaceful tank with colorful shrimp. All this talk of creating new breeds and competing in shows could be the furthest thing from their mind. That's cool, too! Even then, knowing the basics of genetics and breeding helps us appreciate the genetic variations in our tank, maintain bright color(s), and keep a colony healthy.

Side note: The "grade" of a shrimp refers to the quality of its breeding. High-grade shrimp typically have brighter colors and well-defined patterns. Ideally, breeding one high-grade shrimp with another high-grade shrimp from the same line also passes their coloration onto their offspring. Few lines of shrimp produce perfectly matching offspring, but some are better than others. The same is true for sellers, so try to purchase shrimp from well-known breeders or companies for the best chance at high-grade shrimp with consistent offspring.

The rest of this chapter answers the following questions about Neocaridina since that's where most people start out in the hobby (much of this also applies to Caridina though):

- **How was the first cherry shrimp created?**
- **How are new breeds created?**
- **Why does a colony lose its color over time?**
- **Can we keep different colors in the same tank?**
- **How do we maintain good colors in a tank?**
- **What are the natural limitations of possible colors/patterns?**
- **Do we ever need to add new genetics to a shrimp tank?**

After this chapter, we dig deeper into the science of genetics and coloration. Additional information focused on the Caridina Family Tree and unique breeding aspects are included in the Caridina Breeding chapter (p246).

*Vibrant Red Cherry shrimp (*Neocaridina davidi *female) on Java moss.*

The First Cherry Shrimp

None of the commercially available colors of shrimp are present in the wild. That's because the same outstanding colorations that make shrimp so appealing for us also appeal to predators. A bright red, blue, yellow, or orange color is a great advertisement for a tasty treat. To avoid predators, shrimp in the wild typically have darker, muddied colors with a mottled pattern that camouflages them. Shrimp with these natural colorations are called **wild types (WT).**

Wild type Neocaridina eating bloodworm - Source: t.ohashi, CC BY 2.0 via Wikimedia Commons

The transformation from WTs to Red Cherry Neocaridina happened through a process called **selective breeding**. Right around the year 2000, a person (we don't know who) fished through the streams of Taiwan to collect WT Neocaridina. Out of the collected shrimp, this person noticed that some of the brown ones had a slight reddish hue to their colors. The reddest male and female shrimp were put together in a tank to breed and see if any of their offspring had even more red on them. Some of the offspring had more red pigment than others, so the new reddest male and female were selected to breed together, leading to even redder offspring. This process continued for a number of generations as the person continued selecting the reddest offspring, until they eventually bred the first Red Cherry shrimp.

Low-grade Red Cherry shrimp - Source: Rafal Konkolewski 2006 - Wikimedia commons - cropped

The same selection process occurred with Caridina, as well as every other ornamental species in the aquarium hobby. If a trait can be passed down to offspring, then it can be selected for.

How Other Neocaridina Breeds Are Created

The progression from a slightly red shrimp to a very red shrimp is a straightforward origin story but, if we look online, there are MANY other colors and patterns of Neocaridina. The origin of some of them seems pretty straightforward, like how an orange Neocaridina (Pumpkin shrimp) might come from red cherries that had slightly more yellow in their shell. Others, like the Bloody Mary shrimp with a translucent shell and red flesh, are harder to explain. There's a lot of research still to be done to get that answer but, in short, a Bloody

Mary shrimp has a **genetic mutation** in their DNA, causing shrimp to store red pigment inside their flesh instead of their shell.

Orange Rili Neocaridina female

Genetic mutations are one explanation for why some offspring in a tank may look very different from their parents. Another explanation may be that the trait is always hidden in the DNA and the offspring happens to have the right combination of DNA to show the trait. Regardless, if our shrimp reproduce enough, we're going to see some interesting combinations of colors and patterns. If you want to stabilize those patterns and create your own line of shrimp, then go for it! Otherwise, just sit back and enjoy the natural beauty of genetic variation.

Background information: **DNA** and **genes** are like letters and sentences. Letters are meaningless until they are put together in a specific order to form words and sentences that we can understand. Likewise, DNA is made of "letters" that are put in a specific order to create sentence-like instructions for your body to interpret. These instructions are called genes and are responsible for physical characteristics of an organism, such as height, eye color, hair length, etc.

Genetic mutations happen when the DNA is not correctly copied or interpreted. Going back to the letter analogy, it's like if you try copying "shell color is black" but write "shell color is blank" instead. There's still an understandable instruction to follow, but it isn't the "right" one. Organisms have ways of catching and correcting DNA mutations, but these autocorrect functions are much better at catching gibberish like "bpack". They aren't as good at spotting comprehensible errors like "blank" instead of "black".

The Neocaridina Family Tree

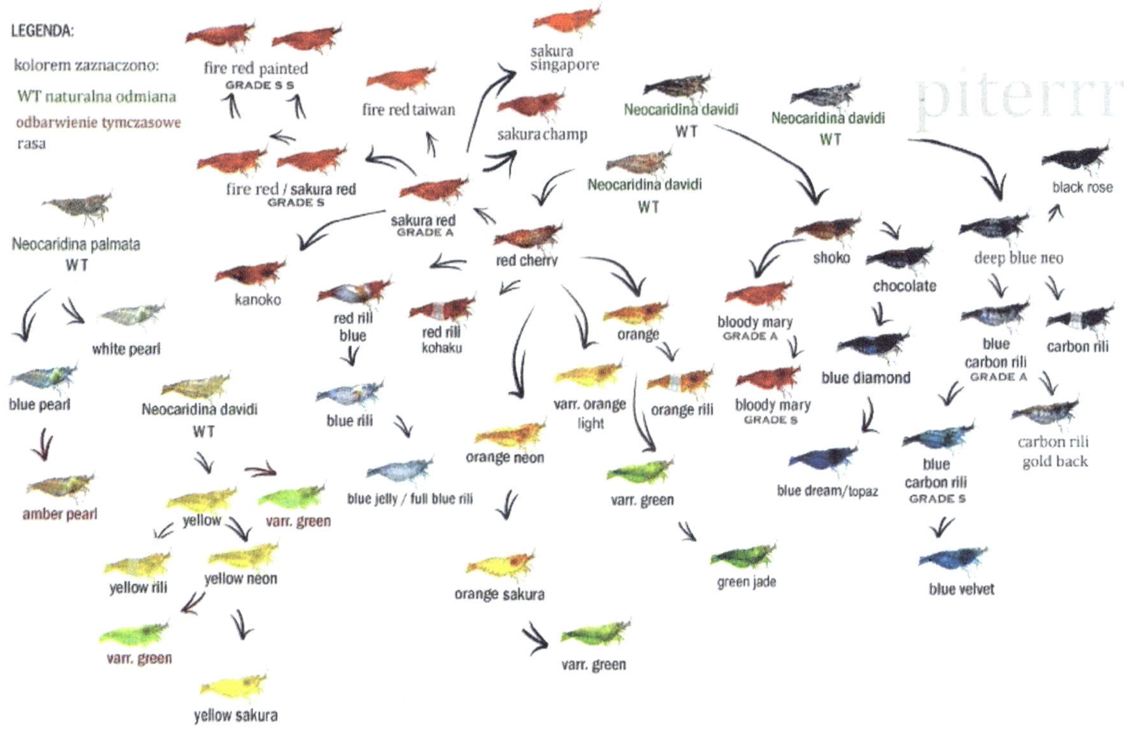

Neocaridina Family Tree - Source: Pethro, CC BY-SA 4.0 via Wikimedia Commons

The family tree may look confusing but is extremely useful once we understand how it works. To explain, we're separating the WT red cherry lineage from the tree for now.

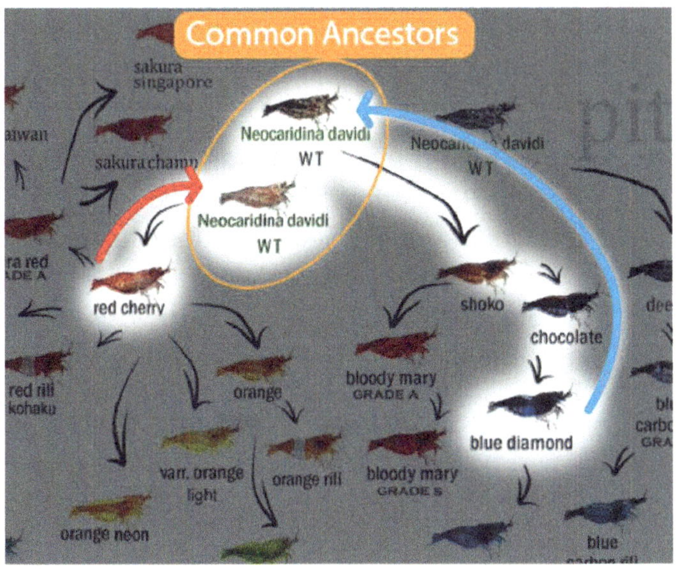

The WT Neocaridina is the **common ancestor** - the lowest common denominator - of every breed. When we mix two different lines in a process known as **crossbreeding**, we reset selective breeding efforts back toward WT shrimp. It's not an immediate change over a single generation though. The first generation of crossbred offspring have very diverse colorations matching any ancestors going back to the common ancestor. For example, if a Red Cherry gets mixed with a Blue Diamond, then the first generation may look like Red Cherry, Blue Diamond, Chocolate, Shoko, or WT. Understanding the basic genetic pathway helps avoid undesirable crossbreeding that may be counterproductive to our goals.

Examples of Selective Breeding Mistakes

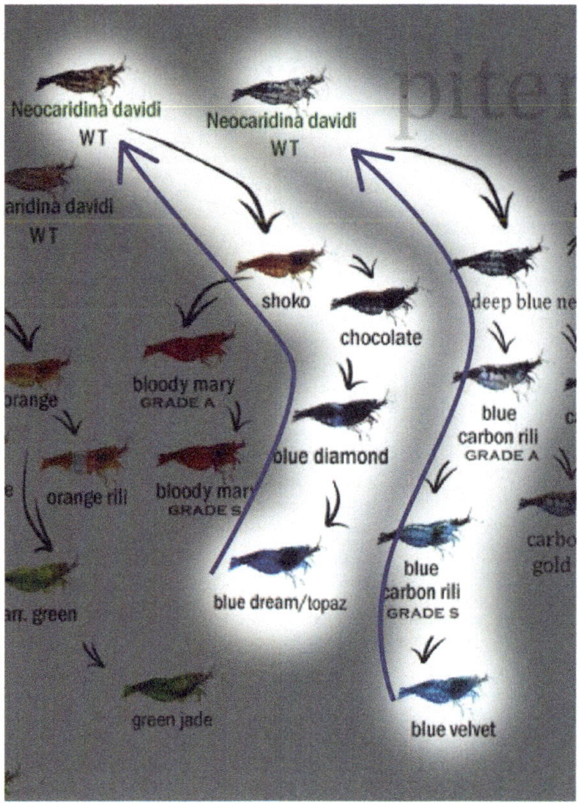

A common mistake is to think that the same color comes from the same line – for example, Blue Dream (BD) and Blue Velvet (BV) shrimp. Without knowing about this family tree, some people mix these while trying to create the bluest shrimp possible and get very disappointed when the offspring show up black (Chocolate), red (Shoko), or WT.

This happens because, if we trace back the ancestry of BDs and BVs, we find their most common ancestor is a *WT Neocaridina*. That means we expect a huge variety of offspring when they breed together—many of which aren't blue. The next generation's offspring now have mixed genes that will produce even more varied offspring, trending toward WT. This is another example of crossbreeding and is the opposite of the selective breeding that created blue Neocaridina in the first place.

Another common attempted combination is trying to breed Bloody Marys (BM) with Fire Red shrimp to make an extra red shrimp. If you remember from earlier in this lesson, BMs have red flesh, whereas Fire Red Cherry shrimp have red shells. The closest common ancestor of these two is a WT Neocaridina so, instead of getting an extra red shrimp, this crossbreeding is mostly likely to result in Shoko and WT offspring.

Please note there are a few different Neocaridina Family Trees available online and they all vary slightly. The original source of some breeds is not clearly documented so there is some disagreement between the tree-makers. With the mystery surrounding different breeds, along with incorrect identification of lines by sellers, it's difficult to predict the exact outcome of crossbreeding. The tree also does not necessarily indicate the most efficient way to get offspring of a certain color.

Why A Colony Gets Less Colorful

While genetic variation is interesting to observe and critical for a healthy colony, it can also lead to less colorful shrimp. This is because all colonies naturally trend toward a brownish WT coloration, which some hobbyists enjoy but others find frustrating. Luckily, we can control this rate of reversion in a variety of ways.

We'll get into what those ways are, but we first want to provide a brief explanation for why this reversion happens. There are three reasons why a colony may trend toward wild type or "impurity":

Reason #1

The accumulation of color pigment requires additional energy (Hill, 1996). It may come at the cost of producing fewer important compounds for immune systems, digestion, and other bodily functions within shrimp. Therefore, shrimp with less pigmentation (WTs) generally require fewer resources and are more fit to survive and reproduce. Without human intervention, the colony slowly becomes less colorful over time.

Reason #2

Shrimp produce offspring that vary in color brightness, with the variability dependent on the quality of the parents – what we call **Type 1** and **Type 2**.

Type 1 has been carefully selected for only the brightest red shrimp for generations, to the point where (nearly) all offspring are high-grade. **Type 2** high-grade shrimp are selected as the best-looking offspring from lower-grade parents (Generation 0) that produce a wide variety of grades. These high-grade shrimp (Generation 1) still have genetics for more variable offspring from Generation 0. This means some of *their* offspring (Generation 2) will be lower grade. The Generation 2 offspring will breed and create a generation (Generation 3) of shrimp that are even more variable. There may still be some higher-grade shrimp but they start to be the exception instead of the rule.

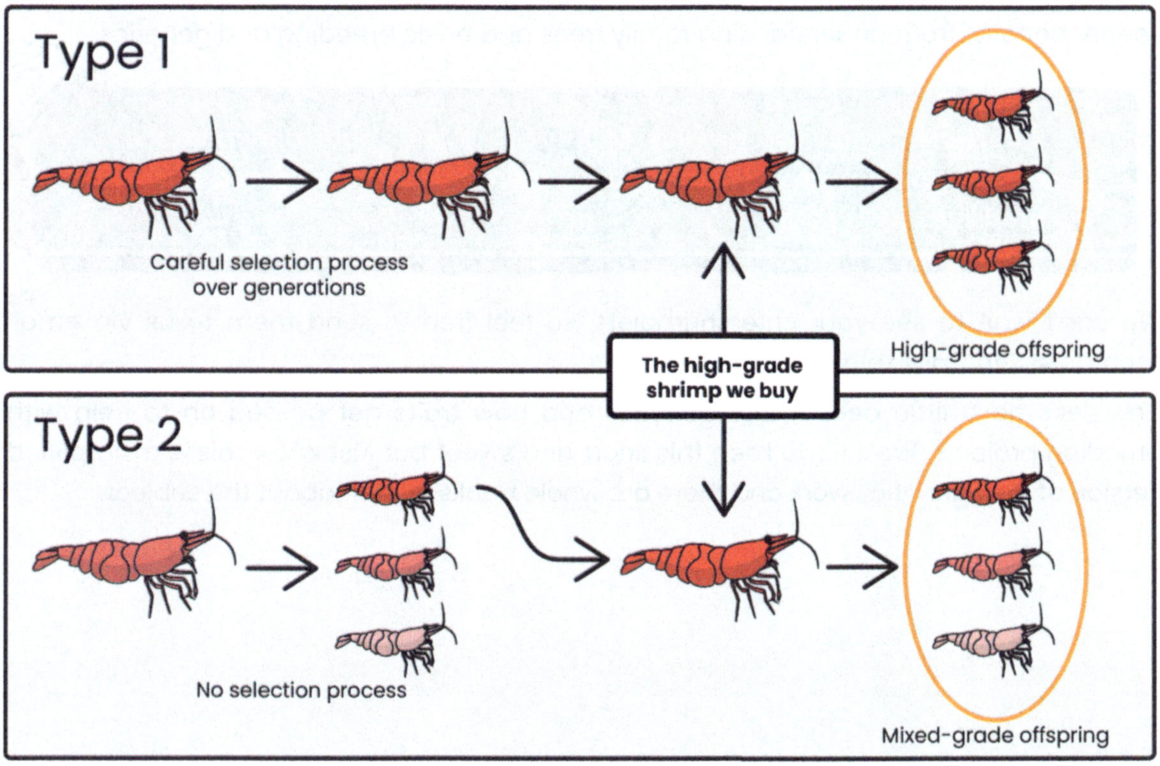

Type 2 high-grade shrimp therefore require more careful breeding and **culling** to maintain a colorful population. Culling is part of the selective breeding process, in which shrimp with

undesired characteristics are separated from the shrimp with the desired characteristics—in this case, low-grade shrimp are separated from high-grade shrimp. Culls are either put into a separate tank, fed to fish, or sold.

Reason #3

The last possible reason a colony trends toward WT is if we mix multiple colors in the same tank. Mixing two different breeds is known as **crossbreeding** and results in more variable offspring, some of which may be WT even in the first generation. The variety of offspring produced depends heavily on how related the breeds are. The Neocaridina Family Tree helps us estimate what the results of crossbreeding might be.

That being said, Neocaridina and Caridina differ somewhat in their reversion to wild-type coloration. Caridina colorations tend to mix "better", maintaining more vivid colorations and interesting patterns even when crossed with very different breeds.

What's Wrong with Wild-Types?

It may seem like we've just been bad-mouthing WTs this whole time and claiming that they are undesirable. For some hobbyists, that is true. Others, however, love the varied colors and stripes associated with WT Neocaridina. If that's you, then use the Neocaridina Family Tree to pick the breeds furthest from each other and see what fascinating offspring occur in the tank (or just buy WTs)! Doing so could result in a cool new pattern or coloration that can be selectively bred to create a pure-bred line of shrimp. Regardless, every shrimp keeper benefits from understanding family trees and basic breeding and genetics.

Side Note: When two parents of the same line mate and produce offspring with a similar color/pattern, then they are considered **pure-bred** lines (or breeds).

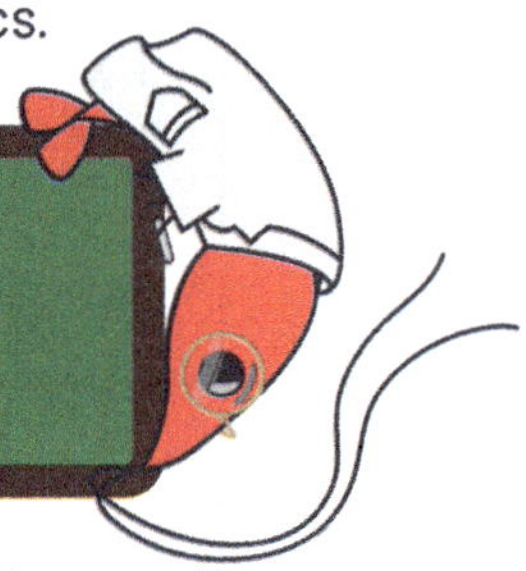

We can't wait to see your cute shrimplets, so feel free to send them to us via email (contact@ShrimplyExplained.com)!

Now, let's go a little deeper into genetics and how traits get passed on to help with breeding projects. We'll try to keep this short and sweet but just know this is a simplified version of how genetics work and there are whole books written about the subject.

The Basics of Genetics

The classic model for understanding genetics is through Gregor Mendel's pea plant experiment (Khan Academy, n.d.). In 1856, Mendel noticed that some pea plants had different characteristics, like some plants were tall while others were short. He found that he could breed tall plants that always produced tall offspring (they were **true-breeding**), or vice versa with short plants. Over eight years, he crossed different pea plants that bred true for a given trait and discovered the basics of how offspring inherit traits from their parents, a concept known as **inheritance**. Here is a brief explanation of his experiment and results:

Mendel had a true-breeding short pea plant and a true-breeding tall pea plant that he bred together, producing offspring that were *all tall*. Parents = P, Offspring = F1 (the first filial generation)

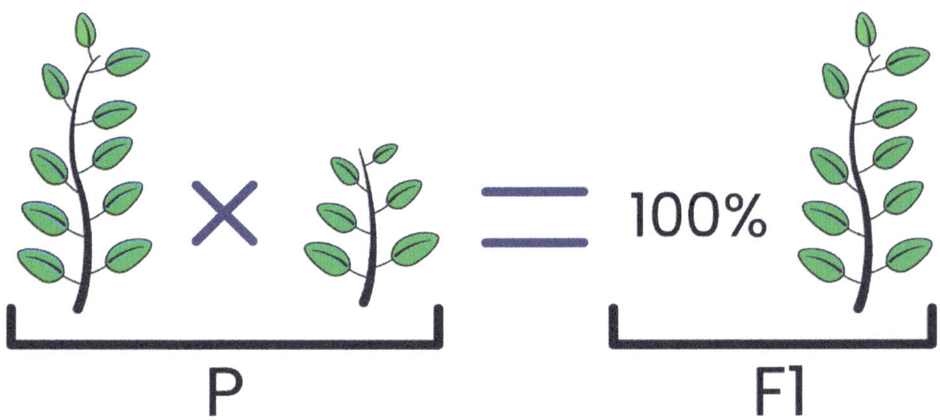

After that, he bred together two F1 plants and found their offspring (F2) were consistently 75% tall and 25% short. To understand why, we have to look at the genes of these plants or, more specifically, the **alleles**.

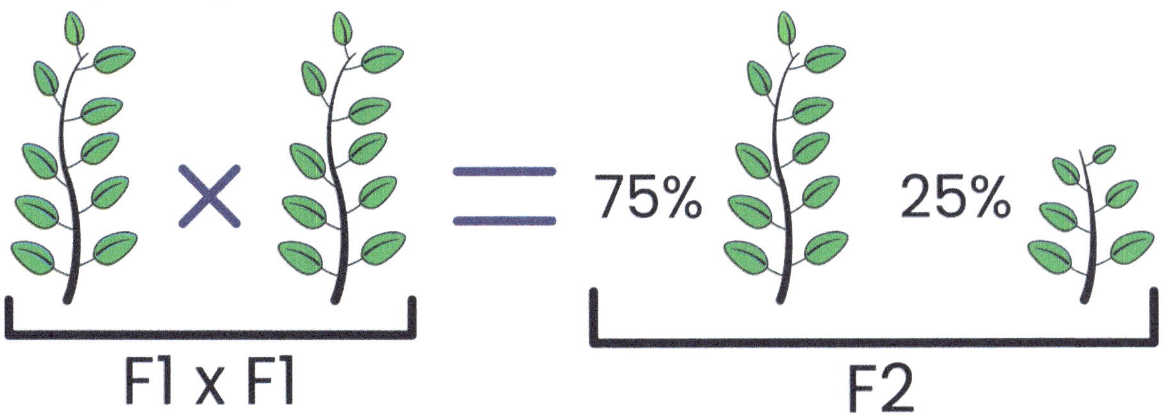

Genes can have one or more "versions", or alleles, in a population of organisms. For example, a gene might encode for a trait like *Height*, while an allele encodes for a specific variation of *Height*, like *tall* or *short*. Different alleles appear due to mutations in the DNA that make up a gene, creating variations in a physical trait like eye color, hair density, or height. In the case of the pea plant, there are two alleles that control height, one of which makes the plant tall (T) and the other makes the plant short (t). Each individual plant has two alleles for the gene (one from each parent), defined as the **genotype.** The **phenotype** is the physical outcome (short vs tall) of the genotype. The following combination of genotypes and phenotypes are possible for *true-breeding* pea plants:

In Mendel's experiment, we learned that crossing a true-breeding short plant with a true-breeding tall plant led to offspring that were all tall (F1), so it seemed like the "short" allele disappeared entirely. Yet somehow a short plant appeared in F2. How did that happen?

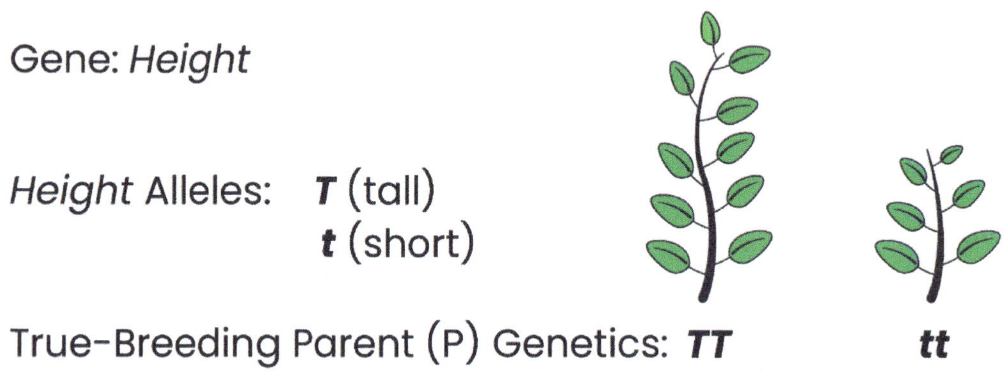

For that, we need to look at a Punnett Square (lower right).

Possible Allele Combinations

A Punnett Square

	T	***t***
T	***TT***	***Tt***
t	***Tt***	***tt***

A Punnett Square is a visual representation of crossbreeding that helps us keep track of how alleles get transferred to offspring. To build one, we start by focusing on one parent, adding a box for each allele the parent has for the trait we're interested in - height, in this case.

That's Step 1.

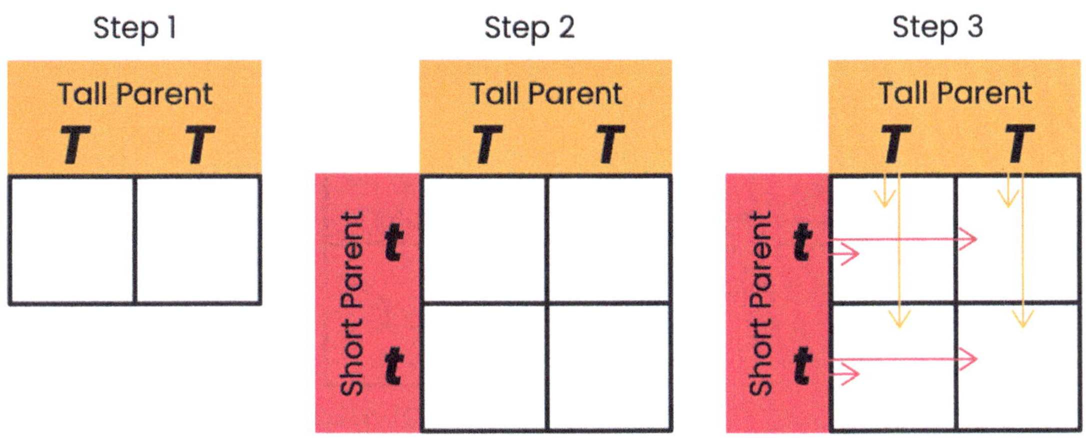

For Step 2, we add a row for each allele from the other parent. Since each plant only has two alleles, this creates a 2x2 grid, with each box representing a potential combination of alleles for offspring. Step 3 is the process of filling in each box with possible combinations of alleles for offspring.

50% of the offspring receive one allele from a parent, while the other 50% get the other. All boxes combined represent 100% of the possible offspring. We can also get the percent chance of each offspring by dividing 100% by the number of boxes - in this case, four boxes. This means offspring have a 25% chance (100%/4) of having the combination of alleles from any given box. For this cross between a true-breeding short and tall plant, all offspring (F1) get the same allele combination (Tt).

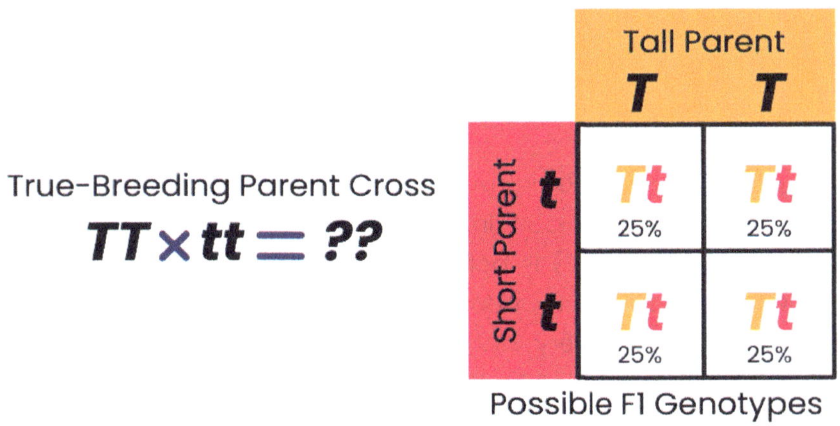

The crossbreeding results of true-breeding parents don't look very exciting – every offspring ends up with the same genotype. It gets more interesting when we breed two plants from the F1 generation. When parents have two different alleles, we see the offspring's (F2) genetics have three different possibilities (TT, 2x Tt, or tt).

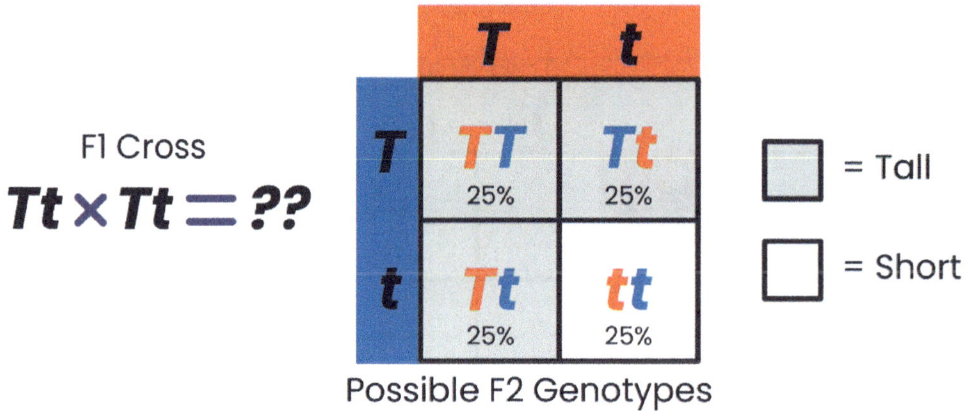

Here, we can see that genotypes with *TT* and *Tt* always produce tall plants, while *tt* is the only combination that produces small plants. Why is that?

Well, this comes down to whether an allele is **dominant** or not. A dominant allele is always indicated by a capital letter (T), while a **recessive** allele is always lowercase (t). **Dominant alleles override the effect of other alleles so only the dominant trait is visible in offspring with at least one copy of the dominant allele**.

To be clear, dominance is the simplest way that alleles interact. There are some that are co-dominant (both traits are partially or fully expressed), there can be more than two alleles per gene, and many physical traits are controlled by more than one gene, leading to complex interactions between different combinations of genes, alleles, and phenotypes.

Applying Punnett Squares to Shrimp

Now, how does all this play out in shrimp? Let's look at the Orange-Eye (OE) trait shared by both Neocaridina and Caridina.

Normal shrimp eyes have central black pupils with surrounding colors ranging from steel-grey to black. On rare occasions, a shrimp with completely orange eyes (OE) is born. When a male and female shrimp that both have OEs are bred together, every single one of their offspring also has OEs. On the other hand, when an OE shrimp breeds with a shrimp that does not carry an OE allele, none of the babies have OEs. This demonstrates the OE trait is recessive and controlled by a single gene. The Punnett Square for the cross looks like the diagram to the right, with E representing the dominant allele (normal eyes) and e representing the recessive OE allele:

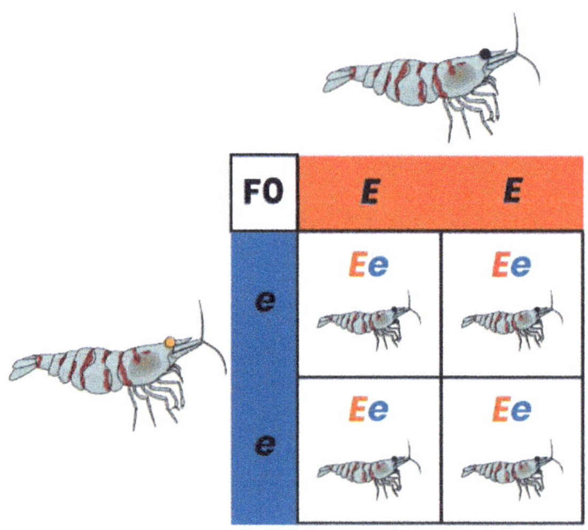

Green Jade Shrimp with normal dark eyes - Credit: Terese Forninge, Sweden

And the Punnett Square looks like the following for a cross between two offspring from the second generation (F1):

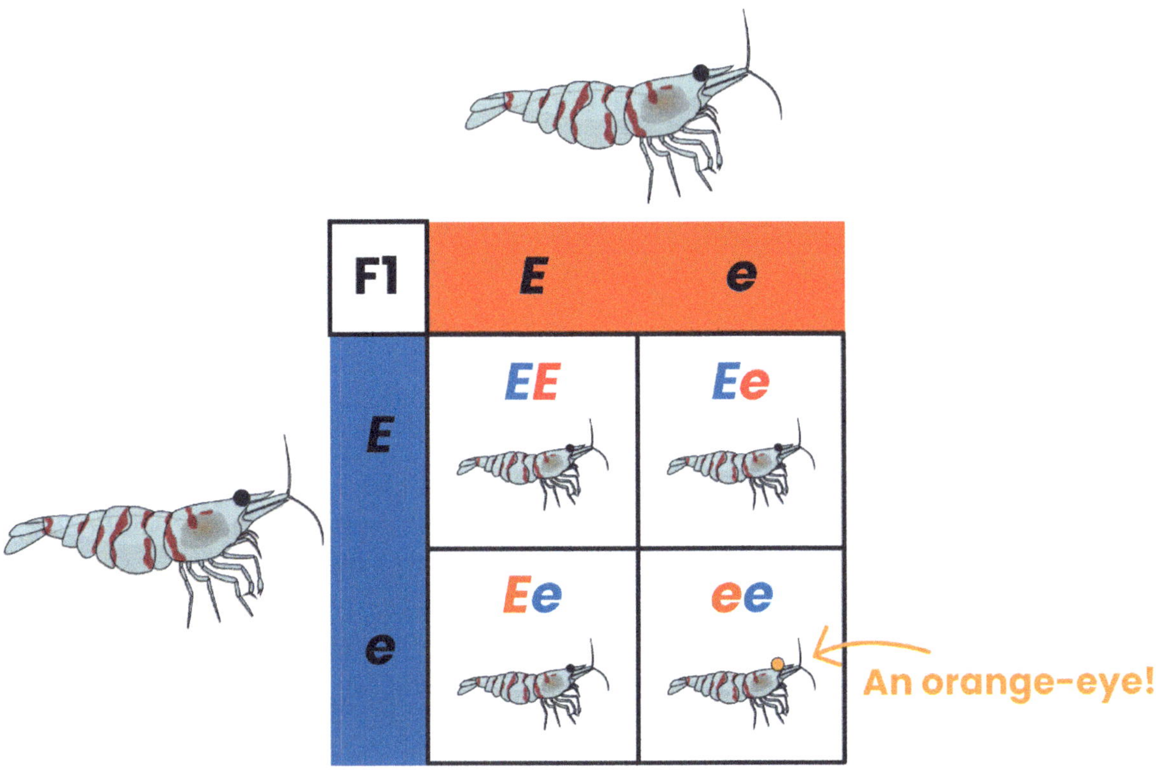

Now, approximately 25% of the F1 offspring (F2 generation) should receive the recessive allele from both parents and have OEs! The F2 shrimp with OEs may then be selectively bred to produce F3 offspring that all have OEs.

Again, many traits are not as simple as OEs since a trait may be controlled by more than one gene. Understanding the basics of how genetics work through Punnett Squares is still useful for visualizing how offspring inherit traits and may help immensely during a breeding project.

OE Green Jade - Credit: Terese Forninge, Sweden

Important Breeding Terms

1. **Crossbreeding**, or **outbreeding**, is the process of breeding two animals from different genetic backgrounds, like a Cherry shrimp with a Blue Velvet shrimp.
2. **Inbreeding** is the process of breeding animals that are closely related. (Yes, it's a weird concept but it's what human food supplies and many pets [dogs, shrimp, etc.] are entirely based around. There are ethical concerns about the practice that deserve their own book.)
3. **Backcrossing** is the process of breeding an animal from one generation with an animal from a previous generation of the same line, like breeding an F3 offspring with an F1 offspring.
4. **Line breeding** is a particular type of inbreeding in which one animal is bred with a related animal that is not a direct relative, like breeding cousins or half-siblings together to maintain greater genetic diversity and reduce inbreeding problems.

Coloration Genetics

In this section, we're delving a bit deeper into how pigments get collected and stored in a shrimp's body (The Pigment Pathway) because this information helps identify what color combinations we can expect when breeding. It also explains why we can't mix blue and yellow shrimp and get green offspring. In addition, we can get hints about what color combinations may be possible in the future. Take a look at the Neocaridina Family Tree below to see how every breed is related.

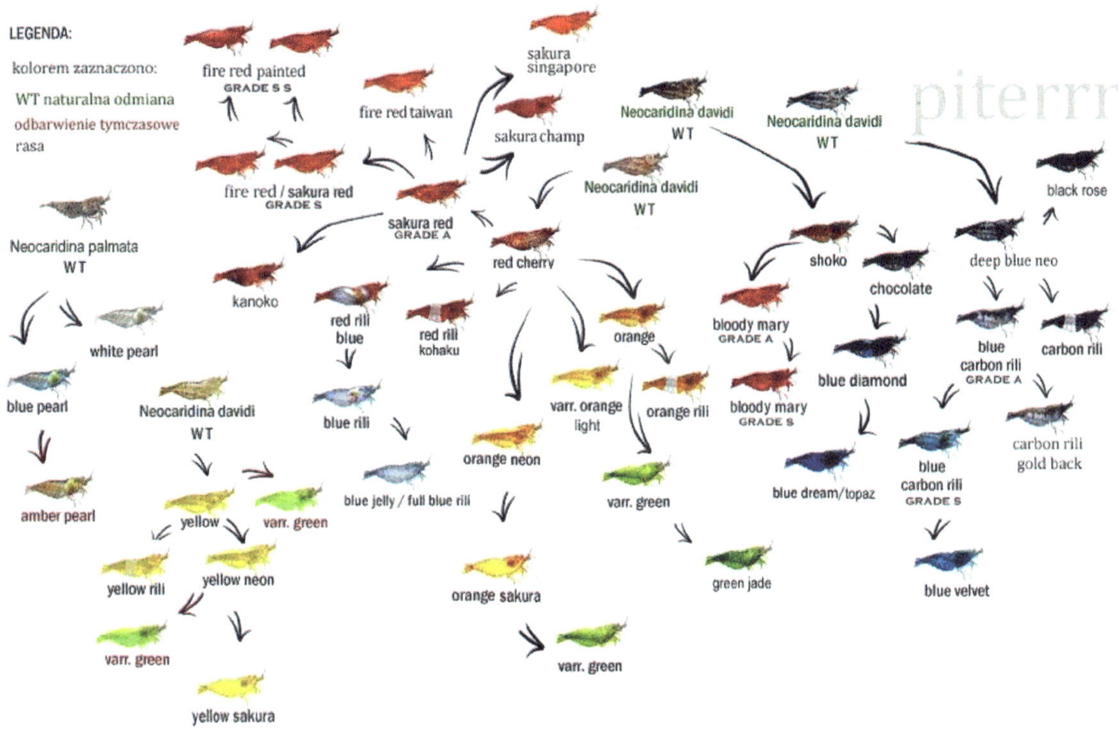

According to the family tree (which may not be completely accurate due to secrecy in the hobby), the first cherry shrimp was bred from a reddish WT over many generations, so breeders slowly selected for the reddest shell over many generations of offspring.

Other breeds, like Bloody Mary shrimp (reddish flesh) are believed to have appeared spontaneously when a mutation affected the way pigments were stored in the body, moving them from the shell to the flesh. This may be a recessive gene that appeared and was then selected for by breeders.

Other spontaneous mutations also happen, causing uniquely colored offspring to suddenly appear - like when orange shrimp randomly have one or more green offspring. Instead of being a slow transition over time, green offspring may pop up in orange populations occasionally. Breeders then took those green offspring, bred them, and created a semi-stable line of green shrimp that produced mostly green offspring. This hints at how green and orange are related. It also explains why we may breed an orange line for many generations and never see any green shrimp. If orange to green was a once in a lifetime genetic mutation, then it may never be replicated by other breeders.

Because green pigments appear spontaneously and do not produce some sort of mixed orange/green shrimp, it may just be a single mutation involving proteins or enzymes that control structural change to a pigment, modifying every pigment in the body from orange to green. The exact mechanism likely won't ever be known (there's no money to be made from research) but **understanding basic genetics is useful because it gives us an idea of what combinations of colors are possible**.

The Pigment Pathway

This is a simplified and theoretical explanation for how pigmentation works in a shrimp body that is useful for understanding the basics of genetics and breeding.

Since shrimp cannot produce their own pigments (Moise, Bandera, & Lintig, 2022), they need to get them from their food and then store it in their bodies, where the pigments may then be modified into different colors (Huang, Zhang, Wang, & Huang, 2022). This process of retrieval, storage, and modification is what we call the Pigment Pathway. Understanding this pathway helps us grasp what happens conceptually when, say, an orange shrimp produces a green baby, which may help with breeding projects. It also explains why mixing different breeds of Neocaridina leads to wild-type coloration and why mixing black crystal caridina with red crystal caridina leads to entirely black crystal offspring.

When a random color appears in a shrimp population, what may happen is a **loss-of-function** of certain genes involved in the Pigment Pathway. Different breeds of shrimp lose separate parts of the pathway, leading them to accumulate, store, or modify pigments in varying ways. When two shrimp from different lines breed (i.e. **Blue Velvet** with a **Red Cherry** shrimp), one often has a functional gene for the part of the pathway the other is missing, leading to their offspring regaining fully functional pathways for all colors and subsequently growing up with WT coloration.

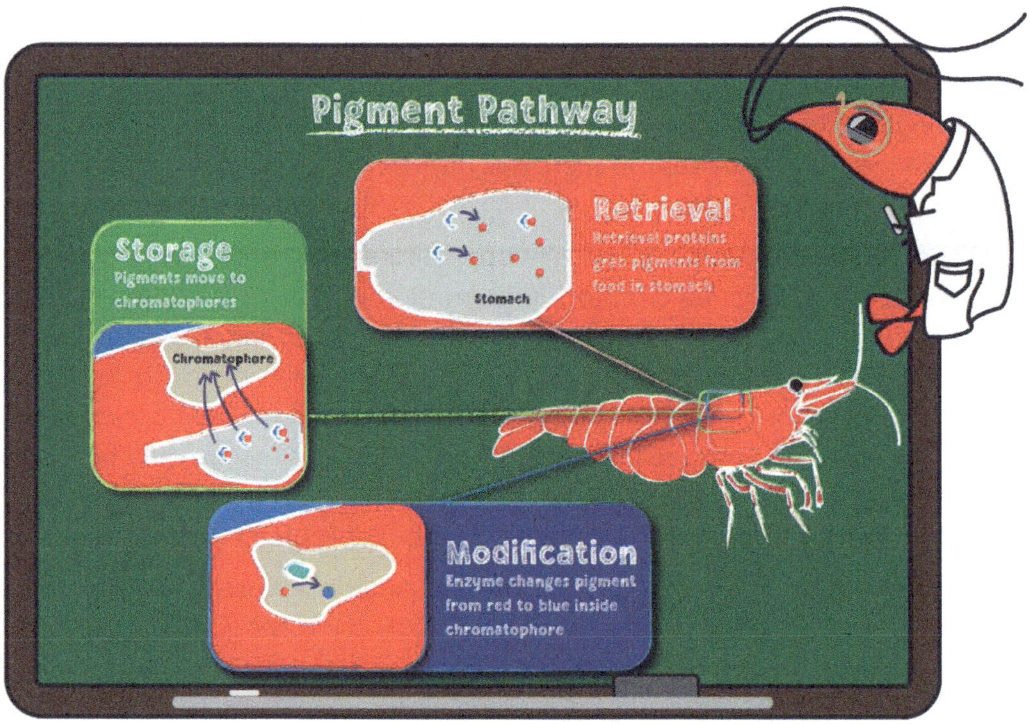

The Three Steps of The Pigment Pathway

To simplify things, let's imagine there are three genes involved in pigment accumulation (there are likely *many* more genes and a much more complex process involved). The first is the **Retrieval Gene**, which encodes for a protein that attaches to pigments in the shrimp's intestines and pulls them into the bloodstream. The second gene in the pathway is the **Storage Gene**, which encodes for a protein that takes the pigment from the Retrieval Gene and moves it to the shell for storage. Finally, the **Modification Gene** encodes for a protein that changes a red pigment to blue or green in certain storage sites.

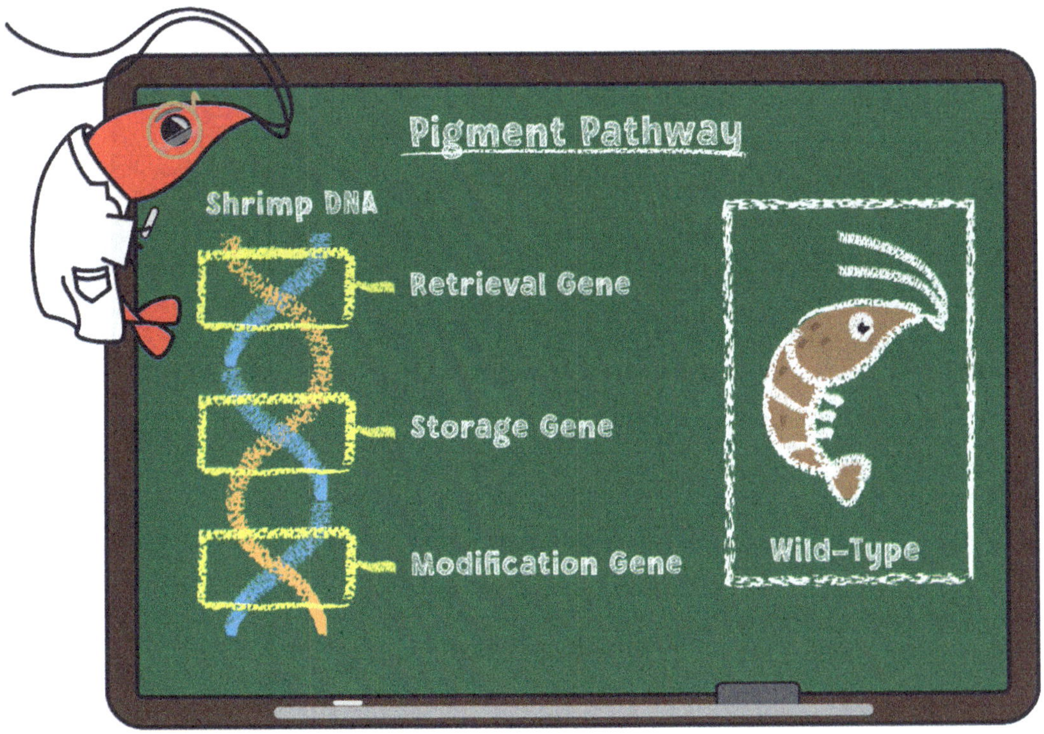

A Crossbreeding Experiment

For a better understanding of The Pigment Pathway, let's look at a common crossbreeding experiment that people try to do: combine Red Cherry shrimp and Bloody Mary shrimp to get ULTRA-RED SHRIMP!!!!

A cherry shrimp (red shell) may have functional Retrieval and Storage Genes but have **non-functional Modification Genes**, leading to entirely red pigment (see image below). In contrast, a Bloody Mary shrimp may have fully functional Retrieval and Modification Genes, but **non-functional Storage Genes**, so pigments never get to the storage sites and therefore can't be modified. Since different parts of the Pigment Pathway are disrupted in the parents, crossbreeding them together results in many offspring with *fully functional* Pigment Pathways, leading to wild-type coloration even though both parents are red.

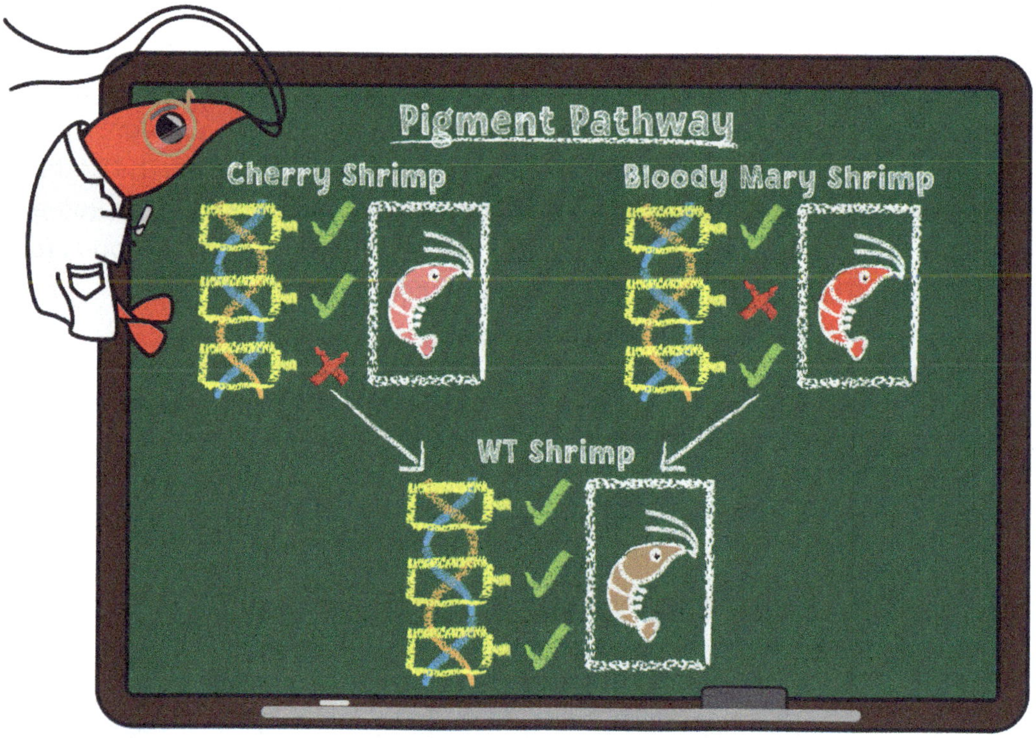

Now, remember that each shrimp has two copies (alleles) of a gene. If the Bloody Mary shrimp has two functional (dominant) Modification alleles, while the Cherry shrimp has non-functional (recessive) Modification alleles, then all offspring get at least one functional Modification allele. Likewise, the cherry shrimp parent has two fully functional Storage alleles and donates one to the offspring and *voila* - the full Pigment Pathway is restored in the offspring, leading to wild-type coloration.

From a genetic perspective, this is why colonies of mixed-color shrimp trend toward wild type coloration. **Different colors have separate disruptions in the Pigment Pathway, so the genetics of each parent "heal" the Pathway disruptions present in the other parent, leading to offspring with fully functional Pigment Pathways**. To be clear, real genetics and Pigment Pathways are complicated, so offspring typically show a variety of colorations.

As another example, green only seems to appear in Neocaridina when a part of the orange Pigment Pathway is fully disrupted, so we will likely never see a Neocaridina shrimp with a mix of orange and green on its shell. You're welcome to try breeding one but the current breeds of shrimp do not seem to have the genetics for it. In contrast, we're still holding out hope for a purple shrimp because there have been offspring that express both red and blue pigment at the same time – a stable breeding line just hasn't been formed as of the writing of this book.

Again, this is a very simplified view of the Pigment Pathway. In reality, MANY genes are involved, and they have varying levels of dominance. There has been some research involving the difference in gene expression in shrimp of different colors but the findings

(arguably) are so specific that they are not useful for hobbyists or breeders. If you want to become a shrimp researcher or love genetics, then they may be interesting so you can check out the following papers:

Additional Genetics Papers

1. Huang, Yongyu, et al. "De novo assembly transcriptome analysis reveals the genes associated with body color formation in the freshwater ornamental shrimps Neocaridina denticulate sinensis." *Gene* 806 (2022): 145929.
2. Lu, Xiqin, et al. "Functional analysis of ABCG2 gene in pigment transport of Neocaridina denticulata sinensis." *Gene* 844 (2022): 146810.
3. Huang, Chang-Wen, et al. "Identification of functional SSR markers in freshwater ornamental shrimps Neocaridina denticulata using transcriptome sequencing." *Marine Biotechnology* 22 (2020): 772-785.

Known Genetic Traits in Shrimp

In general, different lines of Neocaridina seem to have mutations at different points in the Pigment Pathway, so crossbreeding most commonly leads to wild-type shrimp or a reversion to the color of their most common ancestor. In contrast, crossbreeding Caridina shrimp tends to lead to offspring with a combination of the parents' colors. Most Caridina seem to retain their ability to retrieve and modify different pigments, but what changes is if and where these pigments get stored.

Black vs Red Pigment Dominance in Caridina

For example, if we cross a tangerine tiger (*Caridina serrata*) with a crystal red shrimp (*Caridina cantonensis*), then we tend to get offspring with a yellow/orange body and a shell that retains some of the tiger stripes in combination with spots of black and white. The storage locations effectively blend while the colors are retained.

Offspring of Crystal Red (Caridina cantonensis) *x Tangerine Tiger* (Caridina serrata).

You'll notice there's no red in the offspring though. This is because red coloration is a recessive trait (in Caridina) affecting how pigments are modified. Crystal red shrimp have a disruption in their Modification Genes that prevent pigment from turning black. In contrast, tangerine tigers have functional Modification Genes so the offspring of this cross have a fully functional Pigment Pathway, leading to a lack of red pigment.

**Again, the exact genetic mechanisms are not known for shrimp Pigment Pathways. This explanation is useful for understanding the results of crossbreeding but may not use the correct biological mechanisms.*

Unnecessary But Interesting Science

As a quick example of how a mutation might affect the Pigment Pathway, let's think about the Retrieval Gene. In a shrimp with at least one normal (dominant) Retrieval allele, the allele encodes for a Retrieval protein with a **binding site** shaped like a square, which helps it retrieve square-shaped red pigments and pull them into the bloodstream. If the allele gets mutated, it may instead add an extra oxygen or nitrogen to the Retrieval Protein, which changes the binding site to a triangular one (see image below). Even a slight change in the building blocks of a molecule can drastically change its shape. As a result, more triangular (blue) pigments are Retrieved and brought through the Pigment Pathway, leading to a change in coloration. See the picture on the following page for a visual explanation.

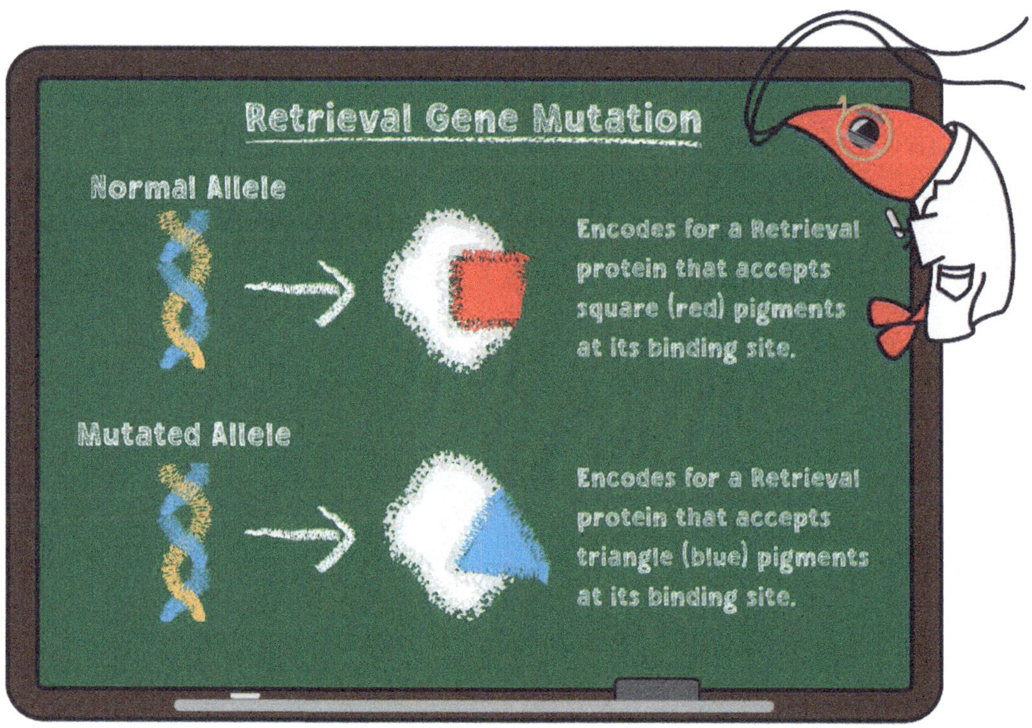

Back Stripes

Both Neocaridina and Caridina can have a lighter strip of colored shell along their back called a **back stripe** or **racing stripe**. The pattern itself seems to be either on or off (possibly a dominant trait) and the exact intensity/shape of the stripe can vary, meaning a shrimp may have the tiniest line of lighter color on their back or it can expand to cover most of the shrimp. Some breeds of Neocaridina have very stable back stripes - like Yellow Goldenback Shrimp - while other breeds have inconsistent back stripes. Bloody Mary shrimp, on the other hand, never have a back stripe. Their original genetic line didn't carry the gene for it, so that's one way to tell the difference between high-grade Cherry Shrimp and Bloody Mary Shrimp.

The back stripe is generally not a trait focused on by Neocaridina breeders (aside from Yellow Goldenbacks) so, if we're looking to breed new types of Neocaridina, selecting for certain back stripe colors/patterns may be an excellent option.

As for Caridina, the back stripe is a staple of many breeds. The "fishbone" pattern seen in some Caridina breeds is a pattern that has been selectively bred for, identified by a solid back stripe and uniformly shaped white tendrils extending down the shrimp's abdomen. Some Caridina breeds may have back stripes from *C. serrata* tiger shrimp, or the pattern may be a result of careful selection from a pinto pattern.

Caridina with "pinto" spotted pattern

Fishbone Back Stripe

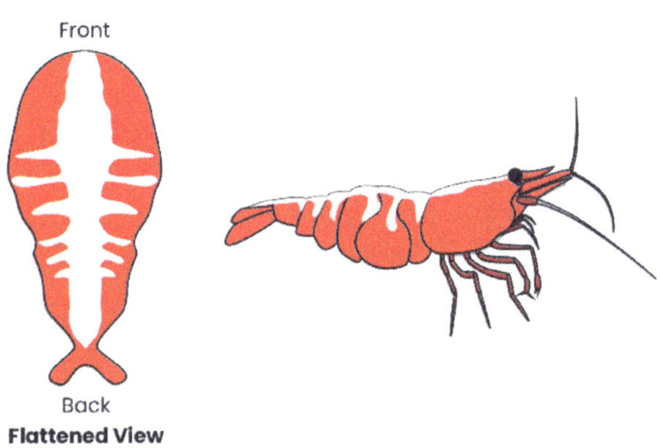

Fishbone Pattern

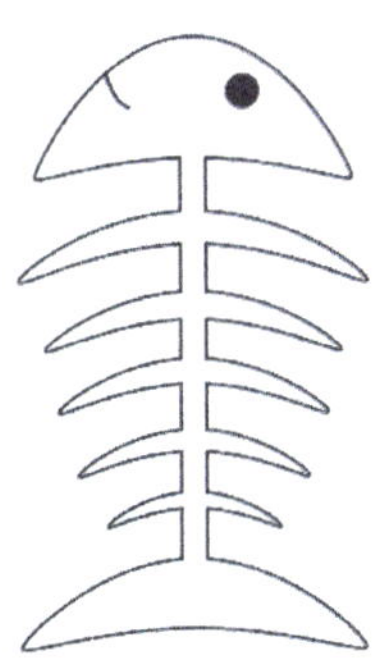

Figure 1. Red BOA Caridina with fishbone pattern.

Orange Eyes

While many shrimp eyes appear to be black, there's actually quite a lot of variation in their eye color, such as hollow-eye or metallic-eye. One of the most striking variations is an eye that isn't black at all and is instead a very light orange color. There are a number of shrimp lines with this trait, such as orange-eye blue tigers (OEBT, Caridina), orange-eye black devils (Caridina), or orange-eye green jades (Neocaridina).

Orange-eye Green Jade - Credit: Terese Forninge, Sweden

The orange-eye trait pops up very rarely but, once you manage to get a male and female with the trait, every one of their offspring have it. This demonstrates that orange eyes are a recessive trait controlled by a single gene which makes breeding them quite simple IF it appears in a population.

Summary

Orange-eyes, back stripes, and red pigment are the traits we're aware of that appear to be controlled by a single gene. Many other color patterns and traits are controlled by multiple genes, the blending of which makes selective breeding a science and an art. Shrimp breeding is an incredibly fun and hobby full of surprises, but there are some issues caused by selective breeding that we need to be aware of.

Inbreeding Issues

If you've ever owned a large dog like a Golden Retriever, then you've likely heard of hip dysplasia. This is a condition where the hips don't form properly, leading to hip joints that dislocate too easily as the dog ages. It affects anywhere from 50-70% of Golden Retrievers (Paster, LaFond, Biery, & et al., 2005) because one of the original golden retriever ancestors carried the genes for hip dysplasia. It likely didn't show the condition, but it passed on these genes to its offspring, some of which did have hip dysplasia. The breeders may not have known about the condition until the offspring showed signs of problems years later – after they had already produced more offspring and spread the trait to other golden retrievers.

This is a common story of what happens when animals that are closely related – like the original Golden Retrievers – are bred together. The selective breeding process leads to desirable purebreds but also causes inbreeding. Inbreeding causes all offspring to have extremely similar DNA, which reduces the genetic diversity in all future offspring. Here's how that plays out when an infection happens:

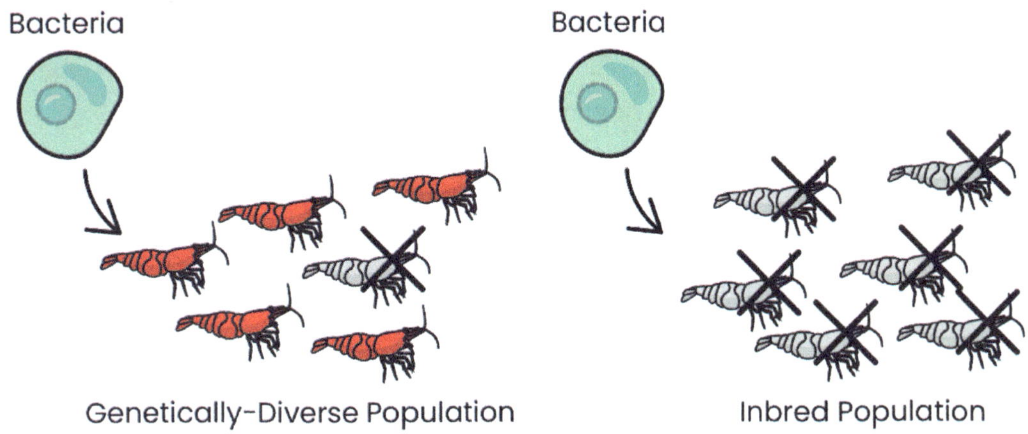

In a normal population with good genetic diversity, each individual has a unique immune system with different vulnerabilities, leading to differences in survival. When many individuals are infected by a bacterial infection, vulnerable individuals may die but others survive because they have greater resistance to that infection.

Inbred populations don't have the same genetic diversity, resulting in very similar immune systems. If the right bacteria comes along that takes advantage of a vulnerability, then it can wipe out EVERY individual.

Side note: This is why "critically endangered" species are still vulnerable even if their population bounces back. When the population becomes too small, then inbreeding can't be helped. Subsequent low genetic diversity leads to poor immune function and the risk of being wiped out by an infection.

The chance that an individual has at least one problematic recessive gene hidden in their DNA is very high. When individuals breed with others that share extremely similar DNA, then offspring are much more likely to get two pairs of a problematic recessive gene, leading to health issues.

Similar problems happen in shrimp, although most of the widely available shrimp breeds have been carefully selected for so many generations that breeders have managed to remove visible problematic traits through a process known as **purifying selection** (This is more difficult to do with dogs because they have a much longer generation time). Regardless, knowing what deformations to look out for can help keep a colony healthy. We've compiled a list of common ones below:

Most Common and Noticeable Inbreeding Issues:

- Small size
- Large head-to-body ratio
- "Short skirt" AKA "Open skirt"
- Decreased lifespan
- Incorrect shape
- Low shrimplet survival rate
- Females who have trouble holding eggs

These problems are common during breeding projects but are relatively rare when purchasing shrimp from established lines. When starting out in the shrimp keeping hobby, it's safe to assume any lifespan issues of adults and offspring, along with any dropped eggs, are likely a sign that their environment has room for improvement. It also takes time to get familiar with shrimp shape, lifespan, offspring survival rate, etc., before we can identify when these health problems appear in a population. It's also difficult to determine if issues like size or offspring survival are caused by tank issues rather than genetic issues.

When To Add New Shrimp

Genetic diversity is incredibly important to maintain a stable population in the wild. It ensures a population has individuals with diverse immune systems to withstand the wide array of diseases faced in nature. Diverse genetics increases the chance that at least some individuals survive a pandemic and re-establish a population. When populations become inbred, they can experience a decline in survival and breeding, known as an **inbreeding depression**. That sounds pretty bad - Do we need to add new genetics to our tank every few years or every few generations to ensure a healthy population?

Unfortunately, there's not much research on shrimp genetics to help determine when it's best to add new genetics, but we can determine when there is a higher risk of genetic issues based on how "fit" a population is. "**Fitness**" is the term scientists use when discussing how likely a population is to survive environmental stress. A more "fit" populations is more resilient to stress and less likely to go extinct. The fitness of a population depends on a number of factors, including but not limited to:

Initial population size

A smaller starting population limits all possible future genetic diversity and decreases the adaptability of future generations. Starting with 2-5 individuals decreases the genetic diversity of a colony so they are less likely to have a healthy long-term population. Starting with at least 10 individuals increases the possibility of genetic diversity and increases the chance of a healthy colony. A larger starting population is always better.

Current population size

The number of shrimp in a colony has a large impact on how fit the colony is. One study on house fly populations found that extinction rate was closely linked to the total population, with smaller populations (<50 individuals) more likely to go extinct sooner than larger populations (>500 individuals) (Reed & Bryant, 2000). The study found that some small populations of 50 individual flies started going extinct after 24 generations due to a lack of genetic diversity. Average generations for population extinction were estimated to be equivalent to the number of individuals, so a population of 50 is expected to survive 50 generations, a population of 500 to 500 generations, and so on.

A more recent study supports the fly study, indicating that a population of 70 individuals is needed to avoid severe inbreeding depression in the short term (Caballero, Bravo, I., &

Wang, 2017). It's difficult to extrapolate the data to shrimp but we can say that **a larger colony of shrimp is less likely to need new genetics to be healthy than a smaller colony**.

What genetics are being added?

The effectiveness of adding new genetics to improve population health depends on what is being added. Are you adding more high-grade Cherry shrimp from the same breeder? If so, that probably won't significantly increase population fitness. Are you adding a blue dream Neocaridina to a red cherry shrimp population? Then that is very likely to improve fitness due to genetic diversity between the lines. What about adding high-grade cherry shrimp from a different breeder than the original high-grade Cherry line? That has a debatable impact because the colonies likely came from the same original line, and all shrimp from this line most likely lack genetic diversity. It *might* help fitness, but the additional money and disease risk associated with adding new shrimp from the same line is such that any genetic benefits may not be worth the costs.

How long has the breeding line been established?

The worst inbreeding depression happens at the start of a new line when parents or siblings are bred together, at which point many unhealthy traits (recessive alleles) pop up. Breeding lines that have been established for a long time have been selectively bred to remove unhealthy traits through a process known as **purifying selection** (Loewe, 2008). For example, a cherry shrimp is much less likely to have genetic issues than a newer line like Green Jade shrimp.

What are our priorities?

Every line of shrimp needs to be at least healthy enough to survive, reproduce, and grow a population. Beyond that, it's up to the keeper to determine their goals. If the goal is to have the hardiest shrimp with the best chance of survival, then adding new genetics from an entirely different line of shrimp (Red Cherry Neocaridina x White Pearl Neocaridina) is likely to create the fittest population. That population is also much more likely to look like wild type shrimp, so this trade-off isn't worthwhile for many shrimp keepers.

Summary

In short, if the goal is to breed a new line of shrimp from a very small gene pool OR if we have a newer line of shrimp (like BOA Caridina), then inbreeding depression is more likely so there are significant benefits to adding new genetics every few generations. Determining when to do so is more of an art than a science.

If the goal is to maintain a single-color colony of shrimp from a well-established line, then adding new shrimp (of the same line and color grade) to an established colony is of questionable benefit and may actually be detrimental due to cost and disease risk. Adding slightly lower-grade shrimp to a colony is more likely to improve genetic diversity and overall fitness than adding shrimp of the same color grade, if we're willing to do some culling to remove any low-grade offspring that are produced.

Low grade Blue BOA Caridina

Breed Bigger Shrimp

Something that surprises a lot of shrimp keepers is that the first batch of shrimp we get often grows the largest, but subsequent batches of shrimp may be much smaller.

That doesn't make much sense, does it? New generations of shrimp never deal with the stress of being sold and shipped across the country; they often don't have to deal with the early mistakes we make with our first shrimp; they have every advantage, so why would they be smaller?

Well, there are at least two science-backed reasons for this effect, and they have to do with population density and reproduction. Hobbyists of all kinds want larger shrimp, but this section is especially important for shrimp breeders and sellers. **Larger shrimp sell better, get graded higher in shows, and reproduce more because larger females carry more eggs**. Let's take a look the research on growing larger shrimp.

How Population Density Affects Growth Rate

The first experiment we'll look at tested the **effect of population density on growth rate and survival** (Vazquez, Delevati-Colpo, Sganga, & Lopez-Greco, 2017). For reference, population density refers to the number of shrimp in a given volume of water, so if we put 10 shrimp into a 2-gallon (7.5 L) tank then we have a population density of 5 shrimp per gallon (4L). This experiment compared three different densities: 2.5, 5, and 10 juvenile

shrimp per liter, which equates to approximately 10, 20, and 40 shrimp per gallon. The experiment lasted for 90 days and weights were measured every 30 days. During this time, all tanks received the same maintenance. Each tank was fed a percentage of the inhabitants' body weight, so food availability was not the issue. Take a look at the results below:

	2.5 shrimp/L	**5 shrimp/L**	**10 shrimp/L**
Mean Weight (Females)	92.40 g	78.47 g	63.68 g

As we can see, there's a large and *statistically significant* difference between the shrimp kept at 2.5 vs 10 shrimp/L after 90 days of growth. The shrimp kept at 2.5 shrimp/L grew approximately 45% faster compared to shrimp kept at 10 shrimp/L. By combining data from another study on offspring (larvae) production of female shrimp (Nur & Christianus, 2013), it is estimated the females kept at 10 shrimp/L (63.68 mg avg. weight) only hatch 20 or fewer offspring, whereas females kept at 2.5 shrimp/L (92.40 mg avg. weight) hatch 25-30 offspring due to differences in body size.

Takeaways

This data shows that shrimp kept at higher densities have slower growth, which in turn affects when they are sold and leads to fewer babies. Maybe someone wants that, but we want MORE SHRIIIIIIIIIIMP!!!!! More shrimp = better person, as shown by the science (Biggums, 2024).

The data also provides support for the use of **grow-out tanks** in breeding setups. Breeders move juvenile shrimp (1-2 months old) from breeding tanks into grow-out tanks so the juvenile shrimp grow up in an environment with a lower population density. This speeds up growth, which in turn lets their colors mature so high-grade shrimp can be added to the breeding population and lower grades can be sold. Any females that breeders choose to add to the breeding population are larger when they reach sexual maturity, so they produce more offspring.

How Reproduction Affects Shrimp Growth

The second paper we'll look at studied how the biological demands of reproduction affected growth rates of female *Neocaridina davidi* (Tropea & Lopez-Greco, 2015). In theory, egg production and care require energy and resources from female shrimp, which may impact their growth rate. Here's how they designed the experiment:

Offspring were collected from a batch of adults and separated into all-male or all-female tanks as soon as they could be sexed. Afterward, twenty replicates of the experimental group were created by placing one virgin "reproductive" female into a tank with two males. The control group had twenty replicates of tanks with one virgin "non-reproductive" female and two smaller females. The two smaller females were placed in the tank to mimic the size and population density of the experimental group, creating a reasonable control group. Each group was observed over a 210-day period and tested for initial & final values of their weight, along with other health factors like lipid and protein concentrations.

Blue Neocaridina davidi - *Credit: c.tiensivu@yahoo.com*

After 210 days, the study found a statistically significant difference between the two groups, with non-reproductive females weighing approximately 17% more than reproductive females kept with males (165 mg vs 141 mg, respectively). The exact number of times each reproductive female reproduced was not provided for these groups (the same study tracked the number of clutches in a separate experimental group, which is interesting but not applicable to this chapter). Molting frequency was tracked throughout the experiment, demonstrating that both groups had the same weight gain per molt, but reproductive females molted less frequently, likely due to longer inter-molt periods while carrying eggs.

Side Note: If you look at the study yourself, you'll see a table that appears to say the exact opposite of what we say: that *reproductive* females were the ones at 165 mg while non-reproductive females weighed 141 mg. We believe this is a labelling error on the table because there are repeated references in the Results and Discussion text mentioning how much larger non-reproductive females were.

An additional study supports the theory that females raised in single-sex tanks grow faster than those raised around males (Tropea & Greco, 2024). At first maturity (~65 days since birth – when ovaries matured in this study), females kept in tanks without males weighed an average of 95mg, while those raised with males weighed an average of 69mg – a difference of around 46%. After approximately 180 days, the females kept in single-sex tanks weighed an average of 140mg, while those kept with males weighed an average of 88mg – **a difference of 66%**. The study observed that females kept in single-sex tanks absorbed their eggs over a period of 2-4 weeks after their ovaries matured, letting the shrimp reuse that energy for growth instead of rearing offspring. Fascinating stuff!

That being said, one confounding factor in this study is that population density was not controlled for. The single-sex aquariums had females at approximately 6 shrimp per gallon (SPG), while the mixed-sex tanks had shrimp at 12 SPG. This almost certainly had an impact on the results but these results, combined with those of the other study, still suggest that keeping females in single-sex tanks has an effect on growth.

Takeaways

This data suggests that keeping females separated from males leads to the highest growth rate, since they can molt more often without the burden of carrying eggs. **If the goal is to breed the largest shrimp for a competition, then it may be best to separate high-grade females into their own grow-out tank at least a few months before the competition**. This gives them time to grow at least 15% larger and possibly focus more energy on shell pigmentation, as well. There are a few questions to consider though:

1. **Does a higher weight lead to a larger size, as opposed to the same size but higher body density?**
   Molting occurs when a shrimp outgrows its shell, so molting should result in a size increase. The non-reproductive females weighed more and molted more often, so we believe conflating weight with overall size is reasonable.
2. **Do reproductive females eventually reach the same size as non-reproductive females, given a longer period of time?**
   It is possible non-reproductive females may reach their maximum size earlier, leading to a plateau in growth. This may give reproductive females a chance to catch up in size later on. We could not find data to answer this question one way or the other.

How Temperature Affects Shrimp Growth

As mentioned in the Parameters chapter, temperature has a large effect on growth rate. It's a bit more complicated than *higher temperatures always lead to faster growth*, despite what we've said both in this book and various other resources. Let's take a look at the research.

Tropea and her team ran a study in 2015 studying the effect of temperature on juvenile growth rate of *Neocaridina heteropoda* (later renamed to *Neocaridina davidi*) at different temperatures (Tropea C, 2015). In two experiments (Experiments 1 and 2 - the differences

between the experiments are irrelevant for our focus), newly hatched shrimp were kept at 24°C, 28°C, or 32°C (75°F, 82°F, 90°F, respectively) and left to grow for 90 days. The researchers weighed the juvenile shrimp at days 0, 30, 60, and 90, so had four data points to compare the rate of growth.

As shown by the graphs on the next two pages, the treatment groups kept at 28°C and 32°C showed equally fast growth while the ones kept at 24°C showed significantly slower growth *at the 30-day mark*. The 24°C group continued to lag behind at 60 days but, interestingly, showed no statistically significant difference in weight by the 90-day mark. By taking a look at Fig 4., we see the juveniles in the 24°C treatment groups showed significantly higher growth rates/increments than the other treatment groups from days 30-60 and 60-90.

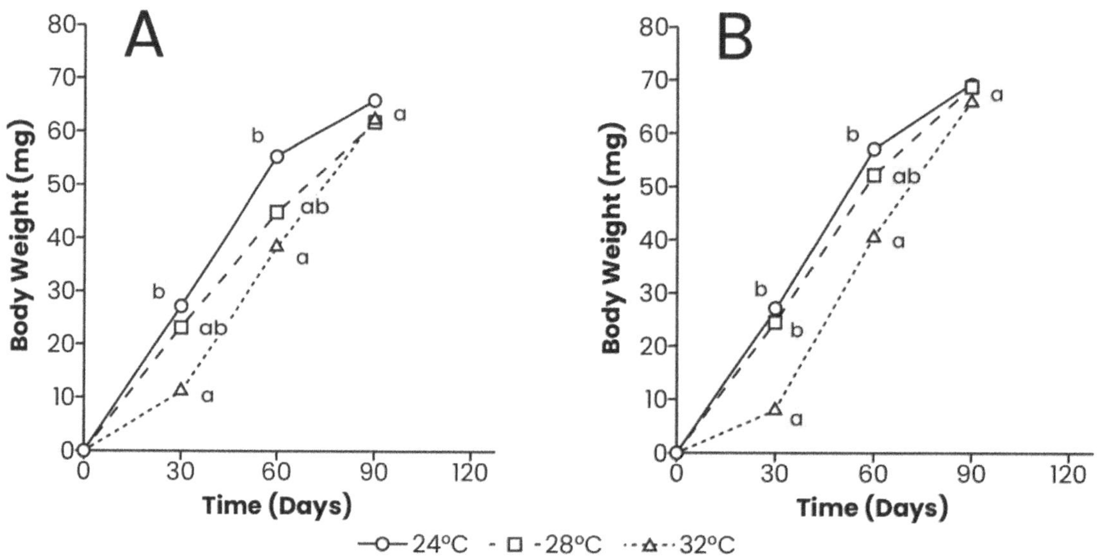

Body weight (mean) of shrimp during the Growth period of Experiment 1(A) and Experiment 2(B). Neocaridina heteropoda heteropoda juveniles were reared at three different water temperatures (24, 28 or 32°C over a 90-day period, and the body weight was recorded every 30 days. Different letters at days 30 and 60 indicate statistically significant differences among treatments ($P < 0.05$): the letter "a" at day 90 indicates the absence of statistically significant differences among treatments ($P > 0.05$). In both experments, body weight of shrimps maintained at 24°C was lower than that of shrimps maintained at 28°C at days 30 and 60, but the difference was no longer significant at day 90 (Tropea C, 2015).

Takeaways

Without more data on growth past the 90-day period, it is difficult to say whether Neocaridina kept at 24°C or 28°C may grow faster over a longer period of time, but it appears raising the temperature to 32°C does not lead to significantly faster growth past the juvenile stage. That being said, 24°C is considered a middling temperature for keeping Neocaridina and Caridina – It can hardly be considered low.

Another paper found that faster growth does happen at higher temperatures for many crustacean species but they have a shorter lifespan that prevents them from reaching the largest size they can reach at cooler temperatures and longer lifespans (Maszczyk & Brzezinski, 2018). As a result, breeders focused on selling shrimp may wish to keep shrimp

at medium or higher temperatures (24°C+), so the juveniles reach a sellable size (60-90 days) slightly faster. On the other hand, breeders focused on winning shows may find that keeping shrimp at lower temperatures slows their growth rate but leads to larger shrimp toward the end of their life.

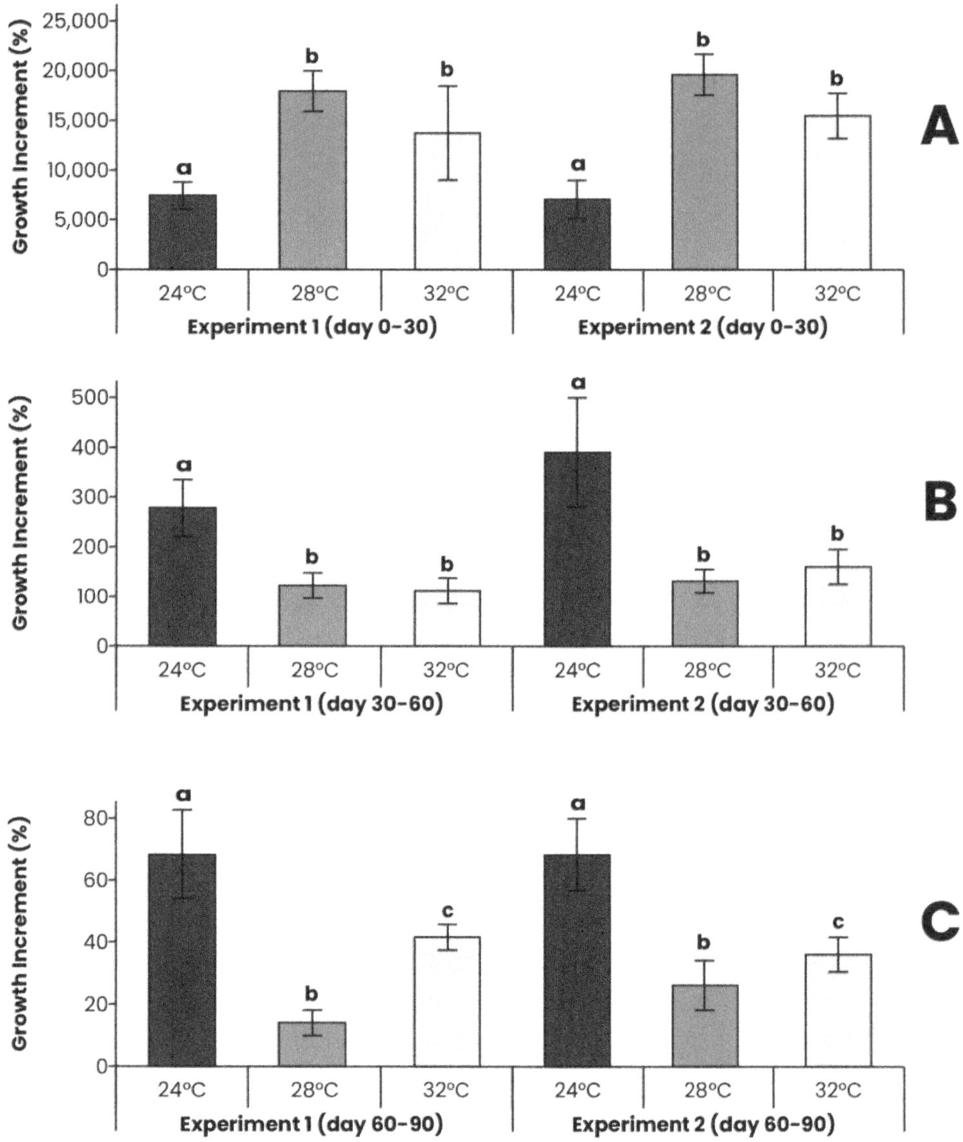

Growth increment (mean±SE) of shrimps during the Growth period in Experiments 1 and 2. Neocaridina davidi juveniles were reared at three different temperatures (24, 28, or 32°C) over a 90-day period and the growth increment was calculated from day 0 to day 30 (A), day 30 to day 60 (B), and day 60 today 90 (C). Different letters indicate statistically significant differences among treatments for each experiment ($P < 0.05$). In both experiments, the growth increment of shrimps maintained at 24°C was lower than that of shrimps maintained at 28 and 32°C at day 30, but this was reversed on subsequent periods. Credit: Tropea et al. 2015.

Another factor to consider is breeding rate, as the highest percentage of Neocaridina were berried at 28°C, while none of the shrimp kept at 32°C became berried. It has been definitively shown that higher temperatures speed up the development of eggs and

reduce the time for them to hatch. On the other hand, higher temperatures increase the risk of disease due to faster bacterial/viral reproduction and compromised immune function (Shields J. , 2019).

As such, our recommendation for Neocaridina breeding is to keep them around 24-26°C [75-78°F] as a good middle ground balancing growth rate, breeding rate, and disease risk. Slightly lower 20-22°C may lead to larger shrimp over a longer period of time.

There is not much scientific research on Caridina but **most breeders prefer cooler temperatures**, finding a good balance of growth, reproduction, and disease risk at 21-24°C (70-75°F).

Chapter Summary

- Larger shrimp are generally more desirable
    - They carry more eggs and perform better in competitions
- Shrimp grow faster when kept at lower population densities
    - Move juvenile shrimp to low density grow-out tank to speed up growth
- Reproduction slows down female shrimp growth
    - Reproductive females put energy into producing and caring for eggs instead of growth
    - Reproduction decreases female molting frequency, which slows growth
    - Female shrimp grow fastest when kept in an all-female tank
- Juvenile shrimp growth varies with temperature
    - Higher temperature speeds up growth for a limited amount of time
        - After 30-60 days, juveniles grow fastest at a medium temperature range
        - Higher temperatures increase disease risk
        - 24-26°C (75-78°F) is a good range to balance growth and disease for Neocaridina
        - 21-24°C (70-75°F) is a good range for Caridina
- Cooler temperatures lead to slower growth but a longer lifespan with more molts
    - This lets shrimp reach a larger size over their lifespan than they otherwise may at higher temperatures

The Different Types of Shrimp Tanks

Everyone has a different reason for setting up a shrimp tank. Some may want a tank that's extremely low maintenance; some are inspired to breed their own lines of shrimp; some are trying to breed massive amounts of shrimp for profit; still others want a beautifully aquascaped tank where shrimp are just an accent piece. Many tanks accomplish more than one purpose. Either way, understanding how our goals affect our approach, equipment, and budget is critical. To get a better perspective on this, we've broken down four different goals and how they affect nine different aspects of tank setup. These aspects include:

- **Goal Priorities**
- **Tank Type(s)**
- **Tank Size(s)**
- **Population Density**
- **Feeding Frequency**
- **Food Type**
- **Plants**
- **Maintenance**
- **Temperature**

A summary of each tank type is provided at the end of each section.

Please note the values given are general guidelines. The important part is understanding why we recommend something so you can adjust the recommendations to fit your goals. For example, we recommend 15-20g for a Caridina breeding setup. This does not mean a 21g or even 25g tank won't work - it just makes shrimp selection more difficult since it can be harder to view and catch shrimp in a larger tank.

Unless otherwise specified, the recommended values are for Neocaridina. Differences in Caridina recommendations are marked by a crystal red shrimp like the one to the right.

Breeding For Profit

Orange Rili Breeding Tank from Master Breeder Mark Shelly

When breeding for profit (BFP), the goal is to produce the largest quantity of shrimp as cheaply as possible.

Priorities: Growth rate, breeding rate, and efficiency

Tank Type(s): Realistically, two tanks are needed for each line of shrimp. One tank is the **Breeding Tank**, where the goal is to start out with a few high-quality shrimp, then let them breed as much as possible and sell the offspring. The second tank is the **Cull Tank**, where lower quality shrimp are moved so they don't mix with the higher quality breeding stock. It's not particularly complicated but maximizing the efficiency and profits from this setup takes awareness.

Ideally, we want to maintain a balanced number of shrimp in the tank. Selling too many shrimp leads to too few shrimp in the tank, which lowers reproduction and profits. In contrast, excess shrimp in the tank lead to slower growth and breeding due to the effects of high population density (p218). Having a ton of shrimp in close quarters also risks disease and deaths due to overpopulation and water pollution. Controlling this depends on sales numbers. Selling too many shrimp leaves us with fewer shrimp to breed so the population recovers more slowly, potentially leading to lower sales. If we're selling too few, then we probably need to work on marketing and should reduce feeding in the meantime...

One additional concern with this setup is that quality eventually declines over time, even when starting with high-grade shrimp. As a result, there needs to be a selection process periodically to reset the tank with only the best shrimp. These shrimp can come from the Breeding Tank or a separate tank where we keep the best shrimp.

Lastly, with the amount of food going into a BFP tank, we need the option to remove waste easily. **A thin substrate or bare-bottom tank makes it easy to vacuum up waste and maintain water quality** without worrying about diseases caused by substrate disturbance.

<u>Tank Size(s):</u>

- Breeding Tank: 40+ g (150+ L)
- Cull Tank: 10+ g (37+ L)

<u>Population Density:</u> In theory, the Breeding Tank has population dynamics like the graph below. The tank starts with the best quality shrimp (3-5 SPG) and is left to breed for a few months to reach the maximum population density. After 3 months, the first generation reaches the age they can be sold, at which point the population drops. This repeats until the offspring quality starts to decline, after which the best shrimp are selected and the rest are sold, then the entire cycle repeats. The time it takes for this quality decline depends on the quality of shrimp you start with, so it is worth paying for the best shrimp initially to reduce how often we need to select breeding stock. Breeding rate also tends to decrease over time, so breeding tanks benefit from a reset every 1-2 years using the best quality shrimp.

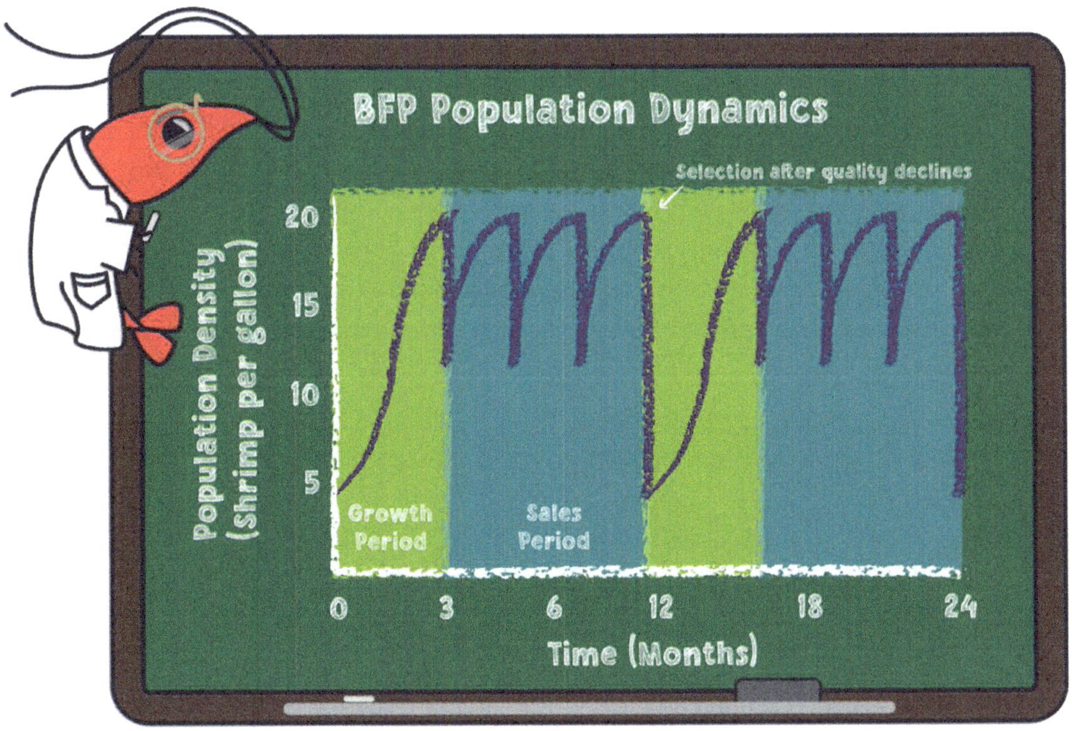

15-20 SPG strikes a good balance between growth rate and water quality for Neocaridina breeding tanks. Since most of the shrimp should be young and small, they don't produce as much waste so more can be kept together without worrying too much about poor water quality (don't forget to test still!). It's possible to reach 20 SPG with sufficient food and more frequent maintenance. Going past this point, it becomes increasingly difficult to maintain water quality so growth and survival rates go down.

To be clear, the timing and population dynamics vary depending on the shrimp species. **Neocaridina can handle higher densities and lower water quality so they follow the above graph fairly well. Caridina, on the other hand, need to be kept at lower densities to maintain water quality**. They also breed slower than Neocaridina so the *Growth* and *Sales Period* are longer while selling frequency is slower. 8-10 SPG is around the maximum for Caridina tanks.

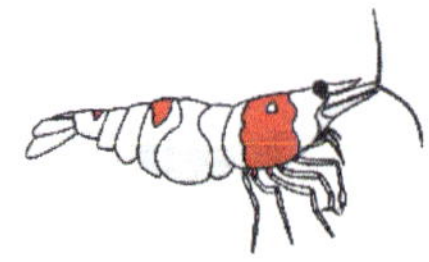

Feeding Frequency: Shrimp need enough food to maximize growth and breeding. Since many breeding tanks have a lot of shrimp, biofilm growth on its own is typically not sufficient to feed every shrimp, nor is it efficient to maximize growth/breeding. As such, frequent feeding (potentially up to twice a day) is recommended.

Neocaridina can be fed more frequently since they are not as affected by water quality - up to twice a day. In contrast, caridina typically do best when fed more sparingly - or when given significantly less food at the same frequency - to maintain water quality.

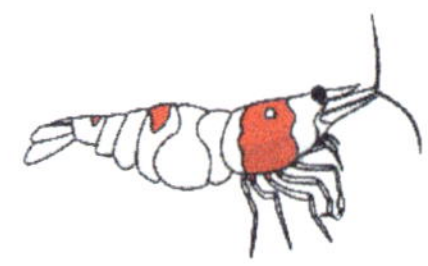

This may seem like a vague recommendation, but setups can vary widely so there is no *perfect* feeding frequency. When starting out, stick with the Feeding Rules of Thumb (p142) to determine how much and how often to feed. See the Food Size section (p144) for more information.

Food Type: When BFP, the goal is to maximize growth and breeding so it's best to feed foods that contain high levels of protein and cholesterol. Shrimp cannot make cholesterol but need it for healthy egg production. Bloodworms and egg yolk are excellent sources of cholesterol, while there are many high protein commercial food products that work well. There is some disease risk with bloodworms (even Hikari triple-cleaned frozen bloodworms), so egg yolk is the safer option. These animal products should be balanced out with high-protein vegetarian foods, as discussed in the Feeding chapter (p137). Higher protein foods produce more nitrogenous waste so frequent monitoring and maintenance are recommended.

For Caridina, opting for higher quality food, a wider variety, and lower quantity of food is typically best to maintain healthy shrimp and a healthy environment.

Plants: At the high densities of shrimp in BFP setups, lots of nitrogenous waste is created. So many plants would be needed to have a positive impact on water quality that they would impact visibility and accessibility. Floating plants work well in other setups to improve water quality but they can be annoying to work around when removing shrimp so they often aren't in BFP setups (except by accident). Instead, most BFP setups rely on consistent water changes to maintain water quality.

That being said, we may improve water quality while making a profit by adding a few clumps of moss, as long as they can be easily removed when visibility is needed.

Maintenance: Maintenance frequency depends on the population density. When it's lower, less feeding and less maintenance is required. As the population increases, more frequent water testing and water changes are needed to eliminate the large amounts of waste

produced by lots of well-fed shrimp. In addition, excess food needs to be removed to prevent water quality issues. Substrate cleaning also needs to be done frequently to remove waste. Shrimp that can survive frequent maintenance also tend to be hardier and more likely to adapt to a customer's tank.

Temperature: We know that higher temperatures increase growth and breeding rate but also increase the risk of disease. As such, an upper-middle range of **72-75°F (22-24°C) for Neocaridina tanks strikes a good balance between growth and disease risk for tanks with high population density**.

Caridina tanks typically do better at slightly lower temperatures so 70-73°F (21-23°C) is a good place to be at

Summary

A summary of the ideal BFP setup is provided below:

SUMMARY	**Breed For Profit (BFP)**
Priorities	Growth Rate, Breeding Rate, Cost
Tank Types	1. Breeding Tank This tank contains mostly female shrimp with a few of the best quality males to breed with them. 2. Cull Tank This is where lower quality shrimp are placed so they don't breed with the main colony.
Tank Size	Breeding Tank: 40+ g (150+ L) Cull Tank: 10+ g (37+ L)
Population Density	Neocaridina Breeding Tank: 15-20 SPG Caridina Breeding Tank: 10-12 SPG Cull Tank: N/A (Keep it healthy)
Feeding Frequency	1-2x per day - rotating food types
Food Type(s)	1x High protein food (Mixed or separate protein sources), 1x Powdered food (optional) Small amounts (<20% of diet) of high cholesterol foods: boiled egg yolk, Hikari Frozen Bloodworms

Plants	No plants are *needed* but a clump of slow-growing moss or larger floating plants (like red root floaters) can be useful. Plants block visibility and accessibility so any plants in the tank should be removeable.
Maintenance	Weekly 30-50% water changes and water testing for Neocaridina. Every 1-2 weeks for Caridina due to lower food volumes. Substrate cleaning may be required if debris builds up in bare-bottom tanks.
Temperature	Neocaridina: 1. 72-75°F (22-24°C) Caridina: 2. 70-73°F (21-23°C)

Breeding For Show

When breeding for show (BFS), the goal is to produce the highest quality shrimp. To do that, we need to be able to clearly see our shrimp to assess their health and select the best ones for breeding. Breeders can be very secretive about their process, so our recommended BFS setup includes some breeder recommendations along with some optimizations based on the available science.

Caridina breeding tank - Credit: Kerric Peyton

Priorities: Shrimp quality, tank health, visibility, accessibility

Tank Type(s): An optimal BFS setup based on the scientific principles discussed in this book includes four separate tanks:

1. **Breeding Tank**

This tank consists of the highest quality females and males at a ratio of 1-2 males for every 10 females to maximize breeding rate and quality. Offspring are typically kept in this tank until they reach 1cm or so before being moved to the Grow-Out Tank.

2. **Grow-Out Tank**
This tank is for separating the offspring from the breeding population for selection. The highest quality shrimp get moved to the Breeding or Show Tank, while the low-quality ones move to the cull tank or get sold.

3. **Show Tank**
This tank is for female shrimp that have been selected for competition. Breeding expends energy and reduces growth rate (Tropea & Lopez-Greco, 2015)so keeping competition-grade females apart from males maximizes their size and overall health. These shrimp are moved over from the Grow-Out Tank as soon as they show the potential for high quality coloration, ideally before they carry their first clutch of eggs. The goal is to leave 10-20 female shrimp that are near their maximum size (takes about a year) so the best 3-4 can be selected for a show.

We have not heard any breeders *say* they have a tank like this but the science shows that virgin females gain significantly more weight (5-10% more) than reproducing females over the same time period (See Growing Bigger Shrimp [p218]). That may not sound like a lot, but every optimization matters if we want first place in a show.

4. **Cull Tank**
This tank is for shrimp that are not of high enough quality for breeding or a show. Breeders may still sell these shrimp as many are still decent quality for hobbyists, but hopefully they are clearly labelled as culls.

Ideally, all these tanks should be set up with the same substrate, plants, filters, and lighting to maintain the same parameters and make it easy to move shrimp from one tank to the other. Neocaridina tanks may have substrate, whereas Caridina tanks need substrate, although it may either be a thick layer on the bottom of the tank or kept entirely in a UGF. Check out the section Comparing Two Caridina Tank Setups (p241) for more information.

Tank Size(s):

- Neocaridina Breeding Tank: 10-20g (38-75L)
- Caridina Breeding Tank: 15-20g (57-75L)
- Grow-Out Tank: 20-30g (75-114L)
- Neocaridina Show Tank: 10-20g (38-75L)
- Caridina Show Tank: 15-20g (57-75L)
- Cull Tank: 10+ g (38+ L)

Larger setups are better for water quality but can make it more difficult to find and select for shrimp. As a result, 10-20g tanks (38-75L) strike a good balance between tank stability and visibility/accessibility. Caridina benefit from larger tanks so we recommend a minimum of 15g (57L) for a more stable environment and a better chance of success.

The Grow-Out Tank needs to have enough space for all the babies produced by the breeding tank so it needs to be slightly larger, ranging anywhere from 20-30g (75- 114L). Anything above this volume is too large to easily observe and catch shrimp – something which needs to be done frequently in BFS tanks.

The Show Tank does not need to keep many shrimp but does need to maintain a stable environment so 10-20g tanks (15-20g for caridina) work well.

In a breeding project, most shrimp end up being culls but they are still desirable for other hobbyists. A 10g (38L) Cull Tank holds a decent number of shrimp but a larger tank may be needed depending on how many babies are produced and how quickly culls get sold.

Population Density: BFS setups are more concerned with overall tank health than BFP setups because breeders don't want to take any risk of losing potentially prize-winning shrimp or their offspring. Maintaining lower population densities in tanks helps ensure adequate food and reduces potential water quality issues. It also promotes the best growth and breeding rates.

For BFS Breeding Tanks, a density of 5 SPG or less (excluding the babies) is optimal for the best growth and breeding rates. There aren't any scientific papers that test such a low density, so the recommendation is based on the experience of breeders we've spoken with. In addition, if we have a higher density in the tank, then that suggests the selection process is likely too loose since only the absolute highest quality shrimp should be in a Breeding Tank producing offspring. Again, the goal is quality over quantity when breeding for a show.

A density of 5-8 SPG for BFS Grow-Out Tanks creates a good balance between growth rate and water quality. **All shrimp in the Grow-Out Tank should be juveniles that have a small bioload and diet, allowing for a higher population density than a Breeding Tank without sacrificing water quality**. Any offspring may be part of a prize-winning group, so this density helps ensure each shrimp gets enough resources to survive. Large amounts of food also aren't necessary to maximize growth rate, so less maintenance is required and better stability is achieved.

For caridina shrimp, slightly lower population densities tend to maintain better water quality and overall health, so **2-3 SPG is recommended in a Caridina Breeding Tank and only slightly higher in the Grow-Out Tank**.

Regardless of species, a Show Tank should have a maximum density of 1-2 shrimp per gallon, or around 10-20 shrimp in total. This optimizes growth (See Grow Bigger Shrimp chapter) but the main reason for such a low number is that the selection process should be extremely rigorous. **Only a small number of shrimp should meet the selection criteria to make it into a Show Tank**.

Feeding Frequency: With the low shrimp density in all BFS tanks, natural biofilm grows back quickly enough to make up most of a shrimp's diet but daily supplementation with a small amount of food is also important for two reasons. First, we don't know exactly what types of biofilms are growing in the tank and it's possible some nutrients may be missing so supplementation increases the chance of having healthy shrimp. Lastly, shrimp gather around the food so it's easier to check their activity level and whether they are getting enough food (based on how quickly they eat what was fed). With this information, we have better idea of when and how much to feed.

Caridina generally do better with less food than Neocaridina, partly because Caridina are kept at lower densities and partly because water quality is more of a concern.

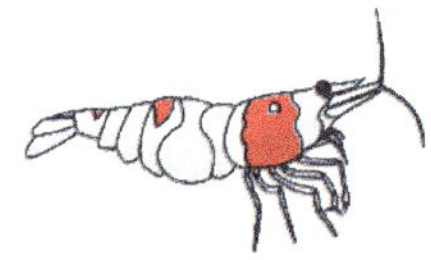

Food Type: Daily addition of *small amounts* of high protein and/or high cholesterol foods help with growth and breeding. Make sure they receive both plant and animal protein. Lastly, there should always be decaying plant matter like Indian almond leaves left in the tank.

Plants: The key is maintaining visibility and accessibility along with a healthy ecosystem. A few slow-growing plants combined with fast-growing floating plants are recommended regardless of tank type in a BFS setup. The fast-growing floaters help remove lots of nutrients to maintain water quality, while the slow growing plants help somewhat with water quality and provide surface area for biofilm growth.

Slow-growing plants like mosses or anubias are typically easy to remove or move around to look at the shrimp, while floaters stay at the top of the tank so they don't block vision from the front of the tank. Be aware that smaller floating plants like duckweed can make it harder to net out shrimp. Larger ones that can be removed easily (like red root floaters) are typically better for BFS setups.

Maintenance: Consistent water changes (20-40% every 1-2 weeks) are recommended for a few reasons. For one, they help maintain water quality. Having a routine also helps ensure the water is remineralized in the same way every time so there are minimal disruptions to the ecosystem. Lastly, shrimp that will be sent to a show need to be able to handle the stress of transit and different water quality that may be in the tanks at the show. Consistent large water changes ensure the surviving shrimp are hardy, although this approach does vary by breeder.

Periodic substrate cleanings are recommended. Rooted plants may take care of debris that builds up in the substrate, but they create issues with visibility and accessibility so the job of maintaining a clean substrate typically falls on the breeder. This approach also varies by breeder. Please see the Comparing Different Caridina Setups (p241) for the pros and cons of different tank setups.

Temperature: Since shrimp are kept at relatively low densities in BFS setups, disease is not as much of a concern so a slightly higher temperature range works when compared to BFP setups. Neocaridina tanks can be pushed to 75-78°F (24-26°C) while Caridina tanks tend to do best around 70-75°F (21-24°C). This temperature bump helps optimize growth and breeding rates.

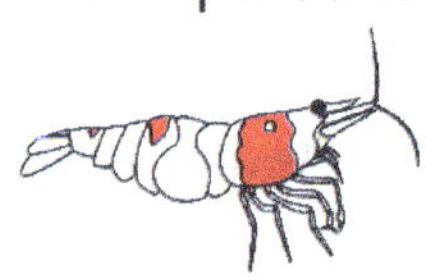

Summary

A summary of an effective BFS setup is provided below:

SUMMARY	Breed For Show (BFS)
Priorities	Shrimp Quality, Tank Health, Visibility, Accessibility
Tank Types	1. **Breeding Tank** This tank consists of the highest quality females and 1-2 of the highest quality males. 2. **Grow-Out Tank** This tank is for separating the offspring from the breeding population to let them grow larger. 3. **Show Tank** This tank is for female shrimp that have been selected for competition. 4. **Cull Tank** This tank is for shrimp that are not breeding or show quality according to the breeder.
Tank Size	Neocaridina • Breeding Tank:10-20g (38-75L) • Grow-Out Tank: 20-30g (75- 114L) • Show Tank: 10-20g (38-75L) • Cull Tank: 10+ g (38+ L) Caridina • Breeding Tank: 15-20g (57-75L) • Grow-Out Tank: 20-30g (75- 114L) • Show Tank: 15-20g (57-75L) • Cull Tank: 10+ g (38+ L)
Population Density	Neocaridina • Breeding Tank: <5 SPG • Grow-Out Tank: 5-8 SPG • Show Tank: 1-2 SPG • Cull Tank: N/A Caridina • Breeding Tank: 2-3 SPG • Grow-Out Tank: 3-4 SPG • Show Tank: 1-2 SPG • Cull Tank: N/A
Feeding Frequency	1-2x per day with rotating food types

Food Type(s)	1x Plant protein source, 1x Animal protein source, 1x Powdered food + Decaying plant matter (Indian almond leaves, cholla wood, etc.) + (optional) Small amounts (<20% of diet) of high cholesterol foods
Plants	1-2 species of removeable slow-growing plants (Java moss, anubias attached to wood, etc.) 1 species of larger fast-growing floating plant (like red root floater)
Maintenance	Weekly testing (minimum) 1. 20-40% water changes every 1-2 weeks 2. Substrate cleaning as needed 3. Top-offs as needed
Temperature	Neocaridina: • 75-78°F (24-26°C) Caridina: • 70-73°F (21-23°C)

The Aquascape

Shrimp are common in an aquascape as a clean-up crew and as aesthetic contrast to the plants or hardscape, but shrimp often take secondary priority to the overall appearance of the tank and health of the plants.

Priorities: Shrimp hardiness, plant health, appearance

A beautiful aquascape with shrimp – Credit: Terese Forninge, Sweden

Tank Type(s): Only a single tank is required for a successful aquascape, although an additional cull tank for shrimp may be useful to maintain strong coloration.

Tank Size(s): Tank size for aquascapes varies widely depending on limitations and goals, ranging from a few gallons to 40+ gallons (150+ L). As always, larger tanks (10+ g) are more stable and more likely to lead to successful shrimp keeping. Shrimp tend to be small and don't stand out as much in larger tanks, so 5-20g aquascaped shrimp tanks are most common - at least if the goal is to show off the shrimp to some degree.

Population Density: Population density does not have a significant impact on an aquascape. In general, we recommend letting the shrimp reproduce and reach the natural population capacity based on the tank and your feeding routine.

Feeding Frequency: Feeding at least once a week is generally a good idea to supplement natural food sources in the tank. Feeding more frequently leads to more shrimp, and vice versa when feeding less. This gives decent control over the total population in a tank.

Food Type: Routinely feeding a high-protein food source is a good supplement in addition to natural food in the tank. Keeping decaying plant matter in the tank, like cholla wood or Indian almond leaves, is also never a bad idea.

Plants: Aquascapes use a variety of plants depending on the hobbyist. Slower-growing plants are often best since they require less frequent fertilization, maintenance, and typically don't require CO_2, leading to a more stable environment that a wider variety of shrimp can handle. You can absolutely keep shrimp with fast-growing plants, but we recommend hardier varieties like Neocaridina or Tiger Caridina.

Maintenance: Aquascape maintenance depends on the plants and fertilization routine. Frequent fertilization requires frequent water changes (often weekly but it depends on the tank). Consistent trimming, substrate cleaning, and algae removal is also often necessary to maintain the desired aesthetic.

Temperature: Any temperature acceptable to all species in the tank. Typically 65-78°F (18-26°C), depending on the shrimp species.

Summary

A summary of the ideal Aquascape setup is provided below:

SUMMARY	**Aquascape**
Priorities	Shrimp Hardiness, Plant Health, Appearance
Tank Types	N/A
Tank Size	5+ g (19+ L)
Population Density	Population density does not matter significantly in an aquascape.

Feeding Frequency	1-3x per week (more frequently if you want more shrimp)
Food Type(s)	-Mixed high-protein food -Decaying plant matter (Indian almond leaves, cholla wood, etc.)
Plants	Various
Maintenance	Maintenance type and frequency depend on fertilization schedule and desired aesthetic.
Temperature	Any temperature acceptable to all species in the tank. Typically in the range of 65-78°F (18-26°C).

The Low Maintenance Setup (LMS)

Many hobbyists want a relaxing shrimp tank that doesn't take too much time to maintain. Here's the setup for an LMS tank.

Priorities: Maintenance time, ecosystem balance

Tank Type(s): A LMS only requires a single tank.

Tank Size(s): A stable tank is critical for a LMS so 10+ g (38+ L) is highly recommended. 20g-long (76 L) tanks strike a good balance between stability, size, and setup cost.

Population Density: The goal of a LMS is to reach a stable population density that can be supported by the tank with minimal feeding. The amount of algae, biofilm, microorganisms, and detritus in a tank can vary greatly between tanks - or even

throughout the lifespan of a tank, so a sustainable density varies in each setup. The natural capacity may vary anywhere from 5-15 shrimp per gallon, with larger tanks trending toward higher densities since they tend to have exponentially more surface area for food growth. **Shrimp breed to the capacity of available resources so they reach population capacity naturally**, assuming lots of additional food isn't being added. It's not particularly important to consider or count density in a low maintenance setup. That being said, large population increases from rapid breeding may increase maintenance requirements.

Feeding Frequency: Ideally, an LMS is given time to mature with plenty of biofilm development so additional feeding is never required. Shrimp absolutely can be healthy living off of naturally produced biofilm. That being said, feeding small amounts of food 1-2 times per week helps balance their diet to mitigate nutrient deficiencies. Doing so also provides some fertilizer to plants.

Food Type: Natural ecosystems have dying animals that provide protein to detritivores like shrimp. Our tanks (hopefully) don't have many dying animals so periodically feeding a Mixed (veggie+animal) protein food source is recommended. The shrimp break it down to produce ammonia and nitrates for plants, reducing the need for supplemental fertilization.

It can help to put a variety of materials in the tank (cholla wood, Indian almond leaves, etc.) that grow diverse biofilm for shrimp to eat.

Plants: Plant diversity benefits low maintenance ecosystems since each plant species takes up nutrients at different rates. Having a variety of plants reduces the need for water changes and reduces the chance of algae growth. Slow-growing plants like cryptocoryne, anubias, and mosses are great options that don't require fertilization or CO_2. Fast-growing floating plants are also good in low maintenance setups since they rapidly remove nutrients from the water to reduce the need for water changes.

Maintenance: When set up properly, some maintenance may be required initially as the tank balances out with organisms finding their niche. Generally, older LMS's require less maintenance as the ecosystem becomes more established. If only small amounts of food and/or fertilizer is being added, then maintenance should consist of little more than topping off the water (with RO/DI water), along with the occasional filter cleaning. A small water change (10-15%) every 1-2 months is often beneficial.

That being said, **wanting an LMS is not an excuse to avoid required maintenance**. As always, maintenance is needed whenever tank parameters like nitrates get near or out of the recommended ranges. Consistent observation and periodic testing is always recommended.

Temperature: Anywhere from 65-75°F (18-24°C) works well for LMS. The shrimp do still grow and breed to reach population capacity but don't breed so fast that they overwhelm the tank. Lower temperatures also slow down the breakdown of waste, so less maintenance is required.

Summary

A summary of the ideal LMS setup is provided below:

SUMMARY	**Low Maintenance System (LMS)**
Priorities	Maintenance Time, Ecosystem Balance
Tank Types	N/A
Tank Size	10+ g (38+ L)
Population Density	Whatever the natural population capacity reaches
Feeding Frequency	1x per week
Food Type(s)	-Small amounts of Mixed high-protein food -Decaying plant matter (Indian almond leaves, cholla wood, etc.)
Plants	Recommended to have at least: • 2x species of slow-growing plants (moss, crypts) • 1x species of fast-growing floating plant More plant diversity is better
Maintenance	Maintenance as required based on periodic testing. Preventative water changes every 1-2 months are also recommended.
Temperature	65-75°F (18-24°C)

Comparing Caridina Setups

Caridina are an interesting conundrum in the hobby. There are so many different breeders claiming that one setup is better than any of the dozens of other ways to set up a tank. Arguably, there is no "best" tank setup – It all depends on our goals, which many recommendations don't consider.

There are a variety of ways to keep them but let's look at two common approaches for Caridina tanks at opposite ends of the spectrum, both of which maintain a healthy tank while accomplishing very different goals. Lessons in this section are true for both Neocaridina and Caridina.

Where Caridina Breeders Agree

One point all Caridina breeders agree on is that caridina need a pH below 7 – Most agree that between 5.5-6.5 pH is optimal.

Another point most breeders agree on is that Caridina tanks require time to mature, just like we've talked about previously in our cycling methods. Giving an ecosystem 2+ months to mature leads to more stable parameters and healthier shrimp, regardless of species. It's especially important for Caridina tanks because this time lets nutrients release from either the substrate or decomposing food/plant matter, which promotes the growth of algae that forms the base of an ecosystem. Microorganisms eat the algae, develop their populations, then in turn get eaten when shrimp are introduced, creating a strong food web to support a healthy tank.

> **Side note**: There are some fast-cycling methods that breeders have come up where they prepare the substrate in a certain way and add specific bacterial powders and other products to their tank. It can work but is highly dependent on products and experience, leading to a low success rate compared to slower tank cycling methods.

As such, basic parameters and cycling recommendations are fairly consistent for Caridina. What differs dramatically is the substrate setup and maintenance recommendations. Let's take a look at two extremely different tank setups that lead to award-winning shrimp.

Full disclosure: This section includes our interpretation of setups of different breeders. None of the information is directly endorsed by any breeders.

Method #1: The KP Method

Named after shrimp breeder Kerrick Peyton – please note this is our interpretation of his method after talking with him.

The first method of keeping Caridina is a very low-maintenance approach focused on keeping a balanced ecosystem with a decent number of plants and a thick substrate layer. The tank setup looks like this:

- 15-20g (57-75L) tank
- Thick akadama substrate layer (2-3")
- Sponge filter
- A few slow-growing plants
- Fast-growing floating plants to soak up nutrients
- Only change water as needed (Elevated TDS, ammonia, nitrates)
- Add small amounts of new substrate periodically
- Daily feeding of very small amounts of food (25-30 mg for 50 shrimp)

One of Kerric's tanks with a thick akadama substrate, plenty of floating plants, and botanicals. Credit: Kerric Peyton

A thick substrate is useful because it creates anaerobic zones toward the bottom of the tank that help with nutrient recycling. This reduces the need for water changes that might otherwise briefly disrupt the ecosystem. Reducing water changes also means the substrate's buffering capacity doesn't deplete as quickly.

The large volume of active substrate maintains a very stable pH and lasts a long time. Deeper parts of the substrate are not exposed to significant water flow, so it takes longer for acid to release fully, further extending the buffering time. This reduces the need for tanks to be reset so a healthy ecosystem can be maintained for much longer than thin soil methods where substrate needs to be replaced every 9-12 months. The lifespan of a thick substrate tank can be further extended by adding new soil periodically, since akadama substrate doesn't leach ammonia. This can extend the tank lifespan up to 2 years before a reset is needed, which is typically determined by when shrimp no longer breed as well as they used to.

Since water changes can have some negative impacts on tank stability and substrate lifespan, this approach reduces the need for them as much as possible. The lack of water changes can be an issue with low-pH systems since nitrifying bacteria and other decomposers process waste more slowly, leading to waste buildup. Kerric Peyton makes two adjustments to help prevent nutrient buildup:

Two of Kerric Peyton's Caridina tanks. Credit: Kerric Peyton

1. **Add floating plants**
   Frogbit, red root floaters, and other floating plants do a fantastic job of soaking up excess nutrients. We recommend larger floating plants because they are more easily removed than smaller floaters like duckweed so don't have as much of an impact on visibility and accessibility.
2. **Feed very little**
   Adding excess nutrients to these systems can be detrimental so the goal in this setup is to let Caridina feed off naturally occurring sources as much as possible. This does limit the population and may slow growth rate but reduces the need for maintenance.

Pros

1. Low maintenance
2. High chance of success
3. Longer tank lifespan

Cons

- More expensive to set up due to thick substrate
- Tanks do not hold very many shrimp (2-3 adults per gallon)

Kerric Peyton's awards for his Cardina shrimp. Upper Left: BOA Category; Upper Right: Red Devil Category; Lower Left: Pure Red Line; Lower Right: Galaxy Pinto Category. Credit: Kerric Peyton

Please note the KP method is mainly based on the experience of Master Shrimp Breeder Kerrick Peyton. He states that those who follow his recommended setup[lxi] to a "T" have a very high rate of success compared to other methods. Based on conversations with him, he seems to be very genuine and does not benefit by selling any recommended products so does not have a clear bias. This contrasts many shrimp sellers who have biases to recommend the various bacterial powders, remineralizers, or special substrates they sell.

Method #2: The UGF Method

This is the method used by award-winning shrimp breeders/sellers like Lawrence Lee (Shrimp Hub Singapore) and Kelvin Wong (KW Shrimps).

The UGF (under gravel filter) Method (outlined throughout this video[lxii]) of keeping Caridina takes a very different approach than the thick substrate method. The stability comes from consistent large (but still slow) water changes with healthy water, rather than environmental filtration used in the KP Method. Here's how to set up a tank with the UGF Method:

- 15-20g (57-75L) tank
- UGF filled with active substrate
- Thin layer of substrate on bottom of tank (optional)
- Sponge filter
- Various additives like bacterial powders, lubao balls, etc. (depends on breeder)
- 1-2 slow-growing plants (clump of moss)
- 30-50% water changes every 1-2 weeks
- Daily feeding of larger amounts of food (150mg for 50 shrimp)

The substrate positioning and quantity are the most important differences between the two methods. Many breeders use UGF filters because the UGF pulls water through the substrate, effectively exposing more surface area of the substrate to release acid and keep the pH low. This is intended to provide a more stable environment but also leads to

faster depletion of the substrate – typically in 9-12 months. At that time, a new tank is set up with new substrate in a UGF and the shrimp are moved over after the new tank matures. These setups are cheaper initially, but the shorter reset time results in larger long-term expenses.

One of the reasons this setup is popular is because it is easy to keep clean. In a thick substrate system, gravel vacuuming and water changes can disturb the substrate, potentially polluting the water or destabilizing the ecosystem. In contrast, with most substrate in the UGF, it's easier and safer to vacuum out any excess waste in the rest of the tank. This lets us feed more to support a larger population.

Shrimp breeders do have different takes on vacuuming. As Darick from Shrimp Affair shows in this video[lxiii], he has substrate-free tanks that he consistently vacuums to remove any debris. On the other hand, Kelvin Wong uses a thin layer of substrate in his tanks and he does not vacuum it. Kelvin's setup may accumulate slightly more waste than Darick's but does not appear to impact Kelvin's shrimp. He also has more filtration in his tanks (double-sponge on top of UGF) which may reduce waste accumulation.

Darick (Shrimp Affairs) uses a single UGF with an otherwise bare floor.

Kelvin Wong (KW Shrimps) uses a UGF (below the moss in the back) with an additional double-sponge filter and thin substrate layer.

In addition, frequent large water changes are done after cleaning out waste to add in healthy new water and dilute any leftover waste. With a good routine, freshly remineralized water is very consistent, so it only leads to temporary pH swings even with large water changes. The combination of UGF + sponge filter also increases water flow to quickly stabilize the pH after a water change.

This method is most often used by for-profit shrimp sellers/breeders for a few reasons:

- It's easier to clean the tanks so food can be fed more frequently, which increases the growth and breeding rate of shrimp. This also enables tanks to hold more shrimp (Higher population capacity/density).

- Shrimp that cannot handle larger water changes are less hardy and more likely to die in a customer's tank. The water change process selects for hardier shrimp and may strengthen the breeder's reputation.
- It is a consistent routine that sellers can do to maintain good water quality.
- This method is cheaper in the short-term because less active substrate is used in UGF compared to the KP Method. It may deplete substrate faster but reducing upfront costs is often better for businesses when starting up.

Pros

- Consistent setup and control of ecosystem
- Selects for hardy shrimp
- Forgiving of errors like overfeeding
- Cheaper setup that uses less substrate
- Higher growth rate and population capacity due to frequent feeding
    a. Leads to more profit per tank

Cons

1. Tanks typically only last 9-12 months before reset
2. More maintenance time
3. May be more expensive in long run due to extra water use and faster substrate depletion

Summary

These two types of setups accomplish different goals but they both hold to the principle of shrimp keeping - creating a balanced environment for shrimp. The KP Method uses thick substrate and minimal feeding, letting shrimp get most of their food from the ecosystem. As a result, he doesn't add enough nutrients to require frequent water changes BUT the ecosystem only supports a limited population. Kerric is focused on breeding high-grade shrimp while saving time with reduced maintenance, so this method works well for him.

In contrast, the UGF Method works for breeders like Kelvin Wong because he wants to breed high-grade shrimp AND make money. As a result, he needs each tank to support a larger population, so he has more shrimp to sell. To increase the population size, Kelvin feeds 5x as much as Kerric. To prevent waste buildup from all that food, Kelvin has a very thin substrate layer that is easily vacuumed. He also performs maintenance much more frequently than Kerric does - again, to remove waste and maintain water quality. Without many plants or a thick substrate layer, the ecosystem doesn't filter out waste as well, so water changes are especially important to maintain healthy water. There's no concern about kicking up debris or disturbing the substrate, so larger water changes come with little risk. Overall, this is a very controlled approach that requires frequent work but produces great results.

Both of these methods prioritize water quality and a healthy ecosystem - they just do so in very different ways and achieve different goals. There is no single way to keep shrimp, so what's important is understanding the fundamentals of water quality,

ecosystems, equipment, and how to adjust a tank setup to achieve the goals we want. The same is true for Neocaridina tanks.

Hopefully, this book helps you achieve your shrimp keeping goals. Let us know if it does! If that doesn't happen, please don't ever contact us.

...Just kidding - we want this book to be a valuable resource for shrimp keepers and need your constructive feedback to improve! Please don't hesitate to reach out to contact@shrimplyexplained.com with any critiques :)

Also, there's still more to learn! We've got one more chapter that dives into the experience of entering a shrimp show, followed by Advanced and Miscellaneous sections with a lot more information. The learning journey is never over.

Shrimp Shows

It all comes down to this. You've spent years selecting for the brightest colors, the best coverage, the biggest bodies, until finally... you're ready to show your project to the world. You sign up for the Aquashella Shrimp Show (great name) and spend the final few months on the most difficult task – choosing the 3-5 highest quality shrimp. Each one needs to match the others in shade, pattern, and size as closely as possible to perform well, so this decision is *critical*. After picking them out, you ship the shrimp over to the venue and hope for the best.

Side note: Many breeders do not actually use their highest quality shrimp due to the risk of losing them from the stress of transport and different tank parameters at the show. Instead, they send their second string to keep the highest quality shrimp breeding for even better offspring in the future.

On the day of the show, you arrive early and make your way to the rows of small, brightly lit display tanks. Each one contains beautiful specimens from dozens of other hopeful breeders. Your nerves build as you go down the rows of competitors, searching for your own shrimp.

At last, near the end of the middle row, you find them. They all survived transit and seem to be active, although two of them have slightly faded carapaces due to stress. Hopefully, the color will return by the time they're judged. There's nothing to be done now, though, so you wander off to visit the booths and say hello to friends.

An hour later, the judges file down the rows of tanks with their coffees and clipboards of scoresheets. Each of these scoresheets looks like the following:

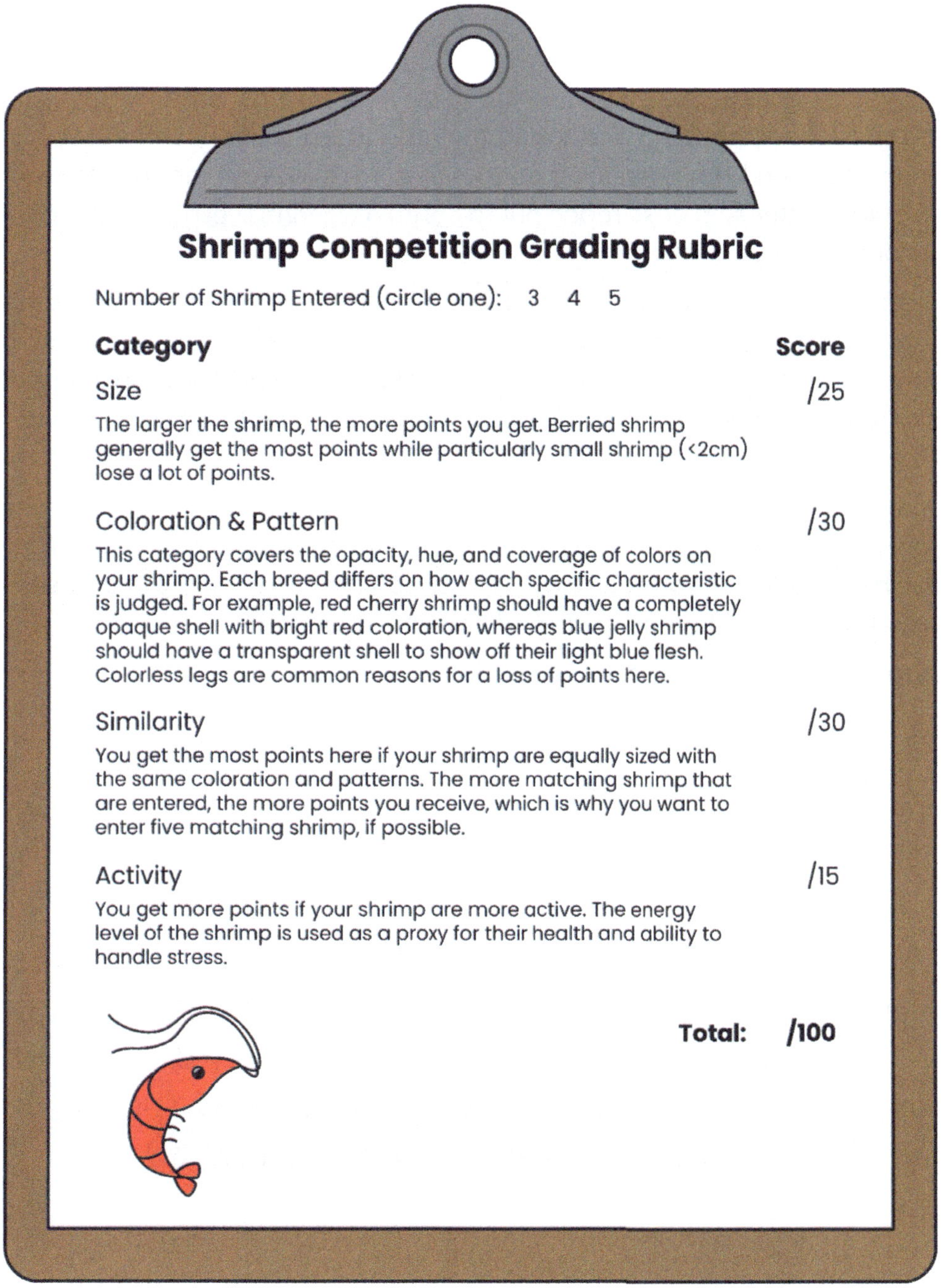

Shrimp Competition Grading Rubric

Number of Shrimp Entered (circle one): 3 4 5

Category	**Score**
Size The larger the shrimp, the more points you get. Berried shrimp generally get the most points while particularly small shrimp (<2cm) lose a lot of points.	/25
Coloration & Pattern This category covers the opacity, hue, and coverage of colors on your shrimp. Each breed differs on how each specific characteristic is judged. For example, red cherry shrimp should have a completely opaque shell with bright red coloration, whereas blue jelly shrimp should have a transparent shell to show off their light blue flesh. Colorless legs are common reasons for a loss of points here.	/30
Similarity You get the most points here if your shrimp are equally sized with the same coloration and patterns. The more matching shrimp that are entered, the more points you receive, which is why you want to enter five matching shrimp, if possible.	/30
Activity You get more points if your shrimp are more active. The energy level of the shrimp is used as a proxy for their health and ability to handle stress.	/15
Total:	**/100**

Figure 2. Rubric based on criteria from Mad Shrimp.

These are the grading criteria from Singapore's Golden Shrimp Awards[lxiv]. Other shrimp competitions may differ in their point distribution.

So, how are your shrimp going to do? That's up to your dedication and patience. The Shrimply Explained Team is rooting for you!

Conclusion

Thank you so much for reading and taking the time to learn about shrimp! We hope you found the information both interesting and helpful. We're always looking for feedback to make the guide better so please reach out to contact@shrimplyexplained.com if you find any errors or have any questions!

Also, there's even more information available in the Supplemental section after this! It contains a variety of topics that may come up in a shrimp keeper's career but isn't necessarily relevant to all shrimp keepers. We didn't want to add too much unnecessary complexity to the main sections of the book but also wanted to be thorough to support you in all aspects of your journey.

Lastly, it's important to remember that science is not static. There's always new research coming out that builds on older studies and may even contradict them. We'll do our best to keep the book updated! You can stay up to date on the latest shrimp info by subscribing to our email list and the Shrimply Explained YouTube channel.

Thanks again for reading and happy shrimpin'!

Rick and Shrimply

P.S. - Enjoyed the book? Help fellow shrimp keepers by leaving a review on Amazon (or sending one to contact@shrimplyexplained.com) and giving the book a shoutout on social media! It also helps support Shrimply Explained so we can continue making valuable content 😊

Supplemental Information

Caridina Shrimp Breeding

Please note this is one of the least scientific chapters in the book. We have done our best to verify information but much of it is hearsay due to secrecy in the hobby. That's why this chapter is in the back of the book.

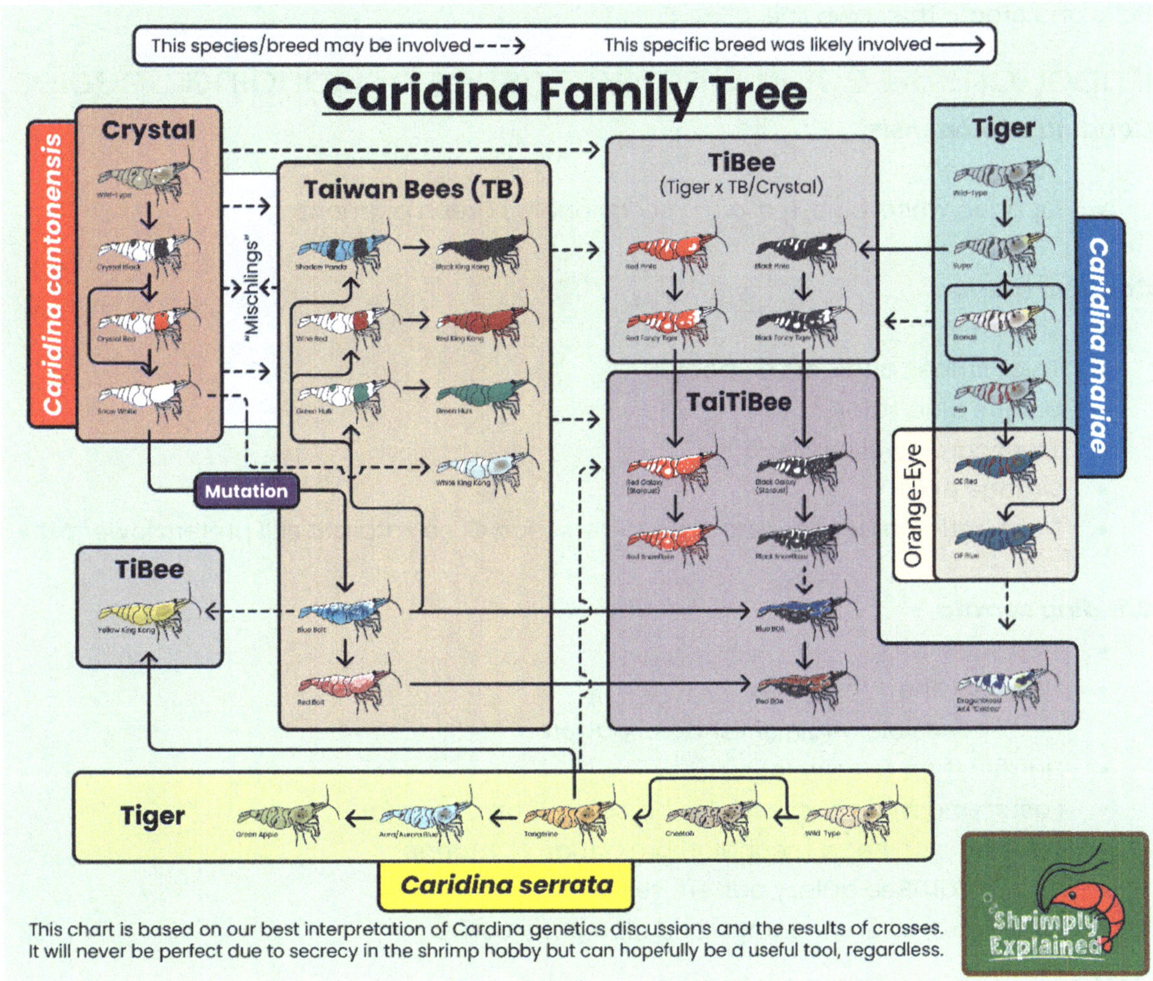

The Trouble with Family Trees

This tree and those we see online are intended to provide *plausible* crosses leading to various colors/patterns. Due to the secrecy of shrimp breeders, we will likely never know what crosses many of them *actually* came from. **Crossing the indicated breeds also is not going to guarantee offspring that look anything like established breeds** and, even if they do, the offspring won't breed true. It takes MANY generations to develop a purebred line and that journey is not a straight line. When creating a purebred line, there will likely be back-crosses to older generations to emphasize certain traits, or the need to reduce inbreeding depression with more diverse genetics before focusing back on the desired

pattern/color goal. The process can take *years*, which is why people are willing to pay so much money for new lines of shrimp developed by master breeders instead of trying to DIY their own breed.

One additional complication with Family Trees is that some lines come from incredibly rare genetic mutations. Multiple sources[1,2] online state that Blue Bolts randomly appeared while breeding a type of crystal shrimp, and that the offspring of the Blue Bolt x crystal cross led to Wine Red and Black King Kong. If it's truly a mutation, then a breeder may need to breed thousands--if not millions--of crystal shrimp to duplicate the occurrence of Blue Bolts and create their own line.

Important Breeding Characteristics of Caridina Species

Caridina cantonensis

- Thick shell
- Opaque white, blue, red, and (occasionally) green pigment

Caridina mariae

- Thin shell
- Thick stripes and dots on body
- Unique blue stripes
- Transparent yellow head
- Orange eyes
- Does better in Neocaridina parameters than *C. serrata* but still prefers lower pH

Caridina serrata

- Thin shell
- Has backline
    - Possible origin of fishbone pattern
- Hardier than *C. mariae*
- Faster and more successful in breeding than *C. mariae*
- More dots on head (cephalothorax) than *C. mariae*
    - TaiTiBee galaxy pattern seen more often
- Unique transparent orange and blue body colors

Basic Breed Descriptions

Crystal Shrimp (Caridina cantonensis)

Crystal Black (CBS)

- This is the first stabilized Caridina breed from wild type coloration
- Has an opaque white shell with black areas

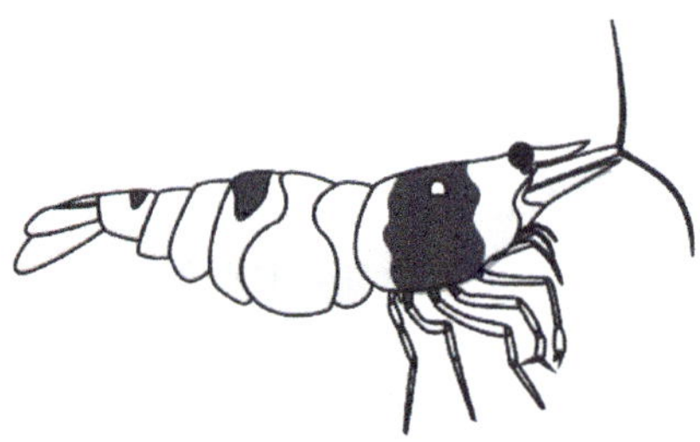

Crystal Red (CRS)

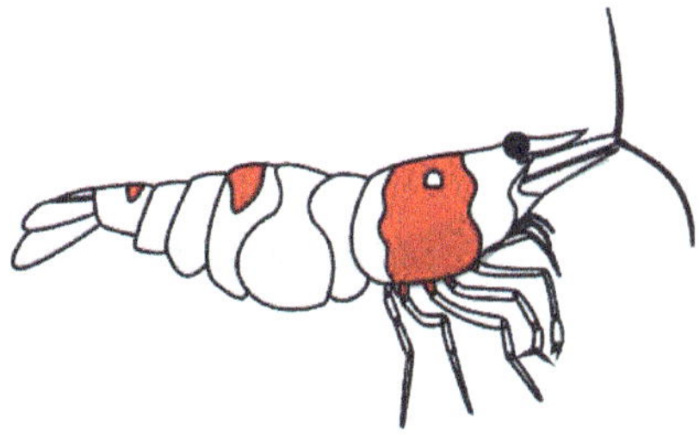

- Has an opaque white shell with red areas
- Originates from selectively breeding CBS
- Red is a recessive gene that takes the place of black pigment
- There are pure lines that only give red and white coloration (pure red line - PRL)
    - Others produce both CRS and CBS offspring

Snow White

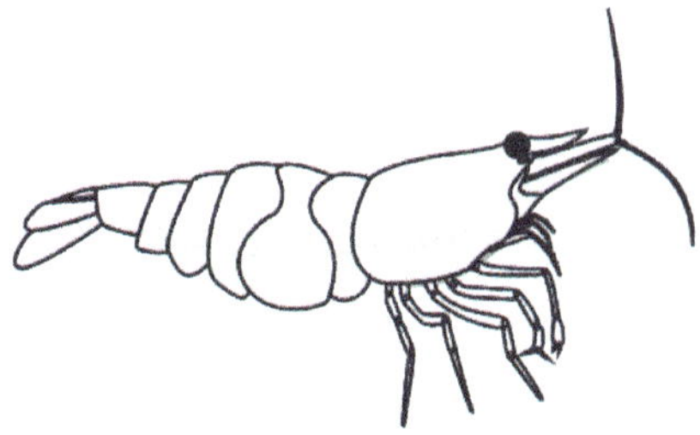

- CBS or CRS taken to the extreme by selecting for all-white shells
- Some have shells with yellowish pigment
    - Referred to as Golden Bee

Mischlings

- Mischlings are Crystal and Taiwan Bee crosses that do not fit into a defined breed

Taiwan Bee Shrimp (Caridina cantonensis)

Blue Bolt (BB)

- White shell with varying amounts and shades of blue
    - Higher grades have little to no white
    - Color tends to have multiple shades speckled in rather than a single blue shade
- Likely bred from a mutated Crystal Shrimp

Red Bolt

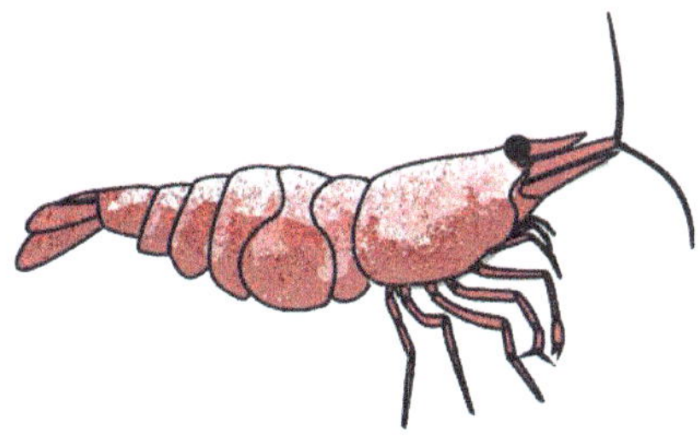

- White shell with varying amounts of light red
    - Higher grades have less white
    - Color is speckled rather than a single red shade
- Likely bred from Blue Bolt
    - Possibly a cross with Snow White/Golden Bee

Shadow Panda

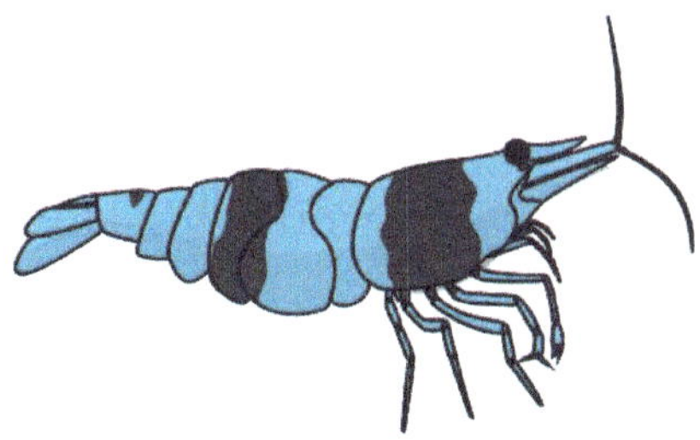

- Has alternating blueish and black stripes
    - Area closest to eyes is blue while rest of head is black
- Blue areas start out light but become darker blue with age
- Likely bred from CBS x Blue Bolt

Black King Kong (BKK)

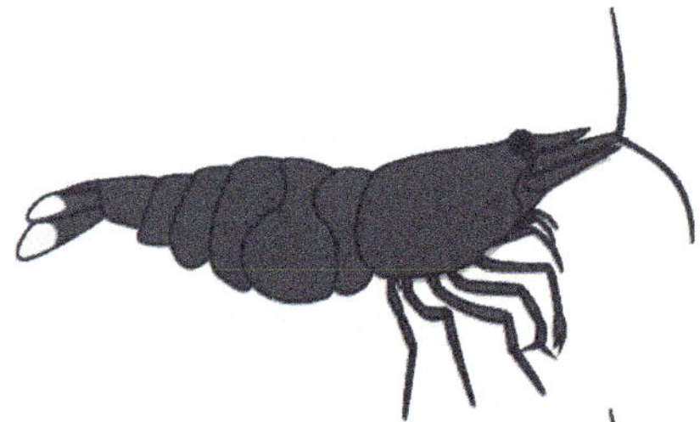

- Mostly black shell with white on the tail
    - Grade depends on the amount of white on rest of body
- Most likely bred from Shadow Panda selected most black coverage
- There are BKKs with orange eyes called OE Black Devils

Wine Red

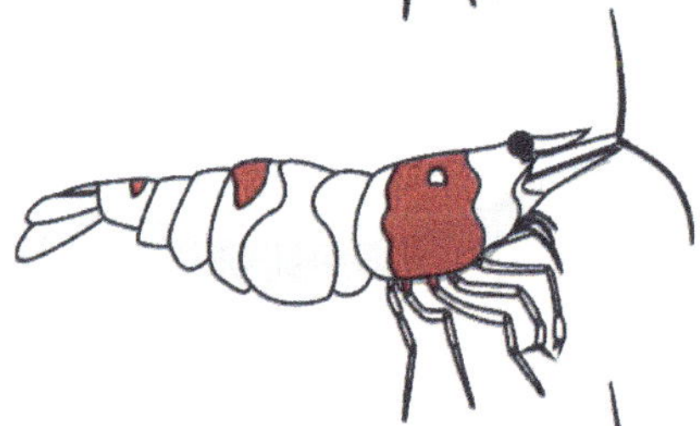

- Mix of white and red striped body
    - Has darker red than CRS
- Most likely bred from CRS x Blue Bolt

Red King Kong (RKK)

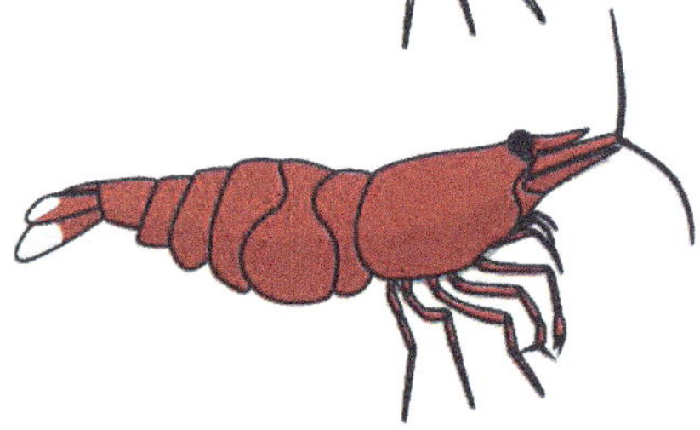

- Mostly red shell with some mottled black + white stripes
    - May have some larger white spots on the tail and smaller white stripes elsewhere
    - Higher grades have less white
- Likely from Wine Red bred for most red coverage

Green Hulk

- Green shell with some white striping
    - The green is more of a turquoise
    - Higher grades have more green area on shell
- The green tends to be unstable so this breed is not widely available
    - Green changes color with age
- Possibly a pigment mutation from Blue Bolts

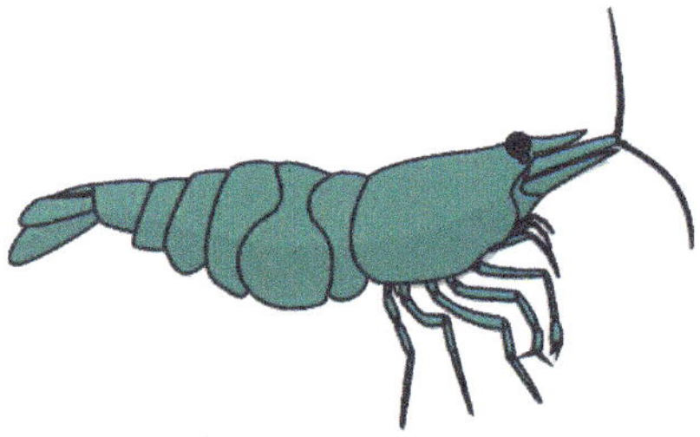

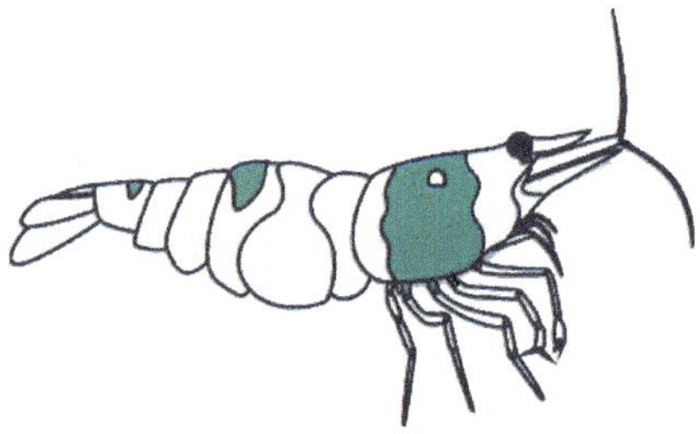

White King Kong (WKK)

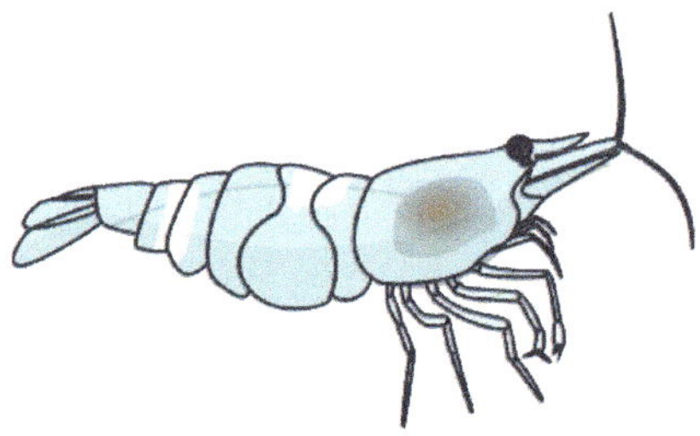

- Mostly clear (transparent) body with white patches
    - Higher grades have less white
- Origin unclear – May have been bred from Snow White but may have Blue Bolt or other TB cross

Tiger Shrimp (Caridina mariae)

Super Tiger

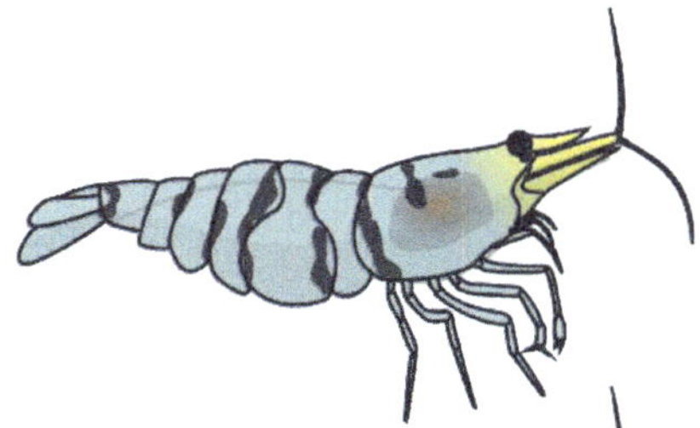

- Transparent shell with numerous black/blue stripes, light blue body, and yellow/orange head
    - May have a yellow/orange tail
    - May have a spattering of white pigmentation
- One of the first tiger shrimp to be crossed with Crystals and/or TBs to create TiBee Pintos

Blonde Tiger

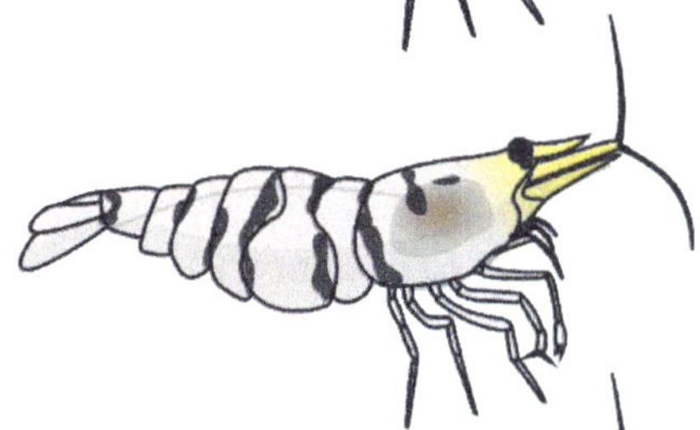

- Like the Super Tiger but with a transparent yellow/white body
- Bred from Super Tiger

Red Tiger

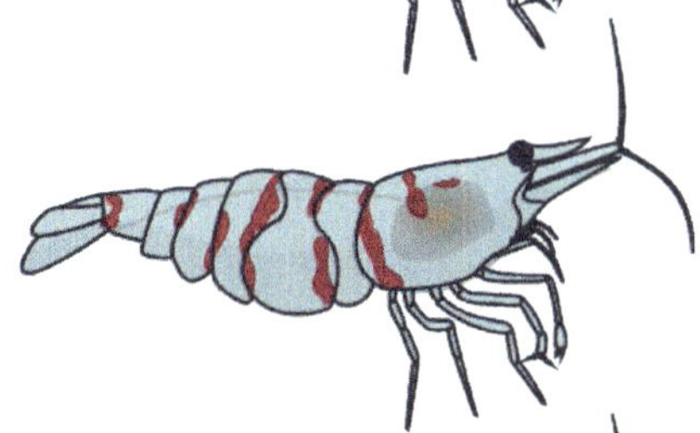

- Super Tiger with red stripes instead of black
    - Transparent body that may be tinted white, yellow, or blue
- Bred from Super Tiger with recessive red stripes

OE Red Tiger (OERT)

- Red Tiger with orange eyes and no apparent pupils
- Orange eyes are a recessive trait found in *Caridina* and *Neocaridina* spp.
    - *Caridina mariae* crosses are the easiest way to introduce OE gene for other breeds
- Bred from Red Tiger

OE Blue Tiger (OEBT)

- Blue body with numerous dark blue/black stripes
    - Higher grades have darker blue bodies that may blend in with the stripes
- Bred from Super Tiger, selected for OEs and blue body

Tiger Shrimp (Caridina serrata)

Cheetah

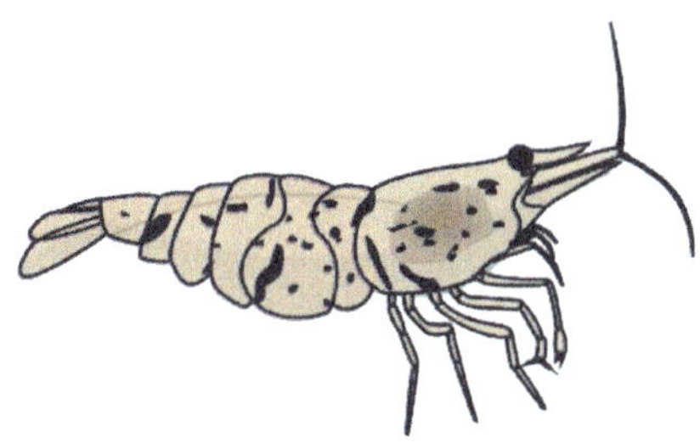

- Transparent brown/yellow bodies with lots of small black spots
    - Far more spots than other types of Tiger Shrimp
- Arguably a Wild-Type (WT) coloration of *C. serrata*[lxv] but may have been bred for more spots

Tangerine Tiger (TT)

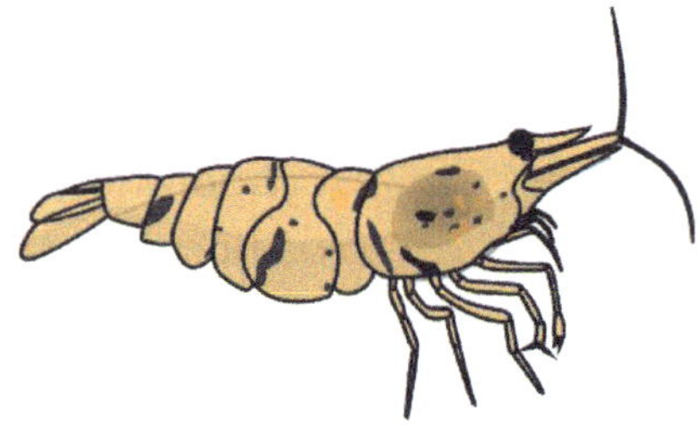

- Transparent orange body with small black stripes and spots
    - Black stripes are typically thinner but more numerous than *C. mariae*
- Bred from WT *C. serrata*
- Hardy – can live and breed in Neocaridina parameters but still prefer Caridina parameters
- Used in crossbreeding to get more stripes and spots than *C. mariae* crosses

Aura/Aurora Blue

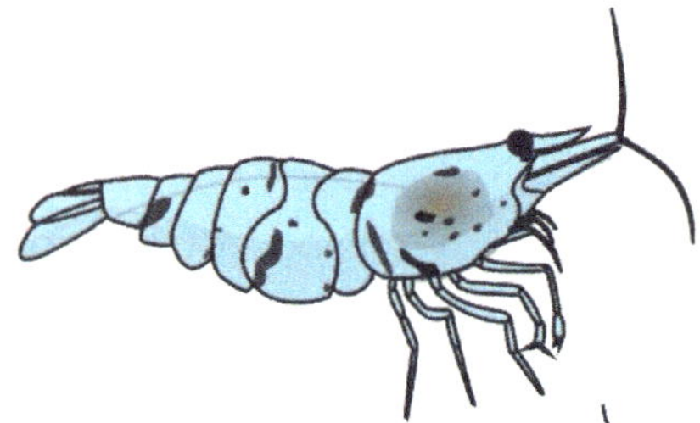

- Transparent blue body with small black stripes and spots
- Bred from TT

Green Apple

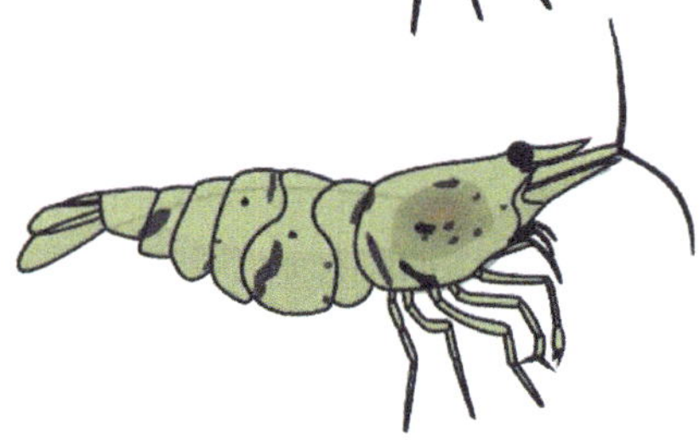

- Transparent green-tinted body with small black stripes and spots
    - Color varies depending on day and condition
- Currently an unstable breed

TiBee

A cross between <u>ti</u>ger shrimp (Typically *C. mariae*) and <u>bee</u> shrimp (Crystal or Taiwan Bee).

Red/Black Pinto

- Opaque black/red body with white spots of varying sizes/patterns
    - Higher grades tend to have more spots
- Specific patterns have their own names like German or Taiwan Pinto
- One of the first stable breeds between Super Tiger and Crystal/TB

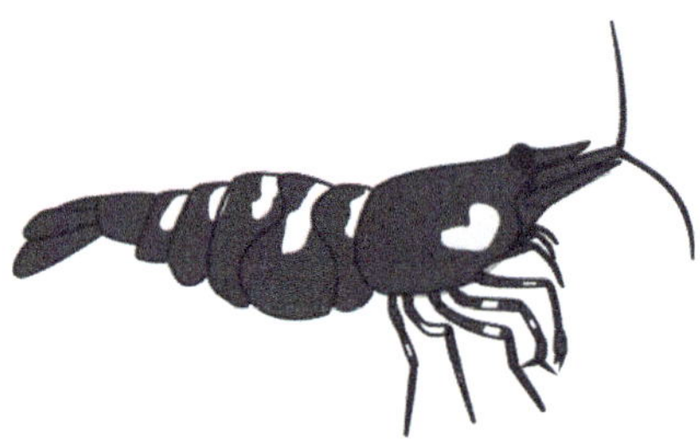

Red/Black Fancy Tiger (R/BFT)

- Opaque red/black body with thick white stripes around the abdomen
- Created by selectively breeding Pintos for defined stripes
  - Additional Super Tiger genetics may be mixed in compared to normal Pinto genetics[lxvi]
- Generally more sensitive to water parameters than Pintos

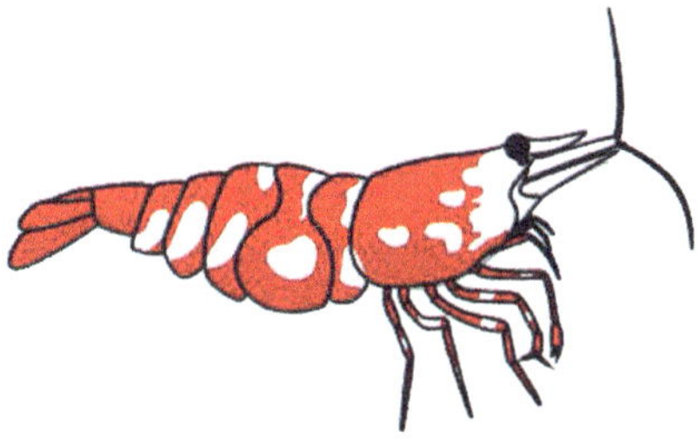

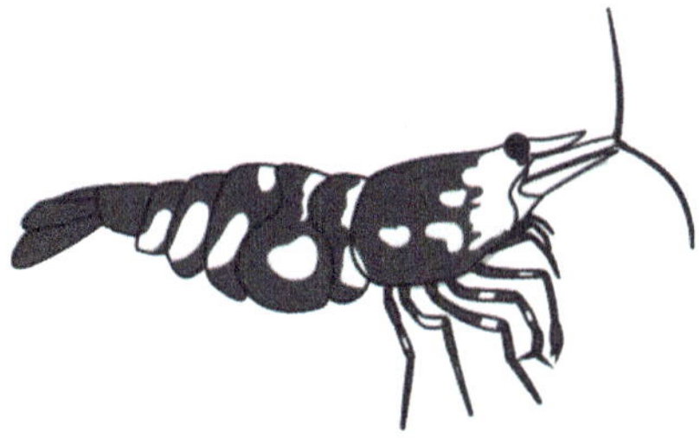

Yellow King Kong (YKK)

- Transparent yellow body
  - May have small opaque white spots on shell
- Believed to be bred from C. serrata and Blue Bolt[lxvii]

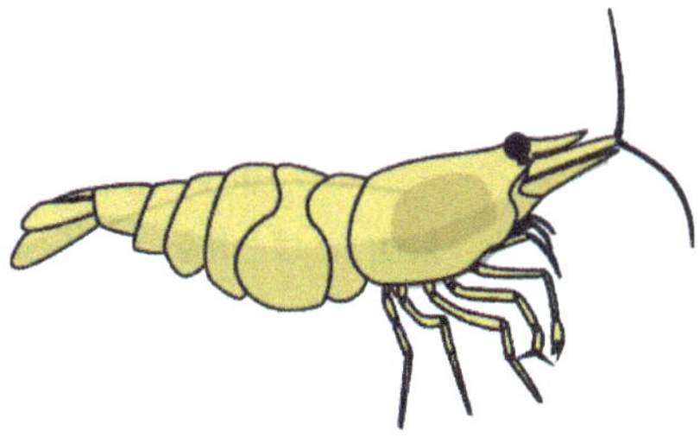

TaiTiBee

A cross between TiBees and Taiwan Bees. Some crosses also have additional tiger shrimp (*C. serrata*) genetics.

Red/Black Galaxy (Stardust)

- Opaque red/black body with white backline and spots across the body
  - "Galaxy" defined by having 5+ spots on the head (cephalothorax)
  - "Stardust" defined by 8+ spots on head
- May have fishbone pattern on back
  - These are called Galaxy Fishbone
- Bred by crossing Fancy Tiger with *C. serrata*
  - C. *serrata* cross provides fishbone pattern, more spots, along with better survival and breeding[lxviii]

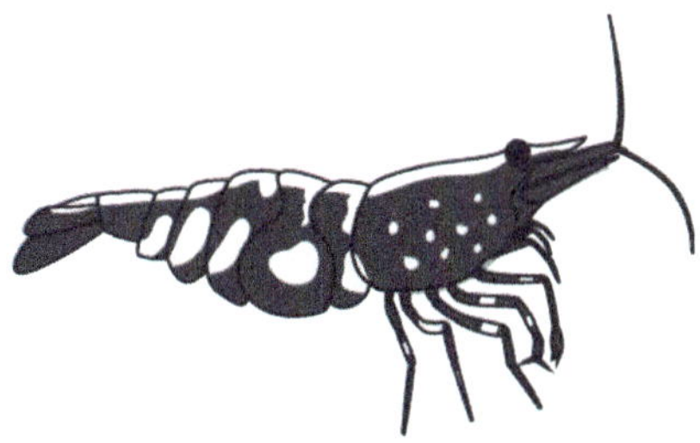

Red/Black Snowflake

- Like Galaxy shrimp but must have fishbone and even more small spots on head
    - Minimum number of spots is undefined
- Bred by selectively breeding Red/Black Galaxy

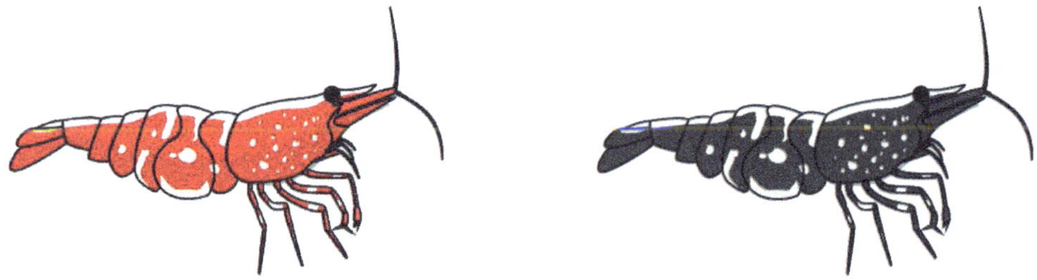

Blue/Red BOA

- Opaque black body with back stripe and spots on body
    - The back stripe and spots have a blue/red-to-white gradient that creates an iridescent or metallic appearance
    - Back stripe may be a cape or have a fishbone pattern
    - Blue/red coloration becomes more visible as shrimp get older
- Named after the iridescent qualities of boa constrictor snakes
- Likely created by crossing TaiTiBee with Blue/Red Bolt
    - Blue/Red Bolts have colored pigments mixed in with their white pigments, giving the gradient seen for white parts of BOAs
    - May have other genetics crossed in
- Even the best BOA lines currently only give 30-50% BOA offspring
- Other colors of BOAs (yellow, orange) are becoming more widely available
    - Some of these lines have red instead of black base color

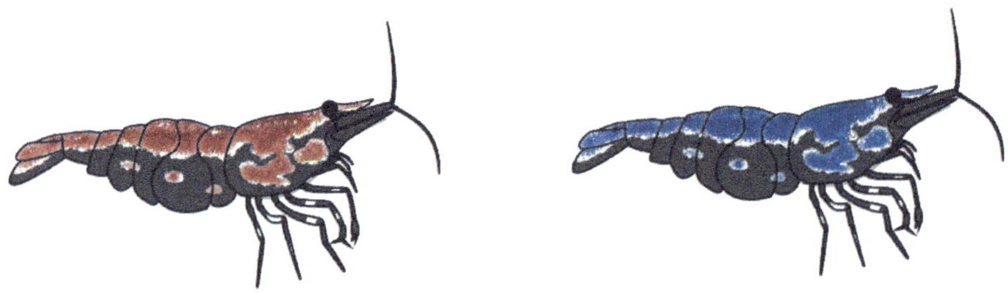

Dragonblood (Red/Blue)

- Transparent body of varying colors with blue/red stripes on abdomen, spots on head/cheeks, and brightly colored organs (yellow/white)
- Cross is unknown but likely a TaiTiBee
  - Thick blue/red stripes suggest cross with blue/red tiger (*C. mariae*)
  - Yellow organs and transparent shell suggest a cross with YKK

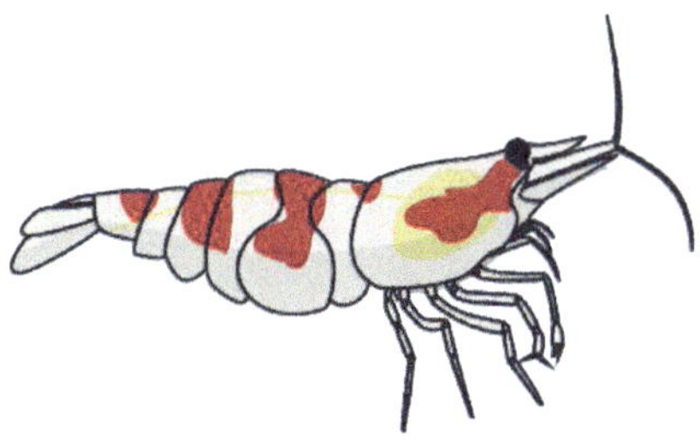

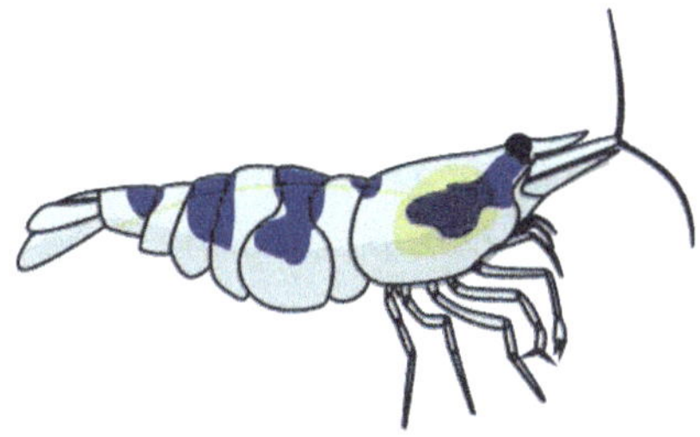

Grading

Standard Competition Requirements

- Size – Larger shrimp earn more points
- Coloration and Pattern – Unique to breed – Discussed below
- Similarity – Shrimp group with close-matching colors/patterns earn more points
- Activity – Higher activity associated with healthier shrimp

Unique requirements

Please note many breeds have even more specific characteristics so we encourage further research before entering a competition.

Multi-Colored Shrimp With Opaque Shells (Crystals, Shadow Panda, Wine Red, Green Hulk, Pinto, Fancy Tiger, Galaxy, Snowflake, BOA)

In addition to standard requirements, higher grades have...

- Well-defined separation between colors
- "Spider legs" AKA legs with alternating stripes of color
- Fully opaque coloration

Multi-Colored Shrimp with Transparent Shells (All Tiger Shrimp)

In addition to standard requirements, higher grades have...

- Well-defined separation of colors
- Spider legs
- Colored spots/stripes that are well-defined, consistent, and opaque

Single-Colored Shrimp With Opaque Shells (Black King Kong, Red King Kong, Snow White, Golden Bee, Blue Bolt, Red Bolt)

In addition to standard requirements, higher grades have...

- Opaque coloration with consistent shading
- Some may have small spots of other colors if those spots are consistent
    - Like the white spots on King Kong tails

Single-Color Shrimp With Transparent Shells (Yellow King Kong, White King Kong)

In addition to standard requirements, higher grades have...

- Completely transparent shell
    - Few to no opaque colored spots on body unless spots are consistent and well-defined
- Consistent coloration along body and across individuals

Parameter Control Methods

pH & KH

KH is easy to change in perfectly controlled amounts. pH is more difficult because every tank has slightly different water conditions so we cannot say exactly how much a certain method will change pH. It requires careful experimentation, so it's best to always be cautious and make changes slowly.

Lowering pH/KH

#1 - Reverse Osmosis/Deionized (RO/DI) Water

RO/DI should produce water with 0 dGH/dKH and neutral pH (pH doesn't really matter for RO/DI water because it shouldn't have enough acids or bases for a meaningful measurement). Mix RO/DI water with tap water to reduce pH (p17), KH (p21), and GH (p23). Adding different water too quickly may cause stress to shrimp so be sure to follow all recommendations for a proper water change (p151).

Spring water is an option for *relatively* pure water, but it is not generally as pure as RO/DI water.

#2 - Buffering Substrate

Buffering substrates are made of volcanic soil that releases acidic compounds into the water. This lowers pH below 7 and removes KH from your tank. The substrate can only remove so much KH throughout its lifetime, however, so they eventually lose the buffering and pH-reducing effect. Adding buffering substrate to water with higher KH depletes the substrate faster, so it is generally used with relatively pure water that has low KH. Once the substrate is depleted, pH must either be stabilized by adding new substrate or adding sufficient KH. An additional problem is that many active substrates are aquasoils that contain ammonia, which is toxic at higher pH.

See the Active Substrate section (p63) for more information on function and precautions.

#3 - Botanicals

Driftwood, peat moss, chola wood, and Indian almond leaves (catappa leaves) leach acidic compounds called tannins, which reduce KH and pH. Peat moss and Indian almond leaves also lower GH. Please note that botanicals are generally best for small changes. They do not have a huge impact on parameters unless we add a LOT – literally dozens of leaves.

In addition to reducing pH, there is some scientific evidence that tannins help protect our aquatic friends from diseases and improve healing after injuries (p82). Dealing with shrimp disease is never fun, which is why we always recommend having at least one item that releases tannins in your tank.

One thing to be aware of is that tannins tint the water slightly brown. Enough tannins create a **blackwater environment** that is desired by certain breeds of fish. If brown-tinted water is not desired, then soak the wood/moss/leaves in boiling water for at least 24 hours before adding them. This also helps sterilize the décor and let it sink. Be aware that there is a tradeoff, as the décor releases tannins into the boiling water so less goes into a tank, thereby reducing the impact on pH, KH, and shrimp health.

Peat Moss NOT Spaghnum Moss

While spaghnum moss and peat moss come from the same plant, they are not interchangeable. They have different harvesting methods that result in very different effects on water quality when added to your tank. Spaghnum moss is harvested directly from the living plant before being dried and used to maintain a moist environment for house or garden plants. In contrast, peat moss dies, sinks to the bottom of a bog in Canada, then gets mixed in with a bunch of other dead plant and animal matter before being harvested and dried. As a result, peat moss is far more acidic and provides additional nutrients to the water, while spaghnum moss has no benefit to an aquarium (although it can be great for paludariums and terrariums). If you're ever confused, just remember this rhyme:

Peat is neat.
Spaghnum is dumb.

Is it an unnecessarily aggressive saying toward regular spaghnum moss? Possibly - but that also makes it easier to remember.

#4 - Water Changes

If the local water is softer than the tank water, then adding dechlorinated tap water to a tank is often the cheapest way of lowering pH and KH. Keep in mind that tap water may introduce unwanted nutrients (ammonia, nitrates, etc.) or metals (from rusted pipes) into a tank, so always test it thoroughly before use and use a dechlorinator to remove heavy metals.

#5 - Acid buffers

Acid buffers are an option to lower pH and KH but **are generally not something we recommend**. Acid buffers are powerful and may cause large pH changes if used incorrectly. It is common to hear about someone adding a few drops of acid buffer to the tank, seeing no change in pH, then adding more buffer. They still see no change, so they add even more. Now the pH drops, and it drops A LOT because the KH was being eaten away each time an acid buffer was added. As discussed before, a tank is vulnerable to

huge pH swings at lower KH, so adding too much acid buffer is a quick way to kill our shrimp friends.

If done right, acid buffers are effective at lowering KH, but we prefer to use the other methods mentioned above because they have a more gradual effect on the water so mistakes are less dangerous.

How to Raise pH and KH

Please note that suddenly adding KH can cause a rapid and stressful pH increase. It's always best to start with small changes and monitor parameters closely.

#1 - Remineralizers

Remineralizers are commercial powders with minerals that dissolve quickly into water. Some raise pH, KH, and GH, while others only raise GH, so be sure to clearly read the product description before purchasing. Remineralizers with RO/DI water (p93) offer complete control over raising pH, GH, and KH, so this is our preferred method.

#2 - Baking Soda

Baking soda is the cheapest and most readily-available option to raise KH in a tank. The stuff we find in our kitchen is made of pure food-grade sodium bicarbonate, so it's completely safe for shrimp tanks. As always, **we recommend adding it to new water outside of the tank, then pouring the water in slowly to avoid sudden pH changes**. The larger the change we're trying to make, the longer we take to add new water .

#3 - Crushed Coral/Aragonite, or Wonder Shells

These products/materials are made of calcium carbonate that dissolves slowly in water to raise pH and KH (and GH). They change the ratio of minerals by an unknown amount, typically by increasing calcium. This may be useful if the tank lacks calcium but may also *cause* molting problems if too much is added.

Each one may be added to the substrate or filter, but we recommend keeping crushed coral in a mesh bag in the filter so it is easily removable. Start with 1 tsp per gallon, wait a few days to observe changes, then adjust accordingly. Wonder Shells do not need to be kept in a mesh bag and should be added according to manufacturer instructions or slightly less than they recommend initially.

#4 - Water Changes

If the local water has higher pH and KH than a tank, then adding dechlorinated tap water may be a cheap solution. As mentioned in Lowering pH and KH (p262), this may introduce unwanted nutrients (ammonia, nitrates, etc.) or metals (from rusted pipes) into a tank, so always test it thoroughly before use and use a dechlorinator to remove heavy metals.

#5 – Substrate Vacuuming

When excess waste builds up in a tank, the decomposition process causes a potentially dangerous decrease in pH and KH. Removing waste through substrate vacuuming (p151) is one potential fix to frequent pH/KH drops.

#5 - Alkaline buffers

As with acid buffers, we are not fans of these. They're expensive and don't have to meet food-grade standards so they are worse than baking soda in every way.

Controlling GH

How to Lower GH In an Aquarium

Please note that all options to lower GH also lower KH.

#1 - Water Changes

The easiest and most controlled option to lower GH is to add softer water to a tank during water changes. The GH goes down in proportion to the difference between the tank water and the new water. For example, if we do a 30% water change by adding water that has 0 GH, then we lower a tank's GH by 30%.

If the tap water is softer (lower GH) than the tank water, then the easiest and cheapest option to lower GH is to do a water change. Many of us don't have soft tap water, but we can purchase RO/DI, distilled, or spring water from the grocery store or the local fish store.

Another option to get softer water is by purchasing a reverse osmosis or distillation system. These are two different methods of purification that create RO/DI water (p88). A brief comparison is provided below, along with recommended products for shrimp keepers.

See our Recommended RO Unit & Distiller[lxix] at shrimplyexplained.com/book-links/.

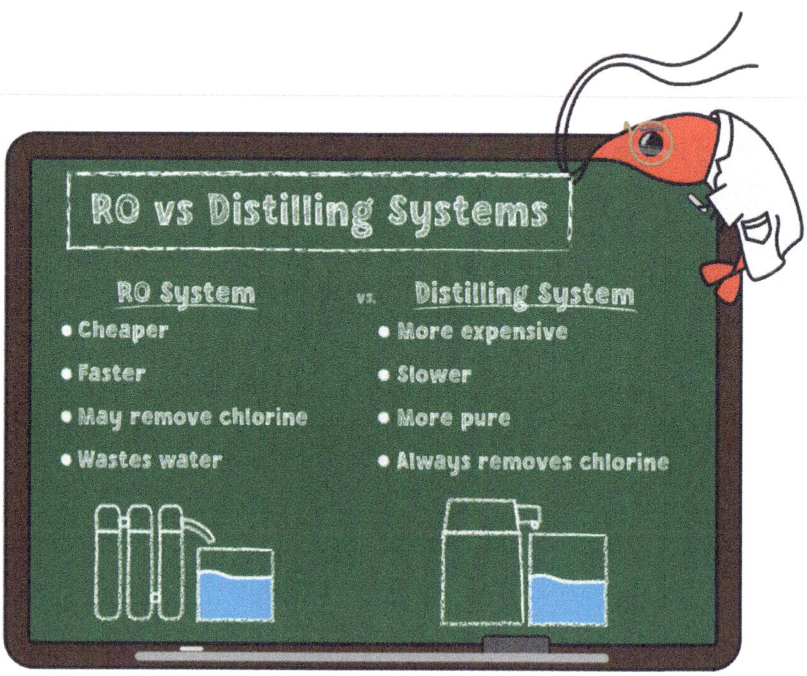

#2 - Peat Moss

Peat moss contains compounds that bind minerals to remove them from the water, thereby lowering GH. This is a biochemical process known as **chelation**. Lowering GH with peat moss is not an exact science due to differences between batches of moss, so water parameters should be carefully monitored during initial use. It also can only hold so many minerals, so it is not sufficient for extremely hard water. Lastly, peat releases tannins which are beneficial in fighting disease but also give water a brown tint (p82).

How To Raise GH in An Aquarium

#1 – Remineralizers

Remineralizers are powders made of calcium, magnesium, and other trace minerals that dissolve into the water. *Do not add these to the tank directly.* Instead, add the powder to new water, mix it in until the desired GH is reached, then add the new water to the tank slowly.

Shrimply Explained™ recommends *shrimp-specific remineralizers* because those ensure the right ratio of minerals. Also, be aware that some remineralizers affect KH, while others do not.

See the Remineralizer section (p93) for an in-depth breakdown of different remineralizers and our recommendations.

#2 – Water Changes

If tap water has higher GH, then simply doing a water change increases GH in a tank. This is the cheapest way of increasing mineral content but is area and time dependent. Water from a single sink can vary significantly depending on the cycles of the water treatment facilities so it is always important to measure parameters before each water change. As we've mentioned before, tap water also does not necessarily have the right ratio of minerals, even when GH is in range. Also, always be aware of GH creep when using tap water during water changes.

#3 - Crushed Coral/Aragonite or Wonder Shells

These products are made of calcium carbonate that slowly leaches into the water over time to increase GH, KH, and pH. They are excellent sources of calcium, but they do not contain any magnesium so molting problems may develop if the water does not already have sufficient magnesium. As such, always start with a small amount, measure the effects over a few days, then adjust as necessary. We recommend putting crushed coral or aragonite into a bag so it is easy to remove from the tank as needed.

A few Wonder Shells cost between $10-30 while a bag of aragonite or crushed coral cost between $20-30 typically. This makes these products among the most expensive options for raising GH, but they last for a few years so they can be a worthwhile investment.

Controlling Ammonia/Nitrites/Nitrates

Lowering Nitrogenous Compounds

The best way to lower ammonia, nitrites, and nitrates is through water changes with water that has lower levels of these compounds. Please note that some tap water has these compounds in it, so always test tap water before adding it into a tank.

Additionally, adding fast-growing plants is another way to consistently reduce nitrogenous compounds. Fast-growing plants, in particular, are very good at soaking up nitrates. Their unlimited access to oxygen lets them absorb it better than fully submersed plants. One way to treat a water source with nitrogenous compounds in it is to fill a bucket with the water and place floating plants in it, then leave it for a few days to let the plants soak up those nutrients. Add the water to the tank once the plants have cleaned it.

As a side note, we've seen hobbyists often detect ammonia in their tank, only to discover that their test is inaccurate. Seeing a false positive of 0.25 ppm ammonia is quite common with liquid test kits. To ensure reliable testing, always wash the test tubes thoroughly. To verify a test kit is working properly, try doing a control test with water that is guaranteed not to have ammonia, like RO/DI, distilled, or spring water from the store.

It's always good to test your tests!

Molting Problems: Cause & Solutions

Molting is critical to the shrimp life cycle but it's also dangerous. Shrimp need the right water parameters, diet, and must be healthy enough to withstand the stress of molting. If anything goes wrong, then a failed molt happens, causing the shrimp to get stuck in their old shell and die. It is extremely frustrating as a shrimp keeper because, once one gets stuck, there is nothing that we can do to help it. Any attempts we make to pull or cut off the old shell are likely to cause additional stress and reduce the chance of recovery.

While there's nothing we can do for an individual shrimp with a failed molt, we can prevent it from happening to other shrimp. To do that, we need to understand causes and symptoms of molting problems.

Understanding Shrimp Molting

Shrimp have a hard shell made of minerals to protect them from predators and their environment. It works great, except that the shell doesn't grow with the rest of their body. To get out of this problem, shrimp go through the molting process repeatedly throughout their life, discarding their old shell and forming a new larger shell. This process is difficult for a few reasons:

1. The act of squeezing out of a restrictive shell is extremely exhausting and stressful. *If you've ever had to take off a tight wetsuit or dress, then you know what this feels like.*

2. They are more exposed to predators and environmental fluctuations while the new shell hardens. *Imagine being left out in the wilderness without any clothes on.*
3. Their body undergoes rapid change as it absorbs minerals from the old shell to soften it before molting. *We don't have a good metaphor for this one.*

Signs of Molting Issues

To identify a failed molt, we first want to understand what a healthy molt looks like. For a healthy molt, a shrimp shimmies its head out of the old shell then flicks backward to release the old abdominal shell, all in just a few seconds. Any longer than that and there is likely a problem. The most obvious signs are when the **shell breaks improperly between the carapace and abdomen** (White Ring of Death) or when **a shrimp gets stuck with the molt on their abdomen or legs**. They may survive anywhere from a few minutes to days afterward, but they eventually die if they can't escape their old shell. Sometimes, shrimp do escape their shell but end up with deformed features, as seen in the picture above. These deformities are often healed after the next molt when the shell softens.

The bent "nose" (scaphocerites and antennae) is a sign of molting complications. Shrimp generally recover from these complications after the next molt.

Identifying the White Ring of Death (WRoD)

The WRoD is commonly misidentified because most sources describe it as simply a break in the shell between the carapace and abdomen. **A small break happens during every molt**, so it's common for new shrimp keepers to see the crack and start worrying – but it's okay. A WRoD is identifiable if one or more of the following characteristics have been present **for more than 24 hours**:

1. **A break larger than 2mm** - the shrimp is either *extremely* close to molting or is experiencing a problem.
2. **Scrunched edges of the shell at the breakpoint** - The shrimp has tried and failed to get the shell off, and likely won't be able to.

Even the WRoD isn't necessarily a death sentence, as roughly half of all shrimp with these symptoms still molt successfully, if it happens within 24 hours of the symptoms appearing. Maybe it needs a new name...

Other Symptoms of Failed Molts

Occasionally, a **shrimp successfully escapes their old molt but dies shortly afterward**. This can happen due to a sudden change in water conditions after molting. The shrimp may also have died from stress caused by molt preparation which could indicate the shrimp was old or unhealthy.

The last sign is not so obvious. It's a combination of **seeing dead shrimp every few days despite seemingly good parameters**, along with **no signs of successful molts in the tank**. You should be able to see at least partial leftover molts if you're checking your tank periodically, as it takes a few days for shrimp to eat their shells. That being said, it's harder to find signs of molting in a heavily planted tank with a lot of places to hide.

Diagnosing The Problem

If we see a shrimp with a molting problem, we need to ask the following questions:

1. Was a large water change done recently? Large changes in water quality (especially hardness) can force a molt to happen before a shrimp is ready.
2. Have there been other successful molts recently? If so, then the problem may be a genetic defect or another issue with the individual shrimp and may not be a tank-wide problem.
3. Does our colony have a varied diet? Molting problems may come from a lack of one or more nutrients, especially cholesterol. More on this later.
4. Are all water parameters in range? If not, then unhealthy water parameters lead to additional stress that may overwhelm a molting shrimp.
5. Are we sure the mineral ratios are correct? When using tap water, we may have the right GH but may not have the right ratio of minerals in the water. More on this later.

After asking these questions, we may suspect that the molting problem is likely a one-off and not a concern to the rest of the tank. If so, observe the tank closely for other successful molts. If we suspect the problem may be tank-wide, then the following sections explain how to assess water parameters and diet.

Short Story: A friend reached out because they started having molting problems in multiple tanks despite the shrimp being fine a few months before. There had been no recent changes in maintenance routine, no new plants or chemicals that might be toxic, and yet 2-3 shrimp were dying from the WRoD each day. After talking with her, we found out the shrimp were mainly getting leftover fish food consisting of frozen artemia and bloodworms. The day she started feeding more diverse foods (Hikari Shrimp Cuisine and Algae Wafers), the molting problems disappeared within a day.

Takeaway: While there is nothing wrong with feeding artemia and bloodworms occasionally (good source of cholesterol), such a limited diet can be an issue. In addition, artemia and bloodworms have very high protein content (40-60%), which may also cause problems. Commercial foods tend to include more diverse ingredients that prevent molting problems and improve overall health.

Molting Problems Due to Water Quality

Molting problems occur due to poor water quality or a lack of the right minerals in the water. If any water parameters are out of range, then check out our Water Parameter Control section (p261). **If all parameters seem to be in the right ranges, then the most likely problem is a mineral imbalance. This is especially likely if we're using tap water**.

"OBJECTION! I have the right GH levels!"

That may be true, but water can still have an unhealthy mineral ratio of minerals for molting. This is because a number of different minerals react with the GH test, including calcium, magnesium, manganese, copper, zinc, cadmium, barium, and others. If GH is entirely from calcium or cadmium, then there are going to be molting problems.

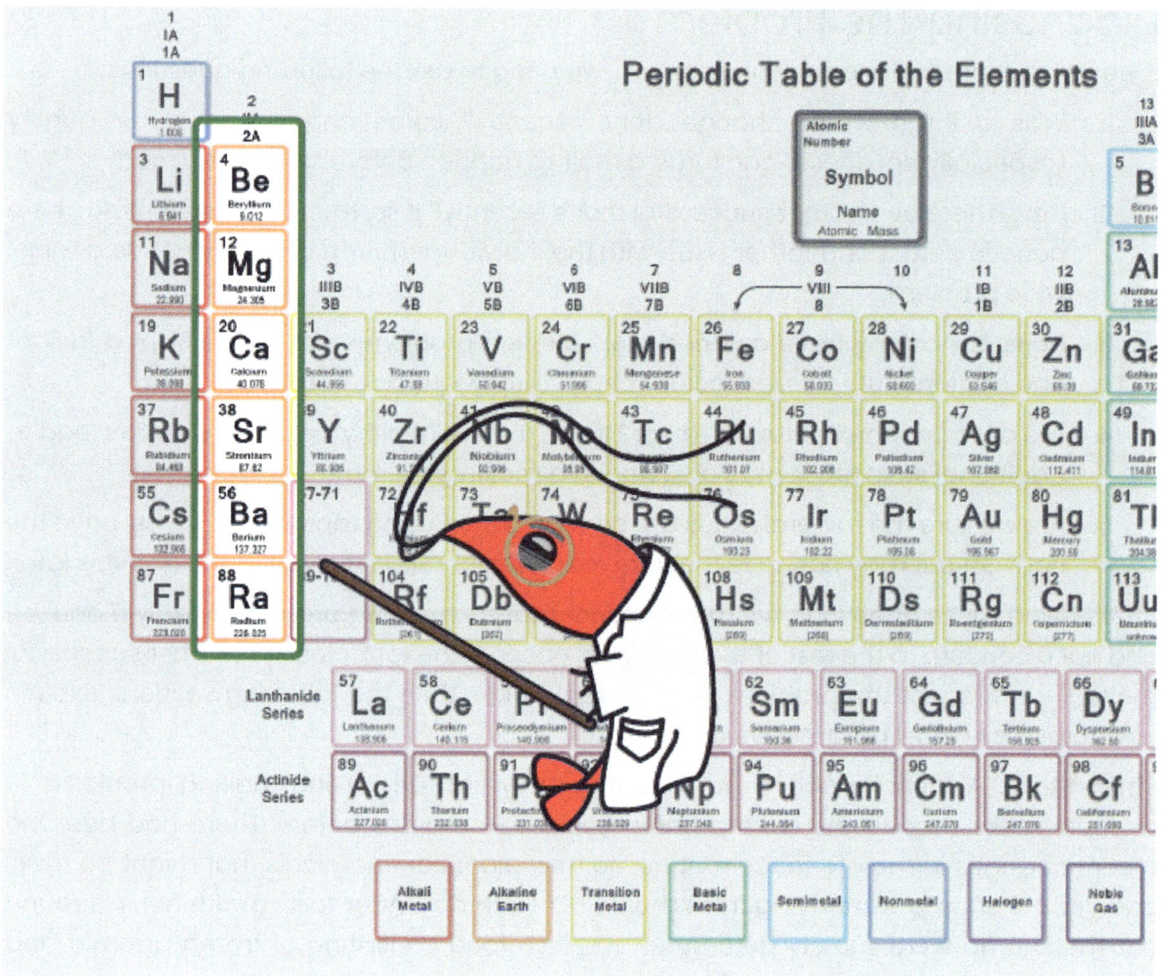

Molting problems due to incorrect mineral ratios are mainly a problem for Neocaridina since they're often kept in tap water. Caridina are typically kept in remineralized RO/DI water, so the most common cause of molting problems is old age or a lack of diet diversity or specifically a lack of cholesterol.

What Are the Right Mineral Ratios? (For Tap Water Users)

When using RO/DI water with a remineralizer like SeaChem Equilibrium or SaltyShrimp GH/KH+, then there should not be an issue with mineral ratios. If molting issues are still happening, then continue onto the Foods To Prevent Molting Problems *section (p274).*

While there can be many different minerals in water, the main two we are concerned with are calcium (Ca) and magnesium (Mg). Ca is critical for proper shell formation because roughly 30% of the shell is made from calcium carbonate (Truong, Hines, & Emerenciano, 2023). Too little Ca leads to soft shells, while too much Ca makes the shell stiff and brittle (Taylor, Gilleard, Allen, & Deheyn, 2015), leading to lower growth and survival (Adhikari, Chaurasia, Naqvi, & Pillai, 2007). Too little or too much Mg prevents proper Ca uptake, leading to issues with molting and reproduction (Tavabe, Rafiee, Shoeiry, & et al., 2015). **Shrimp receive most of their minerals from the water, so the aquatic ratio of these minerals is most important, rather than the mineral ratio in food** (Truong, Hines, & Emerenciano, 2023).

The recommended ratio of Ca:Mg is approximately 2.5-3 parts Ca for every 1-part Mg (by weight), based on a few different pieces of evidence. First, by analyzing water quality data from streams where Neocaridina originate (Han, Hsu, & Fang, 2019), we find their native ratios of Ca:Mg range from 2.3-2.8:1 throughout the year (Hung, Yang, Lai, & Li, 2020). In addition, an experiment with another freshwater shrimp in the Caridea family, *Macrobrachium rosenbergii*, tested different levels of water hardness and varied Ca:Mg ratios (Tavabe, Rafiee, Shoeiry, & et al., 2015). They found that *M. rosenbergii* had the most frequent molts and best offspring quality when they added 38.8 mg/L of Ca and 12.9 mg/L of Mg - a ratio of 3:1. Lastly, remineralizers like SeaChem Equilibrium and Flip Aquatics Shrimple GH include a ratio of 3:1 Ca:Mg by weight in their products (p93). Salty Shrimp remineralizers appear to have a ratio of about 2:1 based on our tests.

The good thing is that, on average, tap water in the US has a 3:1 ratio of Ca to Mg (Pehrsson, Patterson, & Pery, 2008). That being said, this ratio varies depending on location and time of year, with some tap water having no Ca or Mg, others having only Ca, and still others with only Mg. The changing mineral content is another reason why tap water is not ideal for shrimp keeping.

Common recommendations to prevent molting problems include adding crushed coral, wonder shells, seashells, or other forms of calcified rocks. These solutions are made of 95% calcium carbonate so have very little magnesium. It could help shrimp if their problem is a lack of calcium in the water. On the other hand, if calcium levels are already high, then these additives may make a problem worse.

Determining Tap Water Mineral Ratios

The best way to figure out the levels in tap water is to look up the **Water Quality Report** from the local municipal district. These typically include Ca and Mg levels so we can calculate the ratio. From there, we can figure out which minerals, if any, need to be supplemented to get a Ca:Mg ratio of 3:1.

If the Water Quality Report is not available or does not have the right information, then it's also possible to test the mineral ratio in a tank using the instructions and calculator provided on our website[lxx].

Fixing Ca:Mg Ratios

One of the easiest ways to adjust the Ca:Mg ratio in a tank is by adding a remineralizer. Since remineralizers have healthy Ca:Mg ratios, adding them is guaranteed to push a tank's ratios closer to the healthy range. It's okay to add a remineralizer even if it pushes GH out of range. We've tested this by putting Neocaridina in water ranging from 2 dGH to 30 dGH for two months[lxxi]. During this time, they continued to grow, molt, and reproduce normally.

That being said, remineralizers can be expensive and it is wasteful to add unnecessary minerals to a tank. Instead, the following sections explain which ingredients to use to adjust the Ca:Mg ratio.

Adding Ca

If we need to supplement calcium, then there are a couple of options. One is to add crushed coral or wonder shells to the tank. These dissolve over time, releasing calcium carbonate to raise GH and KH. This is an imprecise method, so we recommend adding just a small amount to a tank at first, then monitoring for any changes.

The more controlled option is to add calcium sulfate ($CaSO_4$), otherwise known as gypsum, which is what most remineralizers use as their calcium source. We explain how to calculate the amount of gypsum to add shortly.

Adding Mg

For magnesium, the cheapest and most controlled option is to add magnesium sulfate ($MgSO_4$), AKA Epsom salt, which is what typical remineralizers have in them.

Links to purchase reliable gypsum and epsom salt are available at shrimplyexplained.com/book-links/[lxxii]. There are other options, but the main goal is to avoid any scented salts and additives.

Dosing Ca or Mg

To properly dose Ca or Mg, we need to know three things:

1. Water volume
2. Product molecular weight
3. Quantity of Ca or Mg already present

We also need a scale that measures milligrams, along with all equipment for a water change.

Water Volume

While a tank may be a 10g (37L), at least some of that volume is taken up by substrate, rocks, and other decor. For an accurate estimate of water volume, measure the length of the front (L), the side (W), and the distance from the substrate to the waterline (H). Multiply these together to get the volume (V) in cubic inches or centimeters. For reference, one gallon equals 231 cubic inches and one cubic cm equals 0.001 liters.

If using gallons, convert to liters using the following: 1g = 3.79L

After we have your volume in liters, subtract 10-15% (depending on amount of décor) to be on the safe side. We can always add extra minerals, but it's harder to take them out.

Product Molecular Weight

The portion of the product that is Ca varies depending on the chemical formula. If using **anhydrous** $CaSO_4$, then the molecular weight is 136 g/mol. If using calcium sulfate dihydrate (gypsum), then the molecular weight is 172 g/mol.

Quantity of Ca and Mg Present

Get the concentration of Ca and Mg from the local Water Quality Report. Use this table to help determine what to dose:

	Ca	**Mg**	**dGH**	**ppm**
Quantity Present (mg/L)	?	?	?	?
Max recommended (mg/L)	48	16	10	179
Min recommended	27	9	6	107

If the Ca or Mg values are above the maximum recommended values, then we suggest using purified water when mixing in new minerals. The amount of Ca and Mg changes with the percent of water removed (e.g. 25% water change with pure RO/DI water removes 25% of Ca).

Calculate the total amount of Ca and Mg present using the following formulas:

Total Ca (mg) = Tank Volume (L) * Ca (mg/L)

Total Mg (mg) = Tank Volume (L) * Mg (mg/L)

Calculating Amount to Dose

Once we have the *Total Ca* and *Total Mg*, we can figure out how much Ca or Mg we need to add to the tank.

Adding Ca

1. If your Mg is within the recommended range, then multiply by 3. This gives you the *Desired Ca*.
2. Subtract *Total Ca* from *Desired Ca* to get *Ca Diff*.
3. Divide *Ca Diff* by X to get the weight of product you need to add to the tank.
    a. If using calcium sulfate **anhydrous**, use .29 for X.
    b. If using calcium sulfate dihydrate (gypsum), use .23 for X.
    c. These decimals express the ratio of calcium in the compound based on molar mass.
4. Dissolve that weight of product into water (amount 10% of your tank volume), then do a slow water change to add the new minerals in without stressing out your shrimp.

Adding Mg

1. If your Ca is within the recommended range (27-48 mg/L), then divide it by 3. This gives you the *Desired Mg* in your tank.
2. Subtract *Total Mg* from *Desired Mg* to get *Mg Diff*.
3. Divide *Mg Diff* by .20 to get the weight of magnesium sulfate you need to add to the tank.
    a. The decimal expresses the ratio of Mg in magnesium sulfate based on molar mass.
4. Dissolve that amount of magnesium sulfate into water (amount 10% of your tank volume), then do a slow water change to add the new minerals in while reducing stress.

Foods To Prevent Molting Problems

When looking for solutions to molting problems online, we often find suggestions to feed calcium-rich foods to shrimp. There is at least one research paper supporting this method, which showed feeding snail shells to shrimp (*Marsupenaeus japonicus*) led to improved survival and weight gain (Moss, Ishikawa, Koshio, & et al., 2019). Other papers contrast these findings, showing that supplementing Ca at 1-2% reduced survival rate (Davis & Gatlin, 1993) or had no effect (Sukadi, 1991).

In short, Ca supplementation may or may not benefit shrimp. While these studies were not specifically done on our dwarf shrimp species, a review paper states **that shrimp benefit more from Ca supplementation in the water instead of their diet** (Truong, Hines, & Emerenciano, 2023).

Again, this depends entirely on what's missing from the water. **If the water is already calcium-rich, then feeding a calcium supplement like GlasGarten Mineral Junkie can make molting problems worse**. On the other hand, if the water is magnesium-rich, then it may reduce molting problems. The more we know about our water, the better, which is why shrimp keepers have the most success with remineralized RO/DI water (p93).

Cholesterol-Rich Foods Reduce Molting Problems

In the literature, we found references to what researchers call "Molting-Death-Syndrome" in crustaceans. This syndrome is described as the "inability of the larvae to completely shed their old carapace or integument of the chelipeds or walking legs during molting" (Tian, Yang, Yu, & et al., 2020), which sounds pretty similar to what we see in our dwarf shrimp.

The solution they've found to this Molting Death Syndrome is to supplement **cholesterol** in a shrimp's diet. Adding 0.1-2.0% cholesterol to a crustacean's diet has been shown to improve survival rates, molting rates, and weight gain[1,2]. This happens because cholesterol is one of the main compounds used to create molting hormones. Without the right levels of molting hormones, shrimp can't control their molt timing properly. Therefore, cholesterol supplementation helps ensure they make the right amount of hormones at the right time.

Foods That Include Cholesterol

In the studies we looked at, scientists used pure cholesterol powder to supplement their food in a controlled way. We looked into that option but a 500mg container is around $50, so it's not a reasonable solution for most shrimp keepers.

So, what foods *can* we feed our shrimp that include cholesterol?

Luckily, many commercially available foods already have some cholesterol through ingredients like fishmeal, fish oil, and krill oil. The earlier these additions are on the ingredient list, the more cholesterol the food contains. Some products with them include Fluval Shrimp Granules, Hikari Shrimp Cuisine, Aqueon Shrimp Pellets, Sera Shrimp Food, and others. These foods have fishmeal added to them, which is approximately 0.5% cholesterol (Hjorth, Doncheva, Norheim, & et al., 2022). Unfortunately, most labels don't show the actual percentage of cholesterol on them so it's tough to estimate how much shrimp actually get.

A cheap and readily available option is egg yolk, which has roughly 1% cholesterol content (USDA, 2019). The best way to feed an egg yolk is to hard boil an egg (no salt or anything else in the boiling water), then separate the yolk from the surrounding egg white. After that, feed a small portion of the yolk to see how long it takes shrimp to eat it and to prevent

leftovers from polluting your tank. If they devour the piece quickly, then feed slightly larger pieces going forward. Freeze the yolk to store it for a later date.

Other high cholesterol options include frozen bloodworms and artemia, which have a cholesterol content of 0.5-0.8% (Lytle & Lytle, 1990). These animals also have hormones in them that closely mimic shrimp breeding hormones, so feeding artemia or bloodworms periodically can increase shrimp breeding rate.

Please note we are not sponsored by "Big Worm".

To be clear, **egg yolks and any of the other foods mentioned must be rotated with other foods**. It's especially important that they get both animal and plant protein, as discussed in the Food chapter (p137). Just like humans, **shrimp are healthiest when they eat a varied diet**. Cholesterol just happens to be something they may not always get enough of, so it's worth adding in periodically – potentially a few times a week until molting problems stop.

Shrimp Diseases

Shrimp, like all organisms, are susceptible to diseases that impact their health, including harmful bacterial growth or parasites. As responsible shrimp keepers, it is our job to try to prevent or treat these diseases so our shrimp stay happy and healthy. To do that, we need to know what to look for and how to effectively treat what we find. Let's get into it.

Side note: Caridina tend not to show the diseases in this section due to a combination of factors. First, Caridina tend to be bred in indoor aquariums, so they are not as exposed to natural diseases, whereas Neocaridina are often bred in large, exposed ponds.

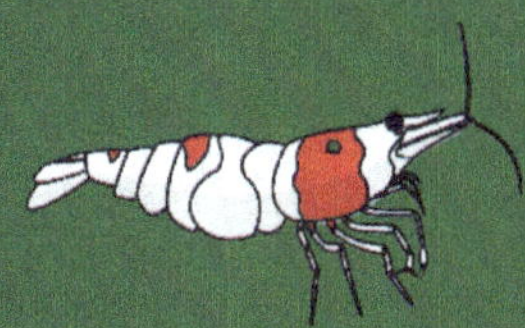

Second, Caridina require cleaner water than Neocaridina and are kept at lower population densities so they are less likely to experience disease. Third, the acidic environment they live in may prevent transfer of disease. Fourth, Caridina are sensitive and may die from poor conditions before showing visible symptoms of disease. Fifth, Caridina tend to have thicker shells that do not show signs of internal diseases.

Organisms That Live on Shrimp

***Scutariella japonica, Vorticella sp., Monodiscus sp., Holtodrilus truncatus*, & similar**

There are a variety of **epibionts** (organisms that live on another organism) that latch onto your shrimp around their rostrum, gills, and legs, making it look like they have little mohawks and beards. These epibionts eat bacteria and other small organisms on surrounding surfaces or in the water. They normally concentrate around the rostrum and gills because these areas are most exposed to water flow, so they offer the highest chance of collecting food. To be clear, we are only talking about short, hair-like epibionts. There are larger worms that look like longer, thicker hairs which we discuss in a different section.

At low densities, the presence of these epibionts is likely not harmful (Maciaszek, Swiderek, Prati, & et al., 2023). When they suction onto a shrimp's shell, they do cause minor abrasions on the surface, but these do not penetrate the shell and are healed after each molt. Epibionts like *S. japonica* and *Vorticella sp.* may even be beneficial because they keep the shell around them clear of potentially harmful organisms. While the impact of these specific epibionts on Neocaridina has not been carefully studied, there is evidence of interactions between crayfish and a species that is closely related to S. japonica, showing that the crayfish grow larger when the epibiont is present (Ames, Helms, & Stoeckel, 2015). As such, it is currently incorrect to call these epibionts parasites.

Monodiscus kumaki *on shrimp rostrum - Credit: R. Maciaszek et al. 2018.*

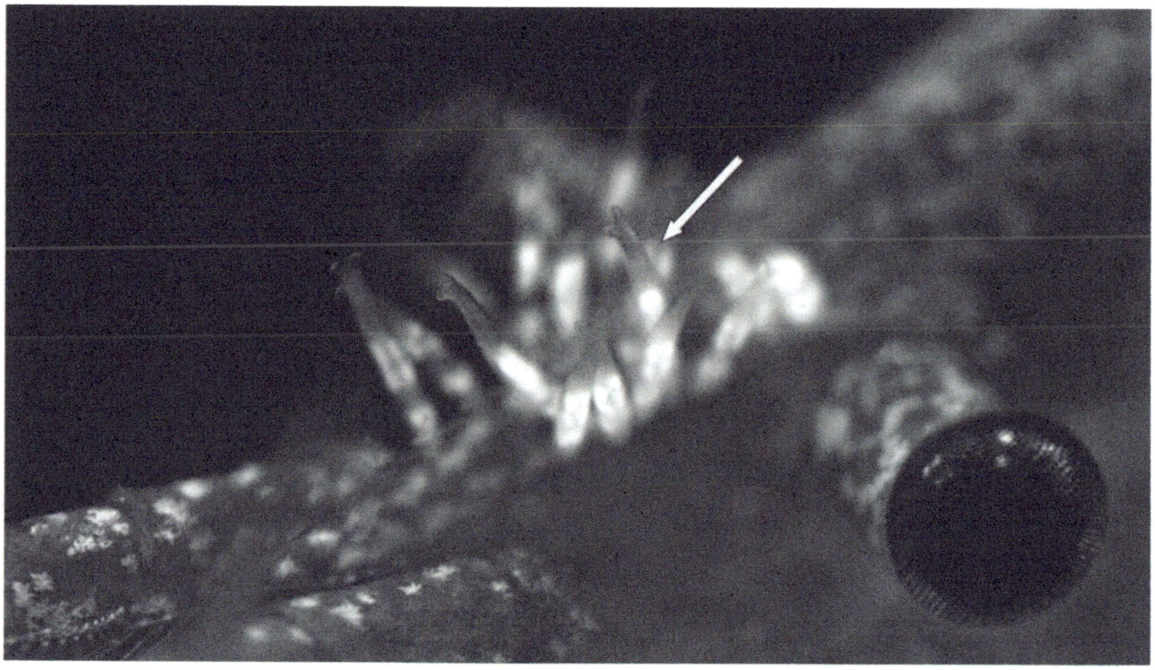

The reason they are listed in this section is because their relationship with shrimp can arguably *become* parasitic at higher densities. For example, *S. japonica* lays eggs in the gills of Neocaridina, which is believed to clog up the gills and prevent them from getting oxygen, potentially leading to death.

Figure 8. Evidence of epibiont activity: eggs of ***Monodiscus kumaki*** sp. nov. removed from the shrimp with its molt and visible melanization of the damaged cuticle (**A**); characteristic melanized damage in branchial chambers caused by egg laying by temnocephalids observed in live shrimp (**B**); round imprints of *Holtodrilus truncatus* on eyes and cuticle of the shrimp (**C**); white scale bar = 100 µm. (Photographs by R. Maciaszek).

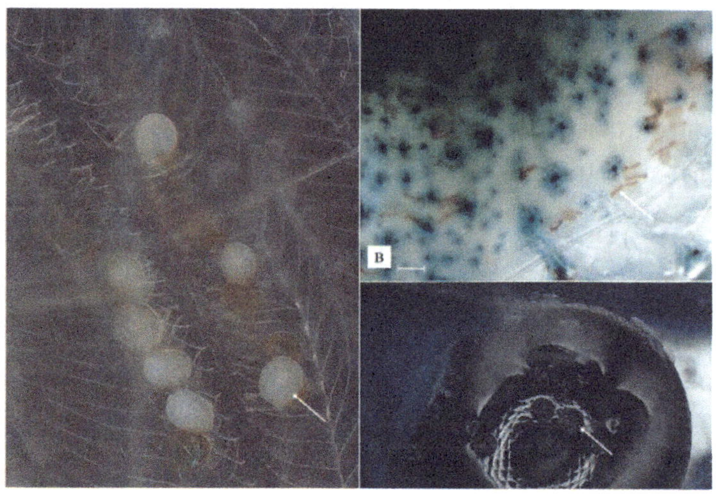

Signs of epibiont activity - Credit: R. Maciaszek et al. 2023.

Our takeaway from the research is that **a few little white hairs here and there are alright but, if we see a lot of them, then it's time to do something**. We might have to treat individual shrimp, but that's just treating a *symptom* of the problem. Treating individual shrimp is like buying an air filter if your house was on fire, instead of calling the fire department. The epibionts are almost certainly elsewhere in your tank, and if only individual shrimp are treated, they will soon multiply to become a problem again.

Instead, we need to understand why the epibionts are multiplying and what we can do to stop them.

Why Epibionts Multiply (Hint: Bad Water Quality)

Epibionts multiply when there's plenty of food available. The main food source of epibionts like *S. japonica* is any smaller microorganism. Smaller microorganisms become plentiful when water quality is poor and there are a bunch of nutrients/food available for microorganisms to eat and multiply. **This means an outbreak of epibionts is a sign of bad water quality**. That's really useful information, since bad water quality may be harming our shrimp!

Poor water quality happens for a variety of reasons, including:

- lack of filtration
- low dissolved oxygen levels
- unhealthy/dead plants and animals
- substrate disturbances
- lack of maintenance
- excessive bioload
- overfeeding

We have to figure out which of these problems apply to our tank. Once we do, our epibiont problem goes away and our shrimp get better conditions. It's a win-win - plus the epibionts are still in the tank and act as lagging indicators if water quality decreases again, which may help save our shrimp.

Solving Water Quality Issues

The first thing to do is stop feeding for a few days so no extra nutrients pollute your tank. Next, consider the list of potential causes for bad water quality above to figure out what might apply to our tank. It never hurts to start cleaning filters more regularly, add an air stone, or add an additional sponge filter to purify and oxygenate the water.

Those little translucent hairs sticking up past the rostrum are M. kumaki

For immediate but temporary treatment, we can also dose hydrogen peroxide in our tank (p295). At the recommended dosage, hydrogen peroxide kills off a lot of unwanted bacteria and microorganisms in the water column without harming our shrimp (Furtado, Serra, Poersh, & Wasielesky, 2014) or disrupting our cycle. Hydrogen peroxide doesn't stick around for long, breaking down quickly into water and oxygen to help aerate our tank.

If any shrimp are absolutely covered in hair-like epibionts, then a 30 second salt dip may also be recommended. Click here for instructions on how to salt dip shrimp safely.

Other Treatment Methods

Many sources online say to studiously watch for molts and remove them, since they may carry eggs from organisms like *S. japonica*. **If we want to completely remove epibionts from our tank, then we do need to remove their eggs**. We would argue that this desire to eradicate them comes from a misunderstanding of what they do to shrimp and ignores the benefits of having them in a tank, if only to act as an indicator of water quality issues. At Shrimply Explained HQ, we see a few *M. kumaki* on a shrimp periodically but never see an outbreak that causes concern.

Shrimp Breeding Theory

If we're trying to breed the largest and healthiest shrimp, we may actually *want* to have some of these epibionts in the tank, based on the evidence that similar species help other crustaceans grow larger (Ames, Helms, & Stoeckel, 2015). *This extrapolated theory is by no means proven* but would be fascinating to experiment with if we are breeding for competition and trying to optimize conditions. The caveat is that all experiments were done in a natural setting where crustaceans were exposed to environmental stressors that may not be present in healthy tanks, so any benefits may be minimal.

Additionally, we'd need to remove the epibionts a few months before the competition using a salt dip and may want to do this before their last molt so there are no abrasions on the shell surface. This probably isn't worth doing but it is an interesting idea.

Epibiont Examples

[lxxiii]*See video examples at www.ShrimplyExplained.com/book-links.*

Singling Out Holtodrilus Truncatus

Holtodrilus truncatus is epibiont that is relatively uncommon in the shrimp keeping hobby but is starting to appear more frequently as of 2026. It has a segmented, transparent white or brown body that reaches up to 5mm in length (Maciaszek, Swiderek, Prati, & et al., 2023). A slight bulge on one end of *H. truncatus* indicates a sucker that lets it attach to shrimp, typically under their legs or on their rostrum. Overall, it looks very much like a small leech – the only differences are that leeches tend to grow much larger and move around the aquarium, while *H. truncatus* tends to stay on its host.

Like *S. japonica* and *M. kumaki*, it lays its eggs inside the gill chamber, but these eggs tend to be 6-7x larger than the other two (0.6-7mm vs 0.1mm diameter, respectively). Another difference is that *H. truncatus* has teeth, which is a little discomforting. While *H. truncatus* hasn't been shown to harm shrimp aside from superficial discoloration and imprints on the shell, they do seem more likely than *S. japonica* or *M. kumaki* to negatively impact shrimp. The size of their eggs may block off water flow through the gills, and their suckers may be strong enough to form small cracks in the shell. That being said, they are a branchiobdellidan like the epibionts studied in Ames, Helms, & Stoeckel, 2015, and their presence is associated with lower rates of other epibionts, so they may still be harmless or beneficial.

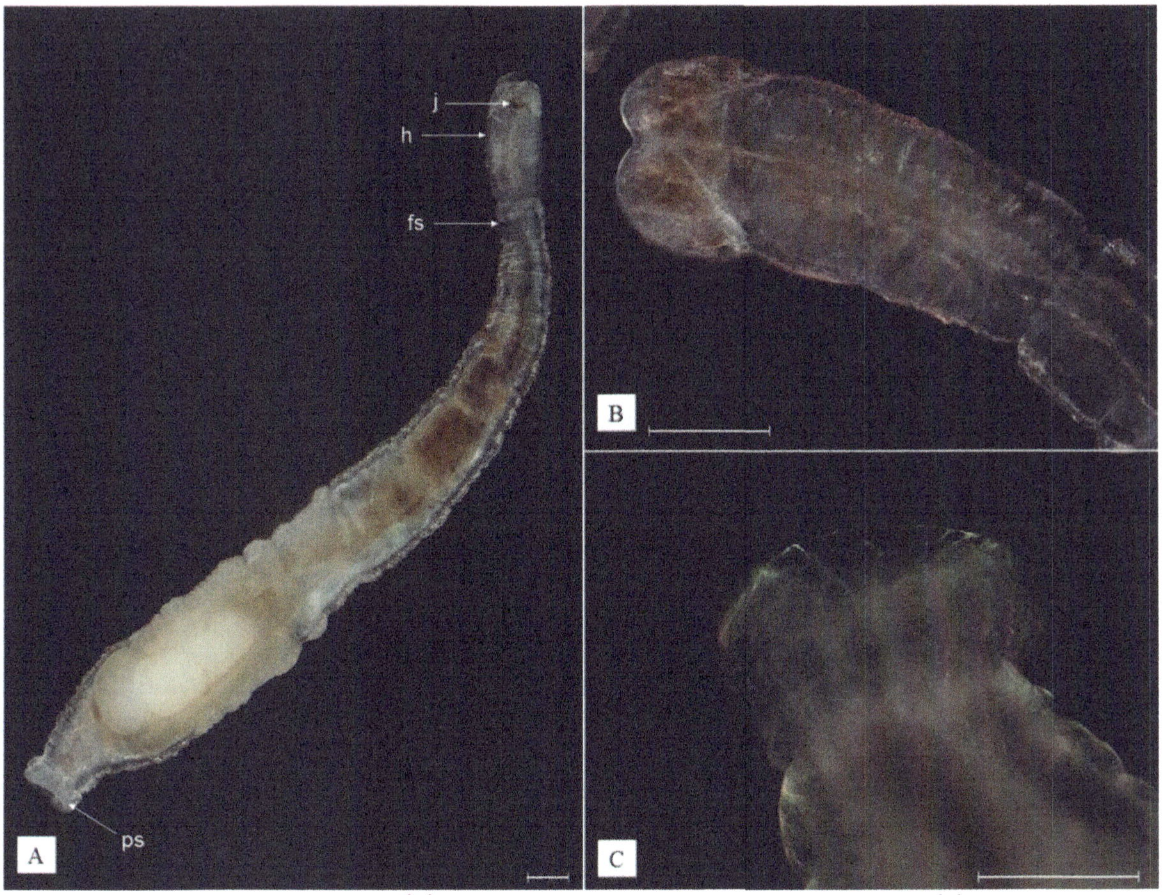

Holtodrilus truncatus: dorsal view (A); lateral view close-up image on the head (B), and lateral view posterior sucker (C); fs—first segment of the trunk region, j—jaws, h—head region, ps—posterior sucker; 100um scale bar. All credit to R. Maciaszek.

From a hobbyist perspective, *H. truncatus* is much larger and more obvious than *S. japonica* or *M. kumaki*, so we understand wanting to get rid of it. They fall off easily with a salt dip (individual treatment) or hydrogen peroxide treatment (entire tank). The eggs are more protected, so an additional treatment 2-3 weeks later is recommended to kill any newly hatched individuals.

H. truncatus Examples

[lxxiv] *See video examples at www.ShrimplyExplained.com/book-links.*

Bacterial Infections

Rust Disease

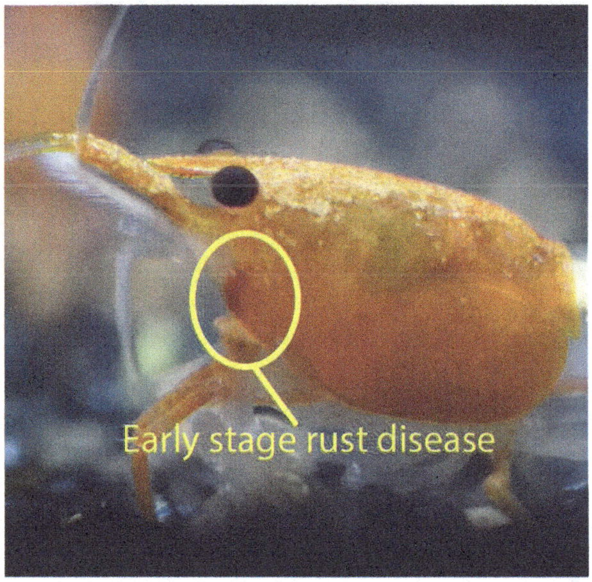

The red rash-like spot at the edge of the "cheek" is an early stage chitinolytic infection. It's often difficult to differentiate from regular color splotches.

Rust disease - also known as a chitinolytic infection, black spot, brown spot, or box burnt disease - is the name for a bacterial infection that eats away at chitin in shrimp shells (Rowley, Coates, & Whitten, Invertebrate Biology, Chapter 15: Bacterial diseases of crustaceans. p412, 2022). It also discolors the shell, so infected shells may have dark, ragged edges or holes that get progressively worse over time. Rust disease is caused by a variety of different bacteria that produce compounds which degrade the chitin in the outer shell, leaving shrimp vulnerable to other infections and health issues (Rowley & Coates, 2023). Affected shrimp may survive for weeks or months without treatment but they are more likely to die.

Since so many different bacteria can cause this infection, it is found all over the world in both natural and farmed crustaceans. That may sound alarming, but here's an example of something similar:

> Our mouth is filled with roughly 700 species of bacteria at this very moment (Deo & Deshmukh, 2019). Under certain conditions, like poor dental hygiene or a high sugar diet, some of these bacteria produce acid that eats away at our teeth and causes cavities. At other times, they are a harmless part of the oral microbiome.

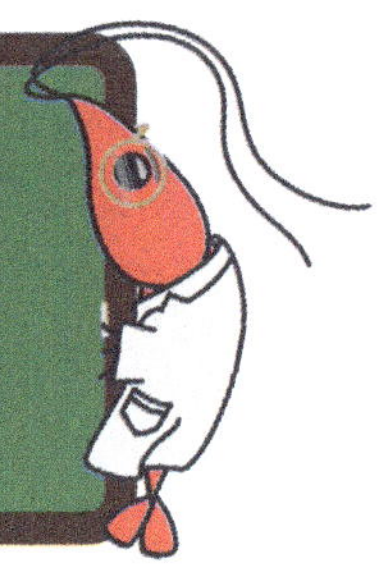

A tank is much like our mouth. Under healthy conditions, the bacteria in our mouth help break down waste. Chitinolytic bacteria perform a similar and fundamental role in marine and freshwater systems. There are many microcrustaceans in the water that die and their leftover shells have resources that may be used and recycled back into the environment. Chitinolytic bacteria are present to break down this leftover chitin and improve the health of the system (Dosta & et al., 2015). Certain environmental factors can change how these bacteria act, however.

According to Chapter 15 of *Invertebrate Pathology*, "disease outbreaks often occur in degraded environmental conditions associated with build-up of waste products, reduced oxygen tension, or pollutants" (Rowley, Coates, & Whitten, 2022) Basically, a poorly

maintained tank. In addition, infections are more likely to occur when shrimp are kept at higher temperatures, fed a poor diet, or as they age and molt less frequently (Rowley & Coates, 2023).

Treatment

Since we know that bad conditions cause chitinolytic infections, we know how to treat them! See our section about How To Improve Water Quality (p279) to learn more.

We may also treat individual shrimp with particularly bad infections using hydrogen peroxide. Please note this is just treating a symptom of a problem and does not protect our other shrimp unless hydrogen peroxide is dosed in the entire tank.

Muscular Necrosis & Milk Shrimp Disease

Muscular necrosis and milk shrimp disease (AKA milky hemolymph syndrome [MHS]) are both internal bacterial infections that cause a shrimp's hemolymph (blood) to turn white (Nunan, Poulos, Navarro, & et al., 2010). The milkiness is caused by *Vibrio spp.* and possibly other bacteria that multiply rapidly inside of shrimp when their immune system cannot keep up. Most of them are low-level infections that stick around like a human cold and take a while to kill shrimp – if they do at all. There are rare examples of infections that are extremely contagious and wipe out a shrimp colony overnight (Nguyen-Phuoc, Doung, Phan, & et al., 2021). Symptoms are easiest to detect in shrimp with transparent bodies. Please note extremely contagious ones are very rare for hobbyists.

Treatment

As with other bacterial infections, muscular necrosis and milk shrimp disease typically occur when shrimp are stressed. This stress may be caused by unhealthy tank conditions or a poor diet. Unlike rust disease, certain bacteria that turn shrimp bodies white can spread, so immediate action is needed to protect a shrimp colony. We recommend a hydrogen peroxide treatment, followed by increased aeration, a water change, and filter cleaning. We also suggest stopping feeding for at least a few days to reduce the amount of nutrients entering the tank. After that, feed a small amount to gather the shrimp in one area and check for other infected individuals. Additional treatments with hydrogen peroxide may be needed if there are more infected shrimp.

With proper measures taken to improve tank conditions, most shrimp should survive. If there are no new infections after a week, then the remaining shrimp should be safe. After that, consider what aspects of your tank setup or care may have led to the problem and adjust accordingly.

Side note: We say "should" because there are still many unknowns about different types of bacterial infections.

A Word on Antibiotics

There are some sources that suggest treating bacterial infections with antibiotics (jayc, 2013). While it may work, we cannot recommend this method. Antibiotic resistance is an enormous problem for humans, as lethal bacterial infections are becoming harder to treat with available antibiotics (World Health Organization (WHO), 2023). This is due to the misuse or overuse in both humans and farmed animals, including shrimp**. The only time to use antibiotics is after consulting with a qualified doctor/veterinarian and *when following the prescription exactly***. When treatment is not followed to completion, bacteria that are resistant to antibiotics are more likely to survive and proliferate, making antibiotics less effective in the future. Even when treatment is completed according to instructions, some percentage of bacteria survive to proliferate their antibiotic resistance.

While it may seem like a small thing to use antibiotics on shrimp tanks, bacteria develop antibiotic resistance, then transfer that resistance to other bacteria (Center For Disease Control, n.d.) in or on humans. Using antibiotics improperly therefore comes with huge risks by bringing up the timeline for when a simple infected scratch may kill us again (Pepi & Focardi, 2021), as it could before antibiotics were found in 1928. To reiterate, ***if you do choose to use antibiotics, please be careful to follow the instructions exactly***.

Water Molds

White filamentous cotton-looking molds are another infection seen in shrimp and fish and are caused by *Saprolegnia* water molds. The mold commonly develops near the base of shrimp swimmerets, an area with very low water flow where debris may be trapped (Maciaszek, Kamaszewski, Struzynski, & Lapa, 2018). When molds do develop, they are often seen on berried shrimp with eggs since eggs obstruct water flow more than normal. "Fanning" behavior is intended to prevent mold but still happens occasionally, especially in poor water conditions. The infection often prevents proper fanning and may cause deaths due to failed molts if not treated. Luckily, most shrimp recover after simple treatment and improvements to water quality.

Treatment

Saprolegnia are "ubiquitous in aquatic freshwater habitats" (Bauer, Jung-Schroers, Teitge, & Adamek, 2021), but potentially harmful infections only happen when water conditions are bad, as these conditions favors the growth of pathogens and makes shrimp more susceptible to diseases (Duan, Feng, Huang, & et al., 2018). A large factor is low water temperature (<10°C or 50°F), which appears to favor *Saprolegnia* growth. Such low temperatures are rarely seen in shrimp tanks but may

happen during shipping, making it all the more important to carefully observe and quarantine new individuals. In addition, affected individuals should be treated with salt dips or hydrogen peroxide, as these are very effective against Saprolegnia (Barnes, Ewing, & Cordes, 1998). Unfortunately, almost all affected eggs are lost (Duan, Feng, Huang, & et al., 2018). Learn more about causes and solutions to bad water quality on p279.

Examples

[lxxv] *See video examples at www.ShrimplyExplained.com/book-links.*

Green "Fungus"

This disease appears as green tufts of feathery growth underneath shrimp, often buried in their swimmerets. It is occasionally found on Neocaridina but very rarely on Caridina, possibly due to the often poor and overcrowded conditions Neocaridina are bred in.

Shrimp with clado - Credit to Reddit u/Hot-cheesecake-905

All researched cases are caused by a parasitic green algae genus called Cladogonium (Maciaszek, Swiderek, Prati, & et al., 2023). There are some sources online claiming that similar-looking infections may be caused by a protist family called *Ellobiopsidae*, a parasitic organism that eats the host's reproductive organs and spreads via sexual reproduction. Some studies show that *Ellobiopsidae* can infect certain species of saltwater shrimp, but recent studies of more than 900 individual Neocaridina identified only Cladogonium infestations. According to the researchers we spoke with, Cladogonium was most likely misidentified early on as *Ellobiopsidae* so the mistake is still being cleared up.

A close-up of a late-stage Cladogonium infection.
Photo Credit: Maciaszek et al. 2023

This is a unique disease that is often misunderstood. There's a decent amount of research, so let's dig into its life cycle and anatomy to help identify and treat the "green fungus".

Cladogonium Lifecycle

Cladogonium initially floats through the water as an asexual spore (technically a zoospore) with a tail (flagellum) to help it move (Maciaszek, Swiderek, Prati, & et al., 2023). Eventually, it finds its way to its host but the exact way the algae attaches to the shrimp is unclear. It may be ingested and move around the body until it finds an area close to the surface where it can punch through the shrimp's shell. It may also enter through previously damaged areas and grow from there. Lastly, it may get stuck in body crevices and burrow into the shell from there.

Regardless of how it gets into the body, the formed structure looks very similar to a plant, as there are rhizoidal cells acting as roots that absorb nutrients from the shrimp's body, followed by basal cells that sprout a stalk-like filament out from the shell. Little sacs called zoosporangium sprout like fruit from this filament and hold developing spores. Once these spores reach maturity, they get released to find a new host.

Unlike the rhizoids, basal cells, and filament, these developing spores soon break off from the rest of the organism, so they need an energy source – in this case, chlorophyll, which is where they get their green color from. The rest of the growth is relatively transparent, which is why infections often aren't caught until they are already well-established. By the time you see green pigment, new spores may already have been released into the environment to seek new hosts.

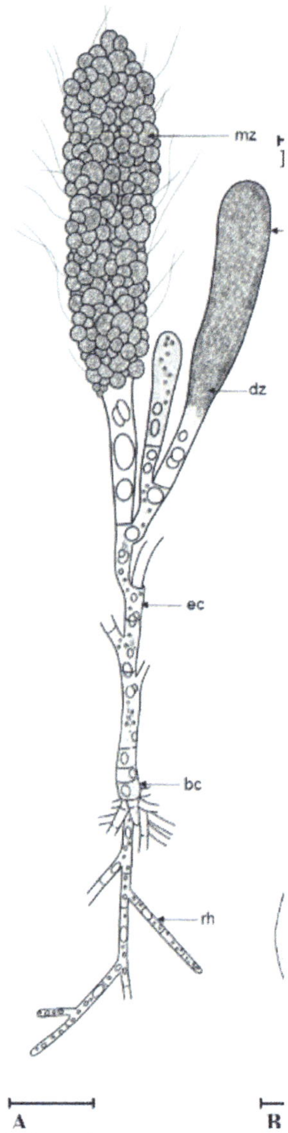

Cladogonium structure - Credit: Maciaszek et al. 2023

How Contagious Is Cladogonium?

There has not been much research on how contagious Cladogonium is so this section is largely based on anecdotal evidence from hobbyists. Regardless, it's worth discussing because it may help alleviate some concerns when found in a tank.

There are very few reports of Cladogonium spreading in a tank once infected individuals are removed. It's likely that crowded conditions from breeders and stress from shipping lead to some individuals being infected before entering a tank. In a healthy tank, Cladogonium may grow on previously infected individuals but it is rare for new hosts to be infected. One of the best reports we have is from a diligent breeder on Reddit[lxxvi]. Though not the most reliable source, they risked their colony to test how contagious Cladogonium can be.

This hobbyist received a shipment of shrimp and found eight of them were infected with Cladogonium. Infected shrimp were quarantined immediately and monitored for 10 months. After that time, the user reported "0 of the 8 shrimp have died, 3 of the cases

cleared entirely on their own within 8 months without treatment, and the remaining 5 have about 1/4th of the original fungus appearance on their abdomens." This supports other evidence that an otherwise healthy shrimp kept in a healthy environment can fight off this infection over time – or at least prevent it from spreading.

The most interesting part is what this hobbyist did with one of the infected shrimp to test how contagious the disease was. They placed an infected shrimp in a colony of around 100 shrimp and left it there for 3 months. ***No other individuals showed visible infections immediately or during the eight months after exposure***.

Again, this is just one account of the disease and needs to be taken with a grain of salt. **It does suggest that, given healthy conditions and individuals, Cladogonium is not a serious danger to uninfected individuals in healthy tank conditions**. Most hobbyists we have discussed this disease with also do not see the infection spread, although reoccurrences on the same individual do happen.

Should infected individuals still be quarantined and exposed individuals be monitored? Absolutely – there's just no need to panic and do something rash.

Where Cladogonium Infects Shrimp

To identify this disease, it helps to understand where infections are most common. Scientists surveyed over 900 Neocaridina shrimp to understand how epibionts like Cladogonium appeared on shrimp (Maciaszek, Swiderek, Prati, & et al., 2023). They did so by counting the number of times each epibiont was found in one of the four areas of the shrimp: rostral, branchial (gills), pereiopodal (legs), and pleopodal (swimmerets). Take a look at the data below to see which areas Cladogonium prefers.

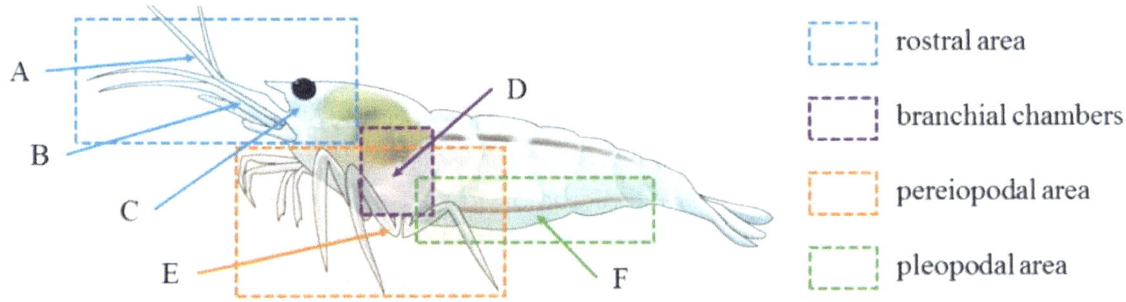

Diagram created by Rafal Maciaszek and shared in Maciaszek et al. 2023.

Variable	Odds Ratio
Rostral area (Head)	0.28
Pereiopodal Area (Legs)	9.46
Pleopodal Area (Swimmerets)	22.62

"Logistic regression output from relative microhabitat preferences (rostral, pereiopodal, and pleopodal areas compared to the branchial area) of epibionts on Neocaridina davidi.*" This table is a recreation of Table 6 from Maciaszek et al. 2023.*

The numbers can be a bit confusing so here's a breakdown. After observing the shrimp specimens for data, the researchers divided the **number of occurrences at a given area** by the **number of occurrences in the branchial area** to compare how much an epibiont favored one area over another. Therefore, if an organism showed up in the branchial area just as much as it did in, say, the pereiopodal (leg) area, then the table shows a ratio of 1.00. If an organism showed up more often in another area compared to the branchial area, then the ratio was >1.00. **This data shows that Cladogonium does occasionally appear on the rostrum and gills but prefers the legs and has an even stronger preference for the swimmerets.**

Treatment

We could not find any scientific papers containing reliable treatment methods, but many hobbyists have experimented with varying degrees of success so we are going to talk about their methods. This information is from many sources like Aquariumbreeder.com, madshrimp.com, and numerous forums.

After quarantining (learn how to set up a quarantine tank on p293), the best treatments come at the disease from two different angles: **External** and **internal**.

External Treatment

Salt baths, hydrogen peroxide, and/or botanicals are all viable options to attack the tendrils growing out of our shrimp. We recommend including botanicals in every quarantine tank and performing salt dips or treating with hydrogen peroxide every few days. Instructions for setting up a quarantine tank (p293) and treatments like salt dips and hydrogen peroxide are found later on (p295).

Internal Treatment

Since Cladogonium digs into our shrimp, treating it internally is also a good idea to avoid recurring infection. Multiple hobbyists have experimented with feeding foods soaked in Kordon Rapid Cure, which uses the antimicrobial properties of malachite green to treat infections. It is regularly used on ornamental fish in the United States but has been banned in some countries due to possible carcinogenic effects (Srivastava, Sinha, & Roy, 2004). It is up to you if you want to use it if you're in the USA.

When Not to Treat Shrimp

In some cases, an infection can progress so far that the host shrimp is unlikely to survive treatment. If the shrimp is very lethargic and does not eat food put in front of it, then the best option may be to humanely euthanize it. Instructions for doing so may be found in this section.

Recommended Cladogonium Treatment Schedule

Day 1

- Set up quarantine tank
- Salt dip infected individuals in separate container
- Treat main tank with hydrogen peroxide
- Prepare food with Kordon Rapid Cure by soaking food and then letting it dry over 24 hours

Every day after

- Salt dip infected individuals, feed small amount of treated food
    - Remove any uneaten food to avoid polluting the water

Every 3-4 days

- Replace salt dip treatment with hydrogen peroxide treatment of quarantine tank
- Repeat until Cladogonium disappears (May take 3-4 weeks)

Worms

Planaria

Planarian - DataBase Center for Life Science (DBCLS), CC BY 4.0 via Wikimedia Commons.

Planaria are translucent white flatworms with a distinct triangular head and two eyespots that they use to slowly glide over surfaces in your tank. They hitchhike in on plants and prey on shrimp after molting. Since they burrow into shrimp, it's often difficult to tell what the shrimp are dying from until we spot a planarian on the glass or substrate - or crawling out of dead bodies...

Despite their predatory nature, planaria are fascinating creatures, partly due to their reproduction. They can reproduce in two ways:

- By sexual reproduction, where they find another planarian and reproduce by laying eggs that hatch in 2-3 weeks (Southern Biological, N.D.).
- By fission, in which they *break apart and clone themselves*. They also do this anytime they get cut in half!

Planaria are commonly reported in shrimp tanks for a number of reasons. First, planaria eggs tend to be very hardy and survive many plant treatment methods, which is part of why they are such a common infestation in shrimp tanks. Second, shrimp-only tanks don't have any predators to eat planaria. Finally, hobbyists also often mistake other flatworms for planaria, so their sightings are overreported.

Planaria vs. Rhabdocoela Flatworms vs. Detritus Worms

Rhabdocoela are harmless flatworms that are commonly mistaken for planaria, due to similar body shapes and movement. The difference is that planaria always have a distinct triangular head that is easily visible. They also have two eyespots in the middle of the triangle, which are visible but can be difficult to see with the naked eye.

Detritus worms almost universally have rounded bodies with comparatively thinner proportions. They can also be differentiated by their movement, as detritus worms tend to wriggle while planaria and other flatworms glide across surfaces.

Examples of Different Types of Worms

[lxxvii]*See video examples at www.ShrimplyExplained.com/book-links.*

Treatment

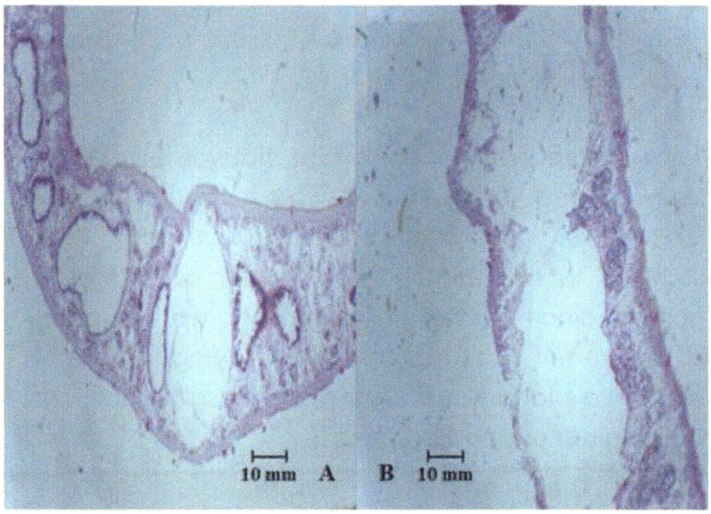

Microscope shots of fluke worms being ripped up by expanding vacuole - Source: Yamson et al. 2019

There are two common medications on the market: benzimidazoles (fenbendazole [dog dewormer], albendazole [poultry dewormer], flubendazole, etc.) and No Planaria (betel/areca nut extract). Both treatments (p297) are safe for shrimp, but they kill planaria and other flatworms, along with snails like Malaysian trumpet, pond, nerite, and mystery snails. At the dosages recommended, many ramshorn and bladder snails survive. Regardless, we recommend removing any snails that you do not want to kill. The more that are left in a tank, the more likely they are to die, decompose, and cause an ammonia spike after treatment.

The two treatments are somewhat different in how they target planaria and related species. Benzimidazoles disrupt the way these animals uptake glucose, reducing their energy supply to lethal levels (Perkins, Yazwinski, & Tucker, 2012). We could not find research on the exact mechanism explaining how No Planaria killed worms but one study compared the effect of No Planaria and a benzimidazole (albendazole) and found both separated the outer layer of skin and expanded vacuoles in the worm's body, effectively ripping it apart (Yamson, Tubalinal, Viloria, & Mingala, 2019).

Benzimidazoles and No Planaria are similar in terms of cost and how quickly they kill planaria (within 72 hours when properly dosed). That being said, No Planaria has been

shown to have a stronger effect on liver flukes (worms related to planaria) than benzimidazoles so it may be slightly more effective (Yamson, Tubalinal, Viloria, & Mingala, 2019). No Planaria is slightly easier to use because it comes with a measuring spoon, whereas we need a scale to weigh out the correct amount of fenbendazole.

There are also other treatment options available, such as planaria traps, but they often are temporary solutions that do not eradicate all planaria.

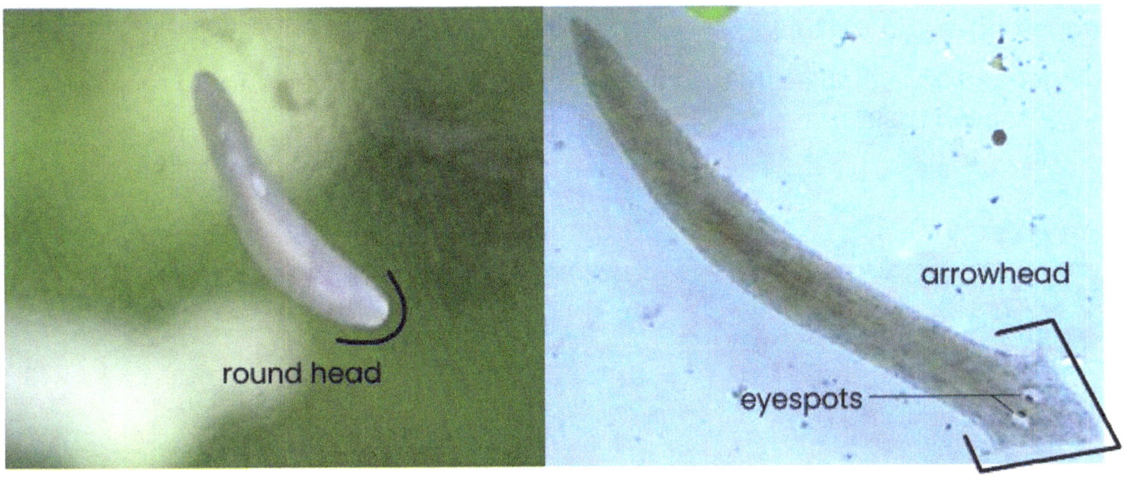

Rhabdocoela (harmless) Planaria (harmful)

Leeches

A much less common predator than planaria is the freshwater leech. These typically have long segmented bodies like earthworms, but they move like inchworms, pulling themselves along quickly instead of gliding like planaria or wriggling like detritus worms. Leeches are relatively uncommon in shrimp tanks but, when you do have one or more, they may be seen suctioned onto the outer part of a shrimp and feeding on it.

Leeches vs. Detritus Worms

Detritus worms are perfectly harmless and help break down nutrients. Leeches and detritus worms are occasionally confused because they are both roundworms of various colors. In general, detritus worms are thin and thread-like, whereas leeches are longer and thicker. They also move very differently, as leeches use their head to grip a surface and curl the rest of their body forward. In contrast, detritus worms move like earthworms, keeping their body straight while expanding and contracting to pull themselves through the substrate.

Examples

[lxxviii] *See video examples at www.ShrimplyExplained.com/book-links.*

Treatment

Leeches, unfortunately, are very difficult to treat. Many do not respond to the treatments used on planaria. **The chemical treatments that do work on leeches typically kill shrimp because they are all copper-based**. This means the only way to eradicate them is to do frequent substrate cleaning and manually pull out any visible leeches. Planaria traps also help catch them. Neither of these methods are guaranteed to eradicate them so the most effective yet radical treatment method is to break down the tank, soak everything in bleach, and restart the tank.

One good thing is that leeches attached to shrimp are relatively easy to remove with a simple salt dip (p295).

Hydra

Hydra are small white/green organisms that look like palm trees and occasionally hitchhike in on plants. They attach to surfaces in your tank with a sucker-like appendage called a peduncle, then extend a trunk into the water with tentacles on the end to catch debris and microorganisms! Hydra are actually closely related to jellyfish, so they have similar stinging cells in their tentacles to incapacitate victims. This sting is strong enough to immobilize baby shrimp, making hydra very unwelcome in a shrimp tank. NOTHING HARMS OUR BABY SHRIMP AND GETS AWAY WITH IT!

Treatment

Hydra population increases when a lot of microorganisms are present, which happens if water quality is low. Please see our section about Water Quality to address this (p279).

We don't want to just reduce the population, though... Unlike *S. japonica*, hydra have no known or potential benefits in a shrimp tank, so we want to eradicate them. To do that, we recommend dosing No Planaria or fenbendazole (p297).

If we don't want to remove snails from the tank, then hydrogen peroxide dosing also works. An effective hydra treatment is to use a syringe to squirt hydrogen peroxide directly on them.

Treating Sick Shrimp

Setting Up a Shrimp Quarantine Tank

There are going to be times in your shrimp-keeping career when you'll need a quarantine tank, whether that's for isolating new shrimp to protect the colony, having a place to isolate old or diseased shrimp, or maybe as a place to move shrimp while treating your fish for disease. Regardless of the situation, always having a quarantine tank ready is ideal.

Realistically though, most aquarists don't have empty tanks sitting around. We may not have the space for them, or we may set one up and then peruse our local fish store and think, "Huh, I want those shrimp – and I already have a cycled tank for them!" And then our quarantine tank is gone... With that reality in mind, this section covers how to set up a new quarantine tank for shrimp in just a few minutes to move our patient(s) into a safe place.

The Tank

To prepare for our patient's arrival, we need a reasonably-sized tank or container – the bigger the better since more water leads to a more stable environment – but realistically anywhere from 1-10 gallons (4-38L) will work. A bucket or large food container works as well as a glass aquarium for a quarantine tank, although clear sides are nice to be able to see our shrimp patient at any angle.

As with any tank, always clean it up before use with vinegar OR hydrogen peroxide, then rinsing with hot water (mixing cleaning agents can be dangerous – vinegar and hydrogen peroxide becomes corrosive, while vinegar and bleach creates chlorine gas).

Side Note: As animals may be stressed during quarantine/treatment, it is a very good idea to cover the top of the quarantine tank so they cannot escape. Shrimp have been known to climb out of tanks occasionally.

What To Put in The Quarantine Tank

1. **Water**

   If possible, the best option is RO/DI (or spring) water remineralized to the same GH and KH. We recommend freshly remineralized water over water from the main tank because purified water is more controlled to ensure healthy conditions. The water in the main tank may be contributing to shrimp sickness, so we want to eliminate potential contaminants. If RO/DI water isn't available, then we recommend fresh tap water matching parameters in the main tank (don't forget to dechlorinate!).

2. **Filtration**

   Our goal with a quarantine tank is to maintain the healthiest conditions possible, which means we need to oxygenate it and take care of waste. A single shrimp has a very low bioload so a simple airstone in a small quarantine tank is often enough. For larger tanks (>5g) or those with more than 1-2 shrimp per gallon, a pre-cycled filter is recommended to eliminate ammonia and increase water agitation/flow.

   Situations where we need a quarantine tank often don't allow for the week or so it takes to cycle a filter though. If we don't have a pre-cycled filter, just stay on top of water testing and do water changes if ammonia/nitrites/nitrates build up.

3. **Botanicals**

   Botanicals like Indian almond leaves and alder cones release tannins into the water, which have been shown to improve immune function in shrimp and other animals (Novriadi, Fadhilah, Wahyudi, & Trullas, 2021). We recommend quickly boiling botanicals for 2-3 minutes to remove any pesticides/contaminants on them. This does remove some of the beneficial tannins but is safer overall. Once botanicals are added to the quarantine tank, the water may take on a brown tint, indicating tannins are being released into the water.

   About 2-3 botanical pieces per gallon should be enough. Adding too many botanicals can cause pH to drop outside of the recommended range for shrimp.

That's everything we need for a quarantine tank!

Treatments

Hydrogen Peroxide Dosing Procedure

Always check compatibility of other plants and animals in the tank. Certain plants like Vallisneria tend to melt, but they do typically recover well afterward.

This is a quick, cheap, and easy method to improve water quality and treat problems like hair algae, external bacterial infections, and hydra. We found out about this method from Mark's Shrimp Tank. His YouTube channel is a great resource for learning about shrimp.

1. Calculate the tank's water volume by measuring the length (L) and width (W) of the tank, along with the height of the water (H) from the top of your substrate to the waterline (L * W * H = V). Multiply those three measurements together to get the water volume (V), then convert to gallons or liters.
2. Multiply water volume by 0.9 for a conservative estimate, considering additional volume may be displaced by décor and plants.
3. Turn off your filtration and lights.
4. Measure out 1.5ml of 3% hydrogen peroxide per 4.5L (1g) of water volume.
    - According to the research, this is approximately half the maximum dosage considered safe for shrimp (Furtado, Serra, Poersh, & Wasielesky, 2014).
5. Add the measured hydrogen peroxide to a cup of tank water, then pour it into the tank. Try to pour it all over the tank instead of in just one spot to disperse it better.
6. Briefly mix the tank water with a net or a spoon.
7. Set a timer and wait 1 hour.
8. After an hour, turn the filter and lights back on.

Salt Dip Procedure

Supplies

- Aquarium salt (Any table salt without iodine is fine)
- Measuring cup
- Small container (Holds about 1C (240g) water)
- Net
- Measuring spoon

Instructions

1. Add 1-3 tsp (6-18g) of aquarium salt to 1C (240g) of clean tank water and stir to dissolve.

a. If shrimp developed symptoms shortly after being added to the tank, then they may be residual effects of the stress of previous shipping or bad water quality from the seller. If symptoms develop 1-2 months *after* being in the tank, then the tank water may be part of the problem. In that case, we recommend remineralized RO/DI water to the same GH and KH as your tank, since this is the most pure and safe water you can get. The other option is dechlorinated tap water.

- The amount of salt depends on the resistance of the unwanted organism to salt. If there is no improvement with salt dips at lower doses, then increase the salt concentration to 2-3 tsp per cup. The amount also depends on the sensitivity of the shrimp, since hardy Neocaridina can handle 3 tsp per cup for the allotted time perfectly well, whereas sensitive Caridina shrimp may require more cautious dosing.

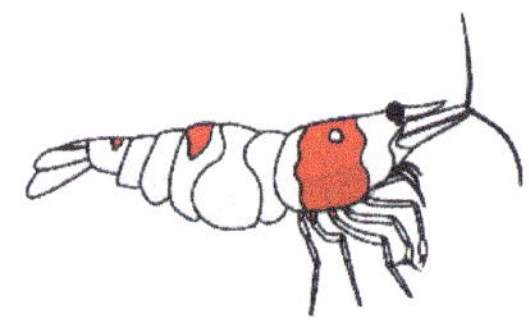

2. Once the salt is entirely dissolved, net the shrimp out of the tank and place it (still in the net) into the cup of salt water.
3. Let the shrimp bathe for 30-60s.
4. Remove shrimp from the salt water and place it back into the tank (main tank or quarantine tank, depending on the situation).
5. Repeat salt dips once daily until symptoms improve.

No-Planaria/Fenbendazole Instructions

For hydra, planaria, and some snail species.

1. Move any snails you don't want to kill into a quarantine tank. Bladder, ramshorn, and pond snails tend to survive the treatment but less hardy snails may be stressed or killed.
2. Remove any carbon filtration temporarily.
3. Dosage:
    a. **No Planaria**

        Day 1 – One spoonful per 13g (50L) (the spoon comes in the bag)

        - Always double-check product instructions for updated information

        Day 2 – ½ of Day 1 amount

        Day 3 – ¼ of Day 1 amount

    b. **Fenbendazole**

        Add 0.1g per 10g (39L) to a cup of tank water and mix well.

        - Always double-check product instructions for updated information

        Add the water to the tank.
4. Stir tank water each time a treatment is added while trying not to disturb the substrate.
5. Wait 3 days (fenbendazole) or 4 days (No Planaria).
    a. Do not feed during these days.
    b. Check ammonia/nitrite/nitrate levels daily - water changes as needed.
6. Gravel vacuum the substrate to remove dead bodies and replace ~25% of the water, following all standard water change recommendations.
    a. If the substrate is not vacuumed and there are a lot of dead planaria/snails in the substrate, then a large ammonia spike may occur
7. Check ammonia levels and perform larger water change, if needed.
8. Add a carbon filter to remove the treatment. Optional but recommended.
9. Do another 20-25% water change.
10. Wait 2-3 weeks, then repeat steps 2-9 to kill any newly hatched planaria that were in eggs during initial treatment.
    a. This step is often not necessary if the tank is only 2-3 months old because planaria need to be 5-6 months old before they can reproduce. This also means we have a few months after the first treatment before we need to do the second treatment.
11. Perform water changes every week until 100% of the water has been replaced, then add snails back in. Observe them and move them back to the quarantine tank if they show any concerning behavior. Wait until another 50-100% of the water has been replaced to try again.

Humanely Euthanizing Shrimp

As pet owners, we always want the best for our pets. Unfortunately, disease or injury happen in life and there are certain times when the peaceful end of a life may be better than the pain of treatment or the suffering caused by the continuation of that life. Since our pets can't talk, the decision to humanely end a life is left up to us as pet owners. We do not take this decision lightly.

That said, there are certain times when humane euthanasia of our pet shrimp may be required, like if they are half eaten, experience severe muscular necrosis, or have a late-stage Cladogonium infection. In these cases, the method we recommend - and the one commonly used by scientists - is to use clove oil to anesthetize, followed by freezing. This offers a pain-free way to end your shrimp's life. Let's go over a little bit of science.

Where Our Numbers Come From

We are basing our instructions off this paper published in 2018 (Li, She, Han, & et al., 2018), which tested the effect of eugenol (the main component of clove oil) on *Palaemonetes sinensis* (Grass/ghost shrimp), a shrimp of similar size to *Neocaridina* and *Caridina*. In it, the researchers found 100-300 mg/L of eugenol led to Anesthesia Stage 2 within 20-40 minutes at temperatures above 16°C (60°F). For reference, Anesthesia Stage 2 is defined by a loss of consciousness and no reaction to stimuli such as movement or light (Li, She, Han, & et al., 2018).

Clove Oil Calculations

Most clove oil products do not say the exact concentration of eugenol in them but, according to Wikipedia, clove oil is approximately 60-90% eugenol. Since our goal is to ensure anesthetization, we will assume it is the lowest concentration to give us the highest calculated dosage. The weight of clove oil is 1 g/mL, so it takes 0.5g of clove oil in a liter to get 300 mg/L, or approximately 0.8 g of clove oil for 500 mg/L.

If you only have a dropper for clove oil, each drop is approximately 0.05 g, so you need between 10-16 drops of clove oil per liter to anesthetize your shrimp within 5-10 minutes.

Humane Shrimp Euthanasia Protocol

1. Add 0.8 g/L of clove oil to a container or bag of water.
    1. If using a container, make sure you have a lid for it.
2. Shake the bag or container to mix the clove oil as best you can.
3. Add your patient (the shrimp) to the bag/container.
4. Let sit for 5-10 minutes.
    1. The patient should no longer react to stimuli. Add a few more drops of clove oil periodically until there are no reactions.
5. Place the bag/container in the freezer and let the bag freeze.
6. Dispose of body in the trash.
    1. Do NOT dispose in the yard/compost because diseases may be carried to local waterways and impact wildlife.

Alternate Methods of Humanely Euthanizing a Shrimp

1. Catch the shrimp.
2. Lay it on a flat surface.
3. Cut its head off with a knife.

This method won't feel good for the person doing the cutting but is the fastest way to put a shrimp out of its misery. What a way to end the book!

Acknowledgments

This book would never be possible without all the impressive research from curious scientists – especially Carolina Tropea and her team. Thank you very much for your hard work. Additionally, we want to give special thanks to Cassandra Tiensivu for spending many hours (Dozens? Hundreds?) in front of her tanks to capture the impressive photos throughout the book. Thank you to Mark Shelly, Kerric Peyton, Kelvin Wong, and everyone else for providing additional photos to support the information in this book and show off the beauty of shrimp tanks! Finally, thank you to beta readers Mimi Wong, Cassandra Tiensivu, and Colleen Noland for going above and beyond to improve the book.

Happy shrimpin' everyone.

Index

Bibliography

@anoxicfiltrationplenums - Dr. Kevin Novak (Director). (2023). *What Is An Anoxic Filtration System?* [Motion Picture]. YouTube.

Adhikari, S., Chaurasia, V., Naqvi, A., & Pillai, B. (2007). Survival and Growth of Macrobrachium rosenbergii (de Man) Juvenile in Relation to Calcium Hardness and Bicarbonate Alkalinity. *Turkish Journal of Fisheries and Aquatic Sciences 7*, 23-26.

Alexander, C., Hindley, J., & Jones, S. (1980). Structure and function of the third maxillipeds of the banana prawn Penaeus merguiensis. *Marine Biology 58*, 245–249. https://doi.org/10.1007/BF00390772.

Ames, C., Helms, B., & Stoeckel, J. (2015). Habitat mediates the outcome of a cleaning symbiosis for a facultatively burrowing crayfish. *Freshwater Biology, 60*, 989-999. https://doi.org/10.1111/fwb.12559.

Andrea D. Shluker. (2003). *State of Hawai'i: Aquatic Invasive Species Management Plan.* Department of Land and Natural Resources.

Anissimov, M. (2024, May 21). *What is Chitin?* Retrieved from All The Science: https://www.allthescience.org/what-is-chitin.htm

Aquarium Co-Op. (2017, March 5). *The Aquarium Nitrogen Cycle.* Retrieved from YouTube: https://youtu.be/PWoiCqCvJco?si=s2EgcSzeAaVoRVLh

Aquarium Co-Op. (N.D.). *What is PAR? Lighting & Brightness Explained for Planted Tanks.* Retrieved from Aquarium Co-Op: https://www.aquariumcoop.com/blogs/aquarium/par-for-planted-tanks

Ayisi, C. L., Hua, X., Apraku, A., & et al. (2017). Recent Studies Toward the Development of Practical Diets for Shrimp and Their Nutritional Requirements. *HAYATI Journal of Biosciences, Volume 24, Issue 3*, Pages 109-117, ISSN 1978-3019, https://doi.org/10.1016/j.hjb.2017.09.004.

Baloi, M., Arantes, R., Schveitzer, R., & et al. (2013). Performance of Pacific white shrimp Litopenaeus vannamei raised in biofloc systems with varying levels of light exposure. *Aquacultural Engineering 52*, 39-44. doi.org/10.1016/j.aquaeng.2012.07.003.

Barnes, M., Ewing, D., & Cordes, R. (1998). Observations on Hydrogen Peroxide Control of Saprolegnia spp. during Rainbow Trout Egg Incubation. *The Progressive Fish Culturist 60(1)*, 67-70. https://doi.org/10.1577/1548-8640(1998)060<0067:OOHPCO>2.0.CO;2.

Bauer, J., Jung-Schroers, V., Teitge, F., & Adamek, M. (2021). Association of the alga Cladogonium sp. with a multifactorial disease outbreak in dwarf shrimp (Neocaridina davidi). *Diseases of Aquatic Organisms 146*, 107-115. DOI:10.3354/dao03625.

Bauer, R., & Caskey, J. (2006). Flagellar setae of the second antennae in decapod shrimps: sexual dimorphism and possible role in detection of contact sex pheromones. *Invertebrate Reproduction and Development 49:1-2*, 51-60. https://doi.org/10.1080/07924259.2006.9652193.

BBC. (2013, October 18). *Mould Time-lapse - The Great British Year: Episode 4 Preview - BBC.* Retrieved from YouTube: https://youtu.be/GY_uMH8Xpy0?si=XliibDYlVwgByl_Y

Bhuij. (2016, May 17). *Everything you need to know about fertilizers in one place.* Retrieved from Fishlore: https://www.fishlore.com/aquariumfishforum/threads/everything-you-need-to-know-about-fertilizers-in-one-place.227835/

Biggums, S. (n.d.).

Biggums, S. (2024). *Are Shrimp Keepers Better Than The Average Person?* Retrieved from Shrimply Explained: https://www.shrimplyexplained.com/shrimp-keepers-better-than-average-person/

Boyd, C. E. (2017, May 8). *How decomposition of organic matter impacts aquaculture ponds.* Retrieved from Global Seafood Alliance: https://www.globalseafood.org/advocate/decomposition-organic-matter-impacts-aquaculture-ponds/

Brazil, T., Caetano, A., Vargas, A., Bozelli, R., & Santangelo, J. (2022). Desiccation increases the hatching of resting eggs of a freshwater calanoid copepod. *Journal of Plankton Research, Volume 44, Issue 2*, Pages 273–277, https://doi.org/10.1093/plankt/fbac008.

Brown, P. (2019, March 26). *Specieswatch: ancient crustaceans still going strong after 450m years*. Retrieved from The Guardian: https://www.theguardian.com/environment/2019/mar/26/specieswatch-ancient-crustaceans-still-going-strong-after-450m-years

Caballero, A., Bravo, I., & Wang, J. (2017). Inbreeding load and purging: implications for the short-term survival and the conservation management of small populations. *Heredity 118*, 177–185. https://doi.org/10.1038/hdy.2016.80.

California Fish and Wildlife Services. (2024, January). *Disinfecting Aquariums and Equipment for the CAEP.* Retrieved from wildlife.ca.gov: https://nrm.dfg.ca.gov/FileHandler.ashx?DocumentID=141750

Calland, N., AlbeckaA., Belouzard, S., & et al. (2012). (2)-Epigallocatechin-3-Gallate Is a New Inhibitor of Hepatitis C Virus Entry. *Hepatology, 55(3)*, 720-729. DOI 10.1002/hep.24803.

Center for Disease Control. (2024, May 10). *Home Pool and Hot Tub Water Treatment and Testing*. Retrieved from U.S. Center For Disease Control and Prevention: https://www.cdc.gov/healthy-swimming/about/home-pool-and-hot-tub-water-treatment-and-testing.html

Center For Disease Control. (n.d.). *How Antibiotic Resistance Moves Directly From Germ to Germ*. Retrieved from Center For Disease Control: https://www.cdc.gov/antimicrobial-resistance/media/pdfs/how-ar-moves-508.pdf?CDC_AAref_Val=https://www.cdc.gov/drugresistance/pdf/threats-report/How-AR-Moves-508.pdf

Ceylan, D. (2019, Feb 19). *Using a Blanching Process to Control Pathogens on Produce*. Retrieved from Merieux NutriSciences: https://na.mxns.com/blog/using-a-blanching-process-to-control-pathogens-on-produce

Chansue, N., & Assawawongkasem, N. (2011). The in vitro Antibacterial Activity and Ornamental Fish Toxicity of the Water Extract of Indian Almond Leaves (Terminalia catappa Linn.). *KKU Veterinary Journal, 18(1)*, 36-45.

Committee, I. A. (2021, January). *Zoonoses Associated with Fish*. Retrieved from Washington State University: https://iacuc.wsu.edu/zoonoses-associated-with-fish/

Coswatte, C., Suraweera, K., Mudalige, A., & Liyanage, N. (2022). Defining a protocol for breeding and rearing crystal shrimp (Caridina cantonensis) using treated municipal water and mineralized reverse osmosis (MRO) water. *IRCUWU.*

Coyle, S., Bright, L., Wood, D., & et al. (2011). Performance of Pacific White Shrimp, Litopenaeus vannamei, Reared in Zero-Exchange Tank Systems Exposed to Different Light Sources and Intensities. *Journal of the World Aquaculture Society, 42*, 687-695. https://doi.org/10.1111/j.1749-7345.2011.00512.x.

Crash Course. (2024). Photosynthesis and Cellular Respiration: Crash Course Botany #5. CrashCourse.

Cronin, T., & Porter, M. (2008). Exceptional Variation on a Common Theme: The Evolution of Crustacean Compound Eyes. *Evolution: Education & Outreach 1*, 463–475. https://doi.org/10.1007/s12052-008-0085-0.

Crump, B., & Koch, E. (2008). Attached Bacterial Populations Shared by Four Species of Aquatic Angiosperms. *APPLIED AND ENVIRONMENTAL MICROBIOLOGY 74(19)*, 5948-5957. doi:10.1128/AEM.00952-08.

Culligan Water of Kansas City. (n.d.). *How Seasonal Changes Can Affect Your Water Quality*. Retrieved from Culligan Water of Kansas City: https://www.culligankansascity.com/seasonal-changes-and-your-water/

Davis, D., & Gatlin, D. (1993). Response of Penaeus vannamei to Dietary Calcium, Phosphorus and Calcium: Phosphorus Ratio. *Journal of the World Aquaculture Society, 24(4)*, 504-515. DOI:10.1111/j.1749-7345.1993.tb00581.x.

Dean, T., & Richardson, J. (1999). Responses of seven species of native freshwater fish and a shrimp to low levels of dissolved oxygen. *New Zealand Journal of Marine and Freshwater Research*, Vol. 33: 99-106.

Deo, P., & Deshmukh, R. (2019). Oral microbiome: Unveiling the fundamentals. *Journal of Oral and Maxilliofacial Pathology 23(1)*, 122-128. doi: 10.4103/jomfp.JOMFP_304_18. PMID: 31110428; PMCID: PMC6503789.

Derby, C. (2021). The Crustacean Antennule: A Complex Organ Adapted for Lifelong Function in Diverse Environments and Lifestyles. *The Biological Bulletin 240:2*, 67-81, https://doi.org/10.1086/713537.

Diaz, A., Sousa, L., Cuartas, E., & Petriella, A. (2003). Growth, molt and survival of Palaemonetes argentinus (Decapoda, Caridea) under different light-dark conditions. *Iheringia. Serie Zoologia 93(3)*, https://doi.org/10.1590/S0073-47212003000300004.

Ding, L., Yu, J., Peng, X., & et al. (2024). Changes in molting frequency and expression patterns of molting-related genes in Macrobrachium rosenbergii with exogenous calcium supplement in water. *Aquaculture, Volume 586*, 740761, ISSN 0044-8486, https://doi.org/10.1016/j.aquaculture.2024.740761.

Donlan, R. (2002). Biofilms: microbial life on surfaces. *Emerg Infect Dis. 8(9)*, 881-90. doi: 10.3201/eid0809.020063. PMID: 12194761; PMCID: PMC2732559.

Dosta, M., & et al. (2015). "Importance and function of microbial communities in aquaculture systems with no water exchange.". *Scientific Journal of Animal Science 4.9* , 103-110.

Duan, J., Feng, Y., Huang, X., & et al. (2018). A Review: Factors Affecting Outbreaks of Saprolegniosis on Aquatic Animals. *The Israeli Journal of Aquaculture*.

ECOS Environmental Consultants Limited. (n.d.). *Wastewater Nitrification: How it works*. Retrieved from ECOS: https://www.ecos.ie/wastewater-nitrification-how-it-works/

Evans, M., & Sober, G. (N.D.). *How Alkalinity Affects Nitrification*. Retrieved from CWEA: https://www.cwea.org/news/how-alkalinity-affects-nitrification/

Fabritius, H.-O., Karsten, E., & et al. (2012). Correlation of structure, composition and local mechanical properties in the dorsal carapace of the edible crab Cancer pagurus. *Zeitschrift für Kristallographie-Crystalline Materials 227.11*, 766-776.

FISHTORY (Director). (2024). *ACTIVE vs. NON-ACTIVE Substrates + ANOXIC vs. AEROBIC Aquasoils* [Motion Picture].

Forster, J., & Hirst, A. (2012). The temperature-size rule emerges from ontogenetic differences between growth and development rates. *Functional Ecology, 26: 483-492.*

Foysal, J. (2023). Host habitat shapes the core gut bacteria of decapod crustaceans: A meta-analysis. *Heliyon*, Volume 9 Issue 6.

Furtado, P., Serra, F., Poersh, L., & Wasielesky, W. (2014). Short communication: Acute toxicity of hydrogen peroxide in juvenile white shrimp Litopenaeus vannamei reared in biofloc technology systems. *Aquaculture International, 22*, 653-659. DOI 10.1007/s10499-013-9694-x.

Gauquelin, F., & Cuzon, G. (2007). Effect of dietary protein level on growth and energy utilization by Litopenaeus stylirostris under laboratory conditions. *Aquaculture, Volume 271, Issues 1-4*, Pages 439-448, http://dx.doi.org/10.1016/j.aquaculture.2006.05.062.

Gong, H., Zhen, Q., Zhao, C., Jitao, L., Zhiqiang, C., Jian, L., & Ping, C. (2022). Effects of Dietary Tannic Acid on Growth, Digestion, Immunity and Resistance to Ammonia Stress, and Intestinal Microbial Community in Pacific White Shrimp (Litopenaeus vannamei). *Fishes, 7(6)*, 327. https://doi.org/10.3390/fishes7060327.

Gorbach, S. (1996). Microbiology of the Gastrointestinal Tract. *Medical Microbiology. 4th edition*, Chapter 95. https://www.ncbi.nlm.nih.gov/books/NBK7670/.

Han, C., Hsu, K., & Fang, L. (2019). Geographical and temporal origins of Neocaridina species (Decapoda: Caridea: Atyidae) in Taiwan. *BMC Genomic Data 20(86)*, https://doi.org/10.1186/s12863-019-0788-y.

Han, C.-C., Hsu, K.-C., Fang, L.-S., & et al. (2019). Geographical and temporal origins of Neocaridina species (Decapoda: Caridea: Atyidae) in Taiwan. *BMC Genetics*, 20:86 https://doi.org/10.1186/s12863-019-0788-y.

Harvey, P. (2002). Evolution: When Females are Brighter than Males. *Current Biology 12:1*, R24-R25, ISSN 0960-9822, https://doi.org/10.1016/S0960-9822(01)00643-1.

Hill, G. (1996). Redness as a measure of the production cost of ornamental coloration. *Ethology Ecology & Evolution, 8(2)*, 157–175. https://doi.org/10.1080/08927014.1996.9522926.

Hjorth, M., Doncheva, A., Norheim, F., & et al. (2022). Consumption of salmon fishmeal increases hepatic cholesterol content in obese C57BL/6 J mice. *European Journal of Nutrition 61(8)*, 4027-4043. doi: 10.1007/s00394-022-02930-y. Epub 2022 Jul 5. PMID: 35788891; PMCID: PMC9596588.

Hoang, T., Barchiesis, M., Lee, S., Kennan, C., & Marsden, G. (2003). Influences of light intensity and photoperiod on moulting and growth of Penaeus merguiensis cultured under laboratory conditions. *Aquaculture 216(1-4)*, 343-354, ISSN 0044-8486, https://doi.org/10.1016/S0044-8486(02)00460-X.

Hong, P., Wu, X., & Shu, Y. (2019). Denitrification characterization of dissolved oxygen microprofiles in lake surface sediment through analyzing abundance, expression, community composition and enzymatic activities of denitrifier functional genes. *AMB Express*, 9(1):129. doi: 10.1186/s13568-019-0855-9. PMID: 31428884; PMCID: PMC6702497.

Huang, Y., Zhang, L., Wang, G., & Huang, S. (2022). De novo assembly transcriptome analysis reveals the genes associated with body color formation in the freshwater ornamental shrimps Neocaridina denticulate sinensis. *Gene 806*, 145929, ISSN 0378-1119, https://doi.org/10.1016/j.gene.2021.145929.

Hung T-C, Stevenson, T., Sandford, M., & Ghebremariam, T. (2018). Temperature, density and ammonia effects on growth and fecundity of the ramshorn snail (Helisoma anceps). *Aquaculture Research*, 49: 1072–1079. https://doi.org/10.1111/are.13556.

Hung, J., Yang, C., Lai, I., & Li, Y. (2020). Rainfall and Human Impacts on Weathering Rates and Carbon-Nutrient Yields in the Watershed of a Small Mountainous River (Kaoping) in Southwestern Taiwan. *Sustainability, 12(18)*, 7689. https://doi.org/10.3390/su12187689.

Ikhwanuddin, M., Moh, J., Hidayah, M., & et al. (2014). Effect of Indian almond, Terminalia catappa leaves water extract on the survival rate and growth performance of black tiger shrimp, Penaeus monodon post larvae. *Aquaculture, Aquarium, Conservation & Legislation, 7(2)*, 85-93.

Ip, Y. K. (2010). Ammonia Production, Excretion, Toxicity, and Defense in Fish: A Review. *Frontiers in Physiology*, 1:134. doi: 10.3389/fphys.2010.00134.

jayc. (2013, September 27). *Shrimp Diseases and Diagnosis*. Retrieved from SKF Aquatics Forum: https://skfaquatics.com/forum/forums/topic/5052-shrimp-diseases-and-diagnosis/

Jha, A., Mishra, A., Kumari, V., & et al. (2011). Softening of Hard Water by Bentonite Mineral. *Asian Journal of Water Environment and Pollution 8(4)*, 93-96. DOI:10.3233/AJW-2011-8_4_12.

Khan Academy. (n.d.). *Mendel and his peas*. Retrieved from Khan Academy: https://www.khanacademy.org/science/ap-biology/heredity/mendelian-genetics-ap/a/mendel-and-his-peas

Krachler, R. F. (2009). "Effects of pH on aquatic biodegradation processes.". *Biogeosciences Discussions 6.1*, 491-514.

Krachler, R. F. (2009). Effects of pH on aquatic biodegradation processes. *Biogeosciences Disussions, 6*, 491-514.

Li, F., Han, S., & Chi, W. e. (2022). Identification and functional study of calcification-related peptide from the oriental river prawn (Macrobrachium nipponense). *Aquaculture Reports 27*, 101354.

Li, X., Lan, H., & Dai, X. (2024). The effects of Ca2+ and Mg2+ in the water column on the growth and molting of Macrobrachium rosenbergii. *Aquacult Int 32*, 8823–8841, https://doi.org/10.1007/s10499-024-01592-3.

Li, Y., She, Q., Han, Z., & et al. (2018). Anaesthetic Effects of Eugenol on Grass Shrimp (Palaemonetes sinensis) of Different Sizes at Different Concentrations and Temperatures. *Sci Rep 8*, 11007 (2018). https://doi.org/10.1038/s41598-018-28975-w.

LibreTexts - Biology. (N.D.). *5.9B: Nitrate Reduction and Denitrification* . Retrieved from Libre Texts: https://bio.libretexts.org/Bookshelves/Microbiology/Microbiology_(Boundless)/05%3A_Microbial_Metabolism/5.09%3A_Anaerobic_Respiration/5.9B%3A_Nitrate_Reduction_and_Denitrification

Liu, X. (2021). Dissolved oxygen disturbs nitrate transformation by modifying microbial community, co-occurrence networks, and functional genes during aerobic-anoxic transition. *Science of The Total Environment*, Volume 790, 148245, ISSN 0048-9697, https://doi.org/10.1016/j.scitotenv.2021.148245.

Loewe, L. (2008). Negative Selection. *Nature Education 1(1):59*, https://www.nature.com/scitable/topicpage/negative-selection-1136/.

Lytle, J., & Lytle, T. (1990). Fatty Acid Composition and Variations in Individual Bloodworms. *Journal of the World Aquaculture Society, 21*, 314-318. https://doi.org/10.1111/j.1749-7345.1990.tb00545.x.

Maciaszek, R., Kamaszewski, M., Struzynski, W., & Lapa, P. (2018). Epibionts of ornamental freshwater shrimps bred in Taiwan. *Annals of Warsaw University of Life Sciences-SGGW. Animal Science*, 57.

Maciaszek, R., Swiderek, W., Prati, S., & et al. (2023). Epibiont Cohabitation in Freshwater Shrimp Neocaridina davidi with the Description of Two Species New to Science, Cladogonium kumaki sp. nov. and Monodiscus kumaki sp. nov., and Redescription of Scutariella japonica and Holtodrilus truncatus. *Animals, 13*, 1616. https://doi.org/10.3390/ani13101616.

Marine Things. (2023, June 25). *Shrimp Molting*. Retrieved from YouTube: https://youtube.com/shorts/sly-ObsGrWE?si=3xBfdEXHC7CzjGrF

Mark's Shrimp Tanks. (2017, January 01). *How To Save Shrimp Eggs*. Retrieved from YouTube: https://youtu.be/Xw3ctPGqWLo?si=ml7WJxJXPhGeqq2Y

Maszczyk, P., & Brzezinski, T. (2018). Body Size, Maturation Size, and Growth Rate of Crustaceans. In *The Natural History of the Crustacea, Volume 5: Life Histories (pp.35-65) Chapter. 2.* Oxford University Press.

Moise, A., Bandera, S., & Lintig, J. (2022). Chapter Fourteen - Carotenoid modifying enzymes in metazoans. *Editor(s): Eleanore T. Wurtzel, Methods in Enzymology, Academic Press 674*, 405-445, ISSN 0076-6879, ISBN 9780323913515, https://doi.org/10.1016/bs.mie.2022.05.005.

Montiel-Leon, J. M., Duy, S., Munoz, G., Verner, M.-A., & et al. (2019). Occurrence of pesticides in fruits and vegetables from organic and conventional agriculture by QuEChERS extraction liquid chromatography tandem mass spectrometry. *Food Control, Volume 104*, Pages 74-82, ISSN 0956-7135, https://doi.org/10.1016/j.foodcont.2019.04.027.

Moss, A., Ishikawa, M., Koshio, S., & et al. (2019). Effects of Different Levels of Marine Snail Shells in the Diets of Juvenile Kuruma Shrimps Marsupenaeus japonicus as a Source of Calcium. *North American Journal of Aquaculture, 81*, 55-66. https://doi.org/10.1002/naaq.10066.

Nakatsuji, T., & Sonobe, H. (2004). Regulation of ecdysteroid secretion from the Y-organ by molt-inhibiting hormone in the American crayfish, Procambarus clarkii. *General and Comparative Endocrinology 135:3*, Pages 358-364, ISSN 0016-6480, https://doi.org/10.1016/j.ygcen.2003.11.001.

Ng, W.-K., Mong, M.-L., & Abdul-Hamid, A.-A. (2023). Dietary montmorillonite clay improved Penaeus vannamei survival from acute hepatopancreatic necrosis disease and modulated stomach microbiota. *Animal Feed Science and Technology 297*, 115581, ISSN 0377-8401, https://doi.org/10.1016/j.anifeedsci.2023.115581.

Nguyen-Phuoc, K., Doung, N., Phan, T., & et al. (2021). Generation and evaluation of polyclonal antibodies specific for ToxA from Vibrio parahaemolyticus causing acute hepatopancreatic necrosis disease (AHPND) in shrimp. *Mol Biol Res Commun 10(1)*, 23-32. doi: 10.22099/mbrc.2020.38774.1561. PMID: 33681394; PMCID: PMC7936389.

Nonwachai, T., Purivirojku, W., Chuchird, N., & et al. (2011). Effects of dissolved oxygen levels on growth, survival and immune response of juvenile pacific white shrimp Litopenaeus vannamei. *Journal of Fisheries and Environment 35(3)*, 1-10. .

Northern Arizona University. (n.d.). *Sexual Dimorphism*. Retrieved from Northern Arizona University: https://www2.nau.edu/~gaud/bio300b/sexdi.htm

Novriadi, R., Fadhilah, R., Wahyudi, A., & Trullas, C. (2021). EFFECTS OF HYDROLYSABLE TANNINS ON THE GROWTH PERFORMANCE, TOTAL HAEMOCYTE COUNTS AND LYSOZYME ACTIVITY OF PACIFIC WHITE LEG SHRIMP Litopenaeus vannamei. *Aquaculture Reports 21*, 100796, ISSN 2352-5134, https://doi.org/10.1016/j.aqrep.2021.100796.

Nunan, L., Poulos, B., Navarro, S., & et al. (2010). Milky hemolymph syndrome (MHS) in spiny lobsters, penaeid shrimp and crabs . *Disease of Aquatic Organisms 91*, 105-112. https://doi.org/10.3354/dao02270.

Nur, F., & Christianus, A. (2013). Breeding and life cycle of Neocaridina denticulata sinensis (Kemp, 1918). *Asian Journal of Animal and Veterinary Advances*, Vol. 8, No. 1, 108-115 ref. 21, DOI: 10.3923/ajava.2013.108.115.

Obaldo, L., & Masuda, R. (2006). Effect of Diet Size on Feeding Behavior and Growth of Pacific White Shrimp, Litopenaeus vannamei. *Journal of Applied Aquaculture, 18(1)*, 101–110. https://doi.org/10.1300/J028v18n01_07.

Ocasio-Torres, M., Giray, T., Crowl, T., & Sabat, A. (2015). Antipredator defence mechanism in the amphidromous shrimp Xiphocaris elongata (Decapoda: Xiphocarididae): rostrum length. *Journal of Natural History, 49(25-26)*, 1493–1506. https://doi.org/10.1080/00222933.2015.1005716.

Palozza, P., & Krinsky, N. (1992). Astaxanthin and canthaxanthin are potent antioxidants in a membrane model. *Archives of Biochemistry and Biophysics 297:2*, Pages 291-295, ISSN 0003-9861, https://doi.org/10.1016/0003-9861(92)90675-M.

Pantaleão, J., Barros-Alves, S., Tropea, C., & et al. (2015). Nutritional Vulnerability in Early Stages of the Freshwater Ornamental "Red Cherry Shrimp" Neocaridina Davidi () (Caridea: Atyidae). *Journal of Crustacean Biology 35:5*, 676–681, https://doi.org/10.1163/1937240X-00002357.

Pantaleão, J., Barros-Alves, S., Tropea, C., & et al. (2015). Nutritional Vulnerability in Early Stages of the Freshwater Ornamental "Red Cherry Shrimp" Neocaridina Davidi (Caridea: Atyidae). *Journal of Crustacean Biology 35:5*, Pages 676–681, https://doi.org/10.1163/1937240X-00002357.

Pantaleão, J., Gregati, R., da Costa, R., Lopez-Greco, L., & Negreiros-Fransozo, M. (2017). Post-hatching development of the ornamental 'Red Cherry Shrimp' Neocaridina davidi (Bouvier, 1904) (Crustacea, Caridea, Atyidae) under laboratorial conditions. *Aquaculture Research, 48*, 553-569. https://doi.org/10.1111/are.12903.

Park In-Seok, Baek Se-Won, & Moon Ki Hwan. (2019). The Sterilization Effect of Methylene Blue, Formalin, and Iodine on Egg and Adult Stage of Marine Medaka, Oryzias dancena. *Development & Reproduction*, 23(3):199-211. https://doi.org/10.12717/DR.2019.23.3.199.

Paster, E., LaFond, E., Biery, D., & et al. (2005). Estimates of prevalence of hip dysplasia in Golden Retrievers and Rottweilers and the influence of bias on published prevalence figures. *J Am Vet Med Assoc*, 1;226(3):387-92. doi: 10.2460/javma.2005.226.387. PMID: 15702688.

Pehrsson, P., Patterson, K., & Pery, C. (2008). *The Mineral Content of US Drinking and Municipal Water.* Beltsville, MD: USDA, Beltsville Human Nutrition Research Center.

Pepi, M., & Focardi, S. (2021). Antibiotic-Resistant Bacteria in Aquaculture and Climate Change: A Challenge for Health in the Mediterranean Area. *Int J Environ Res Public Health 18(11)*, 5723. doi: 10.3390/ijerph18115723. PMID: 34073520; PMCID: PMC8198758.

Perkins, J., Yazwinski, T., & Tucker, C. (2012). Efficacies of fenbendazole and albendazole in the treatment of commercial turkeys artificially infected with Ascaridia dissimilis. *Discovery, The Student Journal of Dale Bumpers College of Agricultural, Food and Life Sciences 13(1)*, 61-66.

Ponpornpisit, A. (2024). "Single-tank sequencing batch bioreactor with a cellulose substrate for simultaneous nitrification and denitrification of an aquarium.". *The Thai Journal of Veterinary Medicine 53.1*, 83-96.

Pourjam, F., Sotoudeh, E., Bagheri, D., & et al. (2024). Dietary montmorillonite improves growth, chemical composition, antioxidant system and immune-related genes of whiteleg shrimp (Litopenaeus vannamei). *Aquaculture International 32*, 5699–5717. https://doi.org/10.1007/s10499-024-01442-2.

Preethi N. (2021, June 29). *Mechanism of Respiration: Common marine prawn, Penaeus indicus*. Retrieved from YouTube: https://youtu.be/tCo9lNvXfDk?si=wHkzGNaw8aQptnV0

Randall, D. (2002). Ammonia toxicity in fish. *Marine Pollution Bulletin*, Volume 45, Issues 1–12, Pages 17-23.

Rasch, J., & Bauer, R. (2016). The functional morphology and role of the thelycum in insemination, and its relation to the mating system in the seagrass shrimp Ambidexter symmetricus (Decapoda: Processidae). *Invertebrate Biology, 135*, 163-173. https://doi.org/10.1111/ivb.12126.

Raz, S., Hamilton, P., Wilt, F., & et al. (2003). The Transient Phase of Amorphous Calcium Carbonate in Sea Urchin Larval Spicules: The Involvement of Proteins and Magnesium Ions in Its Formation and Stabilization. *Advanced Functional Materials 13*, 480-486. https://doi.org/10.1002/adfm.200304285.

Reed, D., & Bryant, E. (2000). Experimental tests of minimum viable population size. *Animal Conservation 3*, 7-14. doi:10.1111/j.1469-1795.2000.tb00082.x.

Rowley, A., & Coates, C. (2023). Shell disease syndromes of decapod crustaceans. *Environmental Microbiology 25(5)*, 931-947.

Rowley, A., Coates, C., & Whitten, M. (2022). *Invertebrate Biology, Chapter 15: Bacterial diseases of crustaceans. p412.* Oxford University Press.

Samocha, T., & Prangnell, D. (2019). Chapter 6 - System Treatment and Preparation. In *Sustainable Biofloc Systems for Marine Shrimp* (pp. Pages 119-131). Academic Press ISBN 9780128180402, https://doi.org/10.1016/B978-0-12-818040-2.00006-X.

Schmidt, M., Chien, H., & Tadesse, T. (2006). Rosette-type tegumental glands associated with aesthetasc sensilla in the olfactory organ of the Caribbean spiny lobster, Panulirus argus. *Cell Tissue Research 325*, 369–395. https://doi.org/10.1007/s00441-006-0163-y.

Schmidt, M., Chien, H., & Tadesse, T. (2006). Rosette-type tegumental glands associated with aesthetasc sensilla in the olfactory organ of the Caribbean spiny lobster, Panulirus argus. *Cell and Tissue Research 325*, 369-395, https://doi.org/10.1007/s00441-006-0163-y.

Schoolman, G., & Arndt, H. (2017). LEAF-LITTER PREFERENCES OF THE INTRODUCED FRESHWATER SHRIMPS ATYAEPHYRA DESMARESTII AND NEOCARIDINA DAVIDI. *Crustaceana, 90(14)*, 1715–1730. https://www.jstor.org/stable/26570872.

Shaikh, S., Rashid, N., Onwusogh, U., & et al. (2023). Effect of nutrients deficiency on biofilm formation and single cell protein production with a purple non-sulphur bacteria enriched culture. *Biofilm, 5*, 100098, ISSN 2590-2075, https://doi.org/10.1016/j.bioflm.2022.100098.

Shields, J. (2019). Climate change enhances disease processes in crustaceans: case studies in lobsters, crabs, and shrimps. *Journal of Crustacean Biology, Volume 39, Issue 6*.

Shields, J. (2019). Climate change enhances disease processes in crustaceans: case studies in lobsters, crabs, and shrimps. *Journal of Crustacean Biology 39(6)*, 673–683, https://doi.org/10.1093/jcbiol/ruz072.

Shrimp Mania. (2023, September 24). *Master Shrimp Breeder Reveals His Top Secret!* Retrieved from YouTube: https://youtu.be/JObLAyBSdn4?si=lYha_DtbF8wK0DWg

Shrimp Mania. (2024, October 04). *Breeding The Most Expensive Shrimp - Inside Singapore's LARGEST Caridina Breeder...* Retrieved from YouTube: https://youtu.be/3rfaYbznv6U?si=RDTEOArKTV_tS18K

Shrimply. (2022, September 29). *What In The World Is Bacter AE?* Retrieved from Shrimply Explained: https://www.shrimplyexplained.com/2022/09/29/what-is-bacter-ae/

Shrimply. (2024, April 13). *A Scientific Dive Into The Cherry Shrimp Lifecycle*. Retrieved from YouTube: https://youtu.be/4aTdsL-t0xY?si=5bHxRXGEnOn8mJ6b

Shrimply (Director). (2024). *Breed More Shrimp With Ice* [Motion Picture].

Shrimply. (2024, May 01). *What Rocks Are Safe For A Shrimp Tank?* Retrieved from Shrimply Explained: https://www.shrimplyexplained.com/2024/05/01/rocks-safe-for-shrimp-tank/

Smith, R. J. (n.d.). *Ostracod Research at the Lake Biwa Museum, Japan*. Retrieved from Lake Biwa Museum: https://www.biwahaku.jp/smith/facts.html/photos_lake_biwa.html

Soares, R., Peixoto, S., Davis, R., & Davis, D. (2021). Feeding behavior and growth of Litopenaeus vannamei fed soybean-based diets with added feeding effectors. *Aquaculture 536*, 736487, ISSN 0044-8486, https://doi.org/10.1016/j.aquaculture.2021.736487.

SoniaBajaj10. (2021, April 08). *Prawn respiratory system,Types of gills,structure of gill plates,working of mechanism,formula of gills*. Retrieved from Slide Share: https://www.slideshare.net/slideshow/prawn-

respiratory-systemtypes-of-gillsstructure-of-gill-platesworking-of-mechanismformula-of-gills/245920255#2

Southern Biological. (N.D.). *Planaria Care Guide*. Retrieved from Southern Biological: https://www.southernbiological.com/planaria-care-guide/

Srivastava, J., Chandra, H., & Kalra, S. (2017). Plant–microbe interaction in aquatic system and their role in the management of water quality: a review. *Appl Water Sci 7*, 1079–1090. https://doi.org/10.1007/s13201-016-0415-2.

Srivastava, J., Chandra, H., Kalra, S., & et al. (2017). Plant–microbe interaction in aquatic system and their role in the management of water quality: a review. *Applied Water Sciences 7*, 1079–1090. https://doi.org/10.1007/s13201-016-0415-2.

Srivastava, S., Sinha, R., & Roy, D. (2004). Toxicological effects of malachite green. *Aquatic Toxicology, 66(3)*, 319-329. https://doi.org/10.1016/j.aquatox.2003.09.008.

Sukadi, M. (1991). Responses of the freshwater prawn, Macrobrachium rosenbergii to different levels of water hardness and dietary calcium. 5095-5095.

Sundberg, C., & Jönsson, H. (2008). Higher pH and faster decomposition in biowaste composting by increased aeration. *Waste Management*, Volume 28, Issue 3, Pages 518-526.

Tanner, C. (1995). Plants for constructed wetland treatment systems — A comparison of the growth and nutrient uptake of eight emergent species. *Ecological Engineering 7(1)*, 59-83, ISSN 0925-8574, https://doi.org/10.1016/0925-8574(95)00066-6.

Tavabe, K., Rafiee, G., Shoeiry, M., & et al. (2015). Effects of Water Hardness and Calcium: Magnesium Ratios on Reproductive Performance and Offspring Quality of Macrobrachium rosenbergii. *Journal of World Aquacult Society, 46*, 519-530. https://doi.org/10.1111/jwas.12217.

Taylor, J., Gilleard, J., Allen, M., & Deheyn, D. (2015). Taylor, J. R., Gilleard, J. M., Allen, M. C., & Deheyn, D. D. (2015). Effects of CO2-induced pH reduction on the exoskeleton structure and biophotonic properties of the shrimp Lysmata californica. *Scientific Reports, 5(1)*, 10608. DOi: 10.1038/srep10608.

Texas AgriLife Extension. (2009, February). *Chapter 1, The Decomposition Process*. Retrieved from Texas AgriLife Extension Service : https://aggie-horticulture.tamu.edu/earthkind/landscape/dont-bag-it/chapter-1-the-decomposition-process/

Tian, H., Yang, C., Yu, Y., & et al. (2020). Dietary cholesterol level affects growth, molting performance and ecdysteroid signal transduction in Procambarus clarkii. *Aquaculture, 523*, 735198, ISSN 0044-8486, https://doi.org/10.1016/j.aquaculture.2020.735198.

Tietjen, T., Vähätalo, A., & Wetzel, R. (2005). Effects of clay mineral turbidity on dissolved organic carbon and bacterial production. *Aquatic Sciences 67(1)*, 51-60. DOI:10.1007/s00027-004-0753-2.

Tlusty, M., Myers, A., & Metzler, A. (2008). Short- and long-term dietary effects on disease and mortality in American lobster Homarus americanus. *DISEASES OF AQUATIC ORGANISMS, 78*, 249-253. doi: 10.3354/dao01867.

Tomas, A., Sganga, D., & Lopez-Greco, L. (2020). Effect of background color and shelters on female pigmentation in the ornamental red cherry shrimp Neocaridina davidi (Caridea, Atyidae). *J World Aquacult Soc, 51*, 775–787. https://doi.org/10.1111/jwas.12660.

Tomas, A., Sganga, D., & López-Greco, L. (2020). Effect of diets on carotenoid content, body coloration, biochemical composition and spermatophore quality in the "red cherry" shrimp Neocaridina davidi (Caridea, Atyidae). *Aquaculture Nutrition, 26*, 1198–1210. https://doi.org/10.1111/anu.13076.

Tong, Z., He, W., Fan, X., & Guo, A. (2022). Biological Function of Plant Tannin and Its Application in Animal Health. *Front Vet Sci*, 803657. doi: 10.3389/fvets.2021.803657. PMID: 35083309; PMCID: PMC8784788.

Tropea C, S. L. (2015). Effect of Temperature on Biochemical Composition, Growth and Reproduction of the Ornamental Red Cherry Shrimp Neocaridina heteropoda heteropoda (Decapoda, Caridea). *PLoS ONE 10(3): e0119468. https://doi.org/10.1371/journal.pone.0119468.*

Tropea, C., & Greco, L. (2024). To invest or not to invest, that is the question: male presence and genetic relatedness as modulators of female reproductive effort in a shrimp. *Canadian Journal of Zoolog*, Volume 102, Number 9.

Tropea, C., & Lopez-Greco, L. (2015). Female Growth and Offspring Quality over Successive Spawnings in a Caridean Shrimp Neocaridina davidi (Decapoda, Atyidae) with Direct Development. *Biology Bulletin 229:3*, 243-254. https://doi.org/10.1086/BBLv229n3p243.

Tropea, C., Stumpf, L., & Lopez Greco, L. (2015). Effect of Temperature on Biochemical Composition, Growth, and Reproduction of the Ornamental Red Cherry Shrimp Neocaridina heteropoda heteropoda (Decapoda, Caridea). *PLoS ONE 10(3):*, 0119468. doi:10.1371/journal.pone.0119468.

Truong, H., Hines, B., & Emerenciano, M. (2023). Mineral nutrition in penaeid shrimp. *Reviews In Aquaculture. 15(4)*, 1355-1373. doi:10.1111/raq.12780.

Türkmen, B., & Karadal, O. (2016). Effects of different aquarium feeds on growth performance and survival rate of red cherry shrimp (Neocaridina denticulata). *Ege Journal of Fisheries and Aquatic Sciences 33(3)*, 217-222. https://doi.org/10.12714/egejfas.2016.33.3.05.

U.S. Forest Service. (n.d.). *Tannins*. Retrieved from United States Department of Agriculture: https://www.fs.usda.gov/wildflowers/ethnobotany/tannins.shtml

USDA. (2019, April 01). *FoodData Central Food Details: Egg, yolk, raw, fresh*. Retrieved from USDA: https://fdc.nal.usda.gov/food-details/172184/nutrients

van Kessel, M., Speth, D., Albertsen, M., & et al. (2015). Complete nitrification by a single microorganism. *Nature*, 528(7583):555-9. doi: 10.1038/nature16459. Epub 2015 Nov 26. PMID: 26610025; PMCID: PMC4878690.

Vazquez, N., Delevati-Colpo, K., Sganga, D., & Lopez-Greco, L. (2017). Density and gender segregation effects in the culture of the caridean ornamental red cherry shrimp Neocaridina davidi Bouvier, 1904 (Caridea: Atyidae). *Journal of Crustacean Biology, 37(4)*, 367–373, https://doi.org/10.1093/jcbiol/rux051.

Vazquez, N., Sganga, D., & López Greco, L. (2022). Effect of different substrates on growth and biochemical composition of the ornamental "red cherry" caridean shrimp, Neocaridina davidi (Atyidae). *Aquaculture Research*, 53, 3001–3009. https://doi.org/10.1111/are.15812.

Vazquez, N., Sganga, D., & Lopez-Greco, L. (2022). Effect of different substrates on growth and biochemical composition of the ornamental "red cherry" caridean shrimp, Neocaridina davidi (Atyidae). *Aquaculture Research, 53*, 3001-3009. https://doi.org/10.1111/are.15812.

Vega-Villasante, F., Martínez-Ochoa, E., García-Guerrero, M., & Arrona-Ortiz, J. (2015). Effect of different light intensities on expression of chromatophores, growth and survival in juvenile Macrobrachium tenellum. *Latin American Journal of Aquatic Research 43(1)*, ISSN 0718-560X. doi.org/10.3856/vol43-issue1-fulltext-22.

Viau, V., Marciano, A., Iriel, A., & Lopez-Greco, L. (2016). Assessment of a biofilm-based culture system within zero water exchange on water quality and on survival and growth of the freshwater shrimp Neocaridina heteropoda heteropoda. *Aquaculture Research 47*, 2528-2542. https://doi.org/10.1111/are.12701.

Wang, Y., Jiang, Z., Lai, Z., & et al. (2021). The self-adaption capability of microalgal biofilm under different light intensities: Photosynthetic parameters and biofilm microstructures. *Algal Research 58*, 102383, ISSN 2211-9264, https://doi.org/10.1016/j.algal.2021.102383.

Weber, S., & Traunspurger, W. (2016). Influence of the ornamental red cherry shrimp Neocaridina davidi (Bouvier, 1904) on freshwater meiofaunal assemblages,. *Limnologica, Volume 59*, Pages 155-161, ISSN 0075-9511, https://doi.org/10.1016/j.limno.2016.06.001.

Wikipedia. (2023, November 12). *Telson*. Retrieved from Wikipedia: https://en.wikipedia.org/wiki/Telson

Wikipedia. (N.D.). *Caridoid escape reaction*. Retrieved from Wikipedia: https://en.wikipedia.org/wiki/Caridoid_escape_reaction

World Health Organization (WHO). (2023, November 21). *Antimicrobial resistance*. Retrieved from World Health Organization: https://www.who.int/news-room/fact-sheets/detail/antimicrobial-resistance

Xu-Feng, X., Hui, G., Guang-Li, L., & Chun-Hua, Z. (2021). Effects of dietary hydrolyzable tannins on growth performance, antioxidant capacity, intestinal microflora and resistance against Vibrio parahaemolyticus of juvenile Pacific white shrimp, Litopenaeus vannamei (Boone, 1931). *Aquaculture Reports, Volume 19*, 100601, ISSN 2352-5134, https://doi.org/10.1016/j.aqrep.2021.100601.

Yaghmaeiyan, N., Mirzaei, M., & Delghavi, R. (2022). Montmorillonite clay: Introduction and evaluation of its applications in different organic syntheses as catalyst: A review. *Results in Chemistry, 4*, 100549, ISSN 2211-7156, https://doi.org/10.1016/j.rechem.2022.100549.

Yam, R., & Dudgeon, D. (2005). Yam, Rita SW, and David Dudgeon. "Inter-and intraspecific differences in the life history and growth of Caridina spp.(Decapoda: Atyidae) in Hong Kong streams. *Freshwater Biology 50.12*, 2114-2128.

Yamson, E., Tubalinal, G., Viloria, V., & Mingala, C. (2019). Anthelmintic effect of betel nut (Areca catechu) and neem (Azadirachta indica) extract against liver fluke (Fasciola spp.). *J Adv Vet Anim Res, 6(1)*, 44-49. doi: 10.5455/javar.2019.e310. PMID: 31453169; PMCID: PMC6702922.

Yanting, Z., Cuilan, D., Xu-xiang, Z., & et al. (2018). Insights into the Gut Microbiota of Freshwater Shrimp and Its Associations with the Surrounding Microbiota and Environmental Factors. *Journal of Microbial Biotechnology*, 28(6), 946–956.

Zheng, J., Mao, Y., Su, Y., & Wong, J. (2020). Effects of stocking density on the survival, growth and physical injury of Marsupenaeus japonicus juveniles in a flowing water aquaculture system. *Aquaculture Research, 51*, 1500–1506. https://doi.org/10.1111/are.14496.

Zhou, Y., Davis, D., & Rhodes, M. (2014). Comparative evaluation of copper sulfate and tribasic copper chloride on growth performance and tissue response in Pacific white shrimp Litopenaeus vannamei fed practical diets. *Aquaculture, Volume 434,* Pages 411-417, ISSN 0044-8486, https://doi.org/10.1016/j.aquaculture.2014.09.004.

Zhu, X., Guo, H., Li, G., & Zhu, C. (2021). Effects of dietary hydrolyzable tannins on growth performance, antioxidant capacity, intestinal microflora and resistance against Vibrio parahaemolyticus of juvenile Pacific white shrimp, Litopenaeus vannamei (Boone, 1931). *Aquaculture Reports, 19*, 100601, ISSN 2352-5134, https://doi.org/10.1016/j.aqrep.2021.100601.

Zhu, Z., Shan, L., Xinyun, Zhang, & et al. (2021). Effects of bacterial community composition and structure in drinking water distribution systems on biofilm formation and chlorine resistance. *Chemosphere, Volume 264, Part 1*, 128410, ISSN 0045-6535, https://doi.org/10.1016/j.chemosphere.2020.128410.

Please visit shrimplyexplained.com/book-links/ for more information on all roman numeral superscript.

Made in the USA
Las Vegas, NV
20 April 2026

45671941R00177